t for twenty or

ſame ſide, during the
inutes: The Motions of
a noiſe very like that
ay hear it two hundred
their Wing grows grea-
ity, and forms a little
Feathers, as big as a
ts Beak are the chief De-
s very hard to catch it
e in open Places, be-
they, and ſometimes
t much Trouble. From

EXTINCT BIRDS

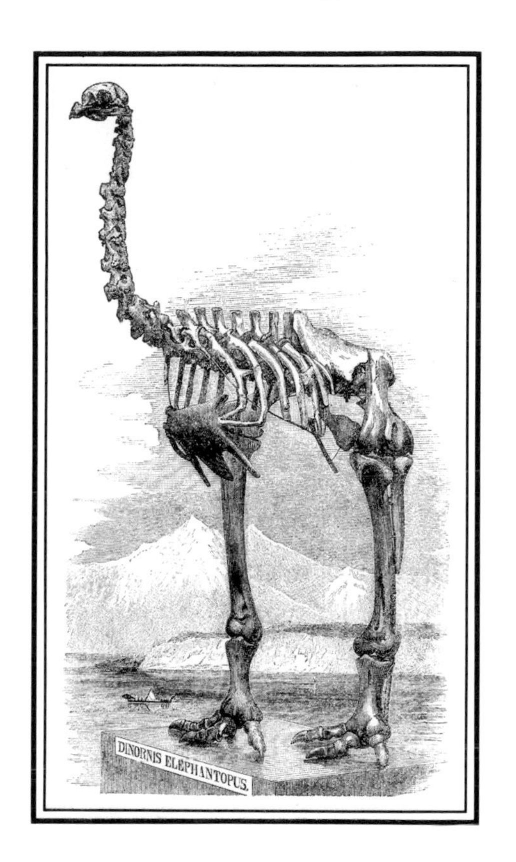

EXTINCT BIRDS

ERROL FULLER

FOREWORD BY

THE HON. MIRIAM ROTHSCHILD

Facts On File Publications
New York, New York • Oxford, England

To the memory of my father, Errol Fuller –
born Calcutta, 1921; died London, 1986.

First published in the United States of America in 1988 by
Facts on File Publications
460 Park Avenue South
New York, New York 10016

Library of Congress Cataloging in Publication Data
Fuller, Errol.
Extinct birds.
Bibliography: p.
Includes index.
1. Birds, Extinct. I. Title.
QL676.8.F85 1988 598′.042 87-9073
ISBN 0-8160-1833-2

This book was designed and produced by
The Rainbird Publishing Group Limited
27 Wrights Lane, London W8 5TZ

Editor: Sarah Bunney
Designer: Lester Cheeseman
Production: Sara Hunt

Text set by Servis Filmsetting Ltd, Manchester
Color origination by Bridge Graphics, Hull
Printed and bound by Graficas Reunidas, Madrid, Spain

10 9 8 7 6 5 4 3 2 1

Endpapers Part of François Leguat's account of the Rodrigues Solitary,
from *A New Voyage to the East Indies* (London, 1708).
Half-title page Skeleton of a moa. Engraving from W.L. Buller's
History of the Birds of New Zealand, Vol. 1 (London, 1887–8).
Frontispiece Hawaii O'os. Hand-coloured lithograph by J.G. Keulemans from
W. Rothschild's *Avifauna of Laysan and the Neighbouring Islands* (London, 1893–1900),
Pl.72. Courtesy of The Hon. Miriam Rothschild.

CONTENTS

LIST OF ILLUSTRATIONS 

COLOUR

BLACK AND WHITE

FOREWORD

The timing of the publication of this volume is perfect. Walter Rothschild's pioneer work, compiled while he was still in his teens although not published until 1907, was a sumptuous production for the specialist or bibliophile. A bound copy of his *Extinct Birds* weighs just under 11 lb and much of the text is concerned with skeletal structures and fossil specimens. Under no circumstances, despite its marvellous coloured plates (some of which are fortunately reproduced in the present volume), can it be described either as a bedside or coffee-table book.

Errol Fuller has realized the deep current interest in the subject but has had the wisdom to begin his study in 1600 – thereby largely avoiding inevitably controversial and relatively dull sections describing species known only from skeletons or bone fragments – and has concentrated on recent extinctions. Furthermore, although well informed, he has adopted a lively, almost anecdotal style that is attractive to the amateur ornithologist as well as the professional, who may feel thankful that such hitherto scattered information is gathered together in one volume and presented in a manner appealing to everyone. The author skilfully arouses the reader's fury at the senseless destruction of many of the beautiful birds he describes. For, like Walter Rothschild, he is emotionally concerned in the unnecessary loss of these unique species, and this is what raises his work above the level of the ordinary bird book.

In the recent past, senseless destruction – especially of defenceless island faunas – and culinary interest linked to financial gain, has been a major cause of extinction, but in present times it is the destruction of their habitat that threatens the existence of many birds. There are frequent appeals for money from the public to 'save' a famous painting by a well-known artist for the nation. But in such an event the work of art merely changes walls, and only a different public can enjoy it. This book brings home to us the melancholy fact that when a bird is exterminated thousands of years of evolution and specialization and unique beauty have gone *for ever*.

Errol Fuller's book is a skilful appeal for the conservation of island faunas, rainforests and other varied habitats, presented with a well-researched and attractive description of the birds we have lost. Let us hope *Extinct Birds* will act as a carrier pigeon and spread the message far and wide.

Miriam Rothschild

Figure i Flocks of a recently extinct bird, the Passenger Pigeon, in North America. Engraving from *The Illustrated Sporting and Dramatic News*, 3 July 1875.

PREFACE

For many years now I have been collecting the materials – pictures, articles, pamphlets, transcripts of diaries etc. – from which this book is assembled. At first, and for a long time, I did it at random with no very definite object in mind; then the collecting became more formal as the idea of making a book slowly took shape.

The beginnings of *Extinct Birds* go back as far as I can remember. I know exactly how and when my attention was first directed to the rare, the curious and the extinct; a large picture book, *Prehistoric Animals* by J. Augusta and Z. Burian, was, I think, totally responsible. I first saw it in a darkened shop window while being dragged along the Charing Cross Road by my parents: they were on their way to see the musical movie *South Pacific* and, being without a babysitter, found they had to take me too! The dust wrapper of *Prehistoric Animals*, as I saw it in the gloom, seemed to me then – and I think perhaps it still does – the most strange and wonderful thing I had ever looked at. Since that day the idea of mysterious monsters, extinct creatures and the South Pacific – from which general area so many extinct birds come – has remained inextricably confused in my mind.

Six months after my first glimpse of *Prehistoric Animals*, my father paid out the then princely sum of 30 shillings for the book and it became my Christmas present. Secure in the knowledge that I possessed the most priceless treasure in the entire history of literature and art, I let Zdenek Burian's paintings become the inspiration for countless dreams of lost worlds and vanished creatures. *Prehistoric Animals* was the only book I had – the only one I wanted – and so it remained for several years.

Extinct Birds really began right then but only much, much later did it truly start to come into being. One day it dawned on me that there was no book to tell me what I wished to know about the subject. I therefore proceeded to make my own. This, then, is a totally self-indulgent book!

There are books on the Dodo, the Passenger Pigeon and a few other extinct birds but, for the rest, very little material is available in book form. No book devoted exclusively to all of the world's recently extinct birds has been produced in English since 1907 when Walter Rothschild published *Extinct Birds* in an edition limited to just 300 copies. Beautifully presented, wildly eccentric, it is now long out of date.

Many species – the Huia, the Passenger Pigeon and the Carolina Parakeet among them – have disappeared since Rothschild wrote his book; information about these birds lies largely uncollected in journals, magazines and miscellaneous writings. In 1958, J.C. Greenway assembled a mass of such material and, with just one colour plate, published *Extinct and Vanishing Birds of the World*. Although dealing as much with rare forms as with those actually extinct, more than a quarter of a century after publication this book remains the single, standard work on extinct birds. Considering how many books on birds are published each year, this neglect is, perhaps, surprising – especially so in view of the intrinsically fascinating nature of many lost species.

As used here, the title *Extinct Birds* may be misleading. A calculation was once made to provide insight into just how many bird species have existed since the first birds appeared some 140 millions of years ago. A vast – albeit hypothetical – number (150,000) was settled upon; assuming it anything like correct, some 94 per cent of bird species are now extinct. Only a comparatively tiny number are featured here. This book deals only with those that have vanished since 1600 – a date chosen fairly arbitrarily as heralding a period during which relatively reliable records accumulated.

Since 1600, some 75 known species have been lost – an approximate number since there can, naturally, be argument over whether an excessively rare bird is actually extinct. Clearly, those featured here are those that in my opinion seem to be extinct; this results in certain eccentricities. Thus, the Cherry-throated Tanager (*Nemosia rourei*), known from a single specimen collected in 1870 and never again located, is listed as *not* extinct, whereas the Passenger Pigeon (*Ectopistes migratorius*) – not then rare – is considered to be so. Such decisions are built around patterns of decline, remoteness of terrain occupied, recent changes to the environment and a whole variety of other elements that may seem relevant in this or that particular case.

Of necessity the book must be incomplete. Many species must have vanished since 1600 leaving no trace at all; others assumed to have disappeared at an earlier date may have lingered until more recently. Species featured here are represented in museums by specimen material – skins, bones or both – although a very few are added whose description rests solely upon unimpeachable testimony. These could be supplemented, according to taste, by a number of 'hypothetical' species and some 'enigmatic' ones; birds belonging to these categories are considered together in the last section of the book.

Accounts of species are variable in length and in the kind of information they contain; about many of these birds very little is known. Where standard pieces of ornithological information are not given, it may be assumed that reliable records have not been located.

The arrangement of the book largely follows that set out by Richard Howard and Alick Moore in their *Complete Checklist of the Birds of the World* (1980). In appropriate places, major bird groupings are reviewed. The purpose of these reviews is to make some inclusion of the extinct races of otherwise extant species. They also give scope to clarify reasons for omitting species that some might consider extinct and to point out which species may vanish in the next few decades.

Each species is illustrated in colour unless it is known only from skeletal material. Should any omission on these grounds be felt a loss, some conjectural pictures are given in Rothschild (1907) and Hachisuka (1953), as, too, are illustrations of several 'hypothetical' extinct birds. Because I am a painter, it may be wondered why I have not produced a completely new set of illustrations for this book. Here, therefore, I wish to place on record the fact that I am not a painter of wildlife subjects and the few paintings I have produced for this volume have been done from sheer necessity, there being no other pictures that I could reproduce. For a book of this type it makes sense to use illustrations made when the featured birds were still extant and, by good fortune, it often happens that such pictures were painted by giants of wildlife art – men such as Joseph Wolf, John James Audubon, Edward Lear and, in my opinion the very greatest of all ornithological illustrators, John Gerrard Keulemans. Any attractions that this book has are largely through their efforts and I wish, above all, to acknowledge them.

Naturally, there are others who I should thank. They are: Peter Blest, Sarah Bunney, Raymond Ching, Rosemary Crane, Peter Dance, Alena Elgie, Linda Foord, Aldwyth Fuller, Celia Hammond, Mike Latter, Mike Lyster, David Medway, Pat Morris, Ikena Muoma, S.L. Olson, Shane Parker, Nick Peters, S.D. Ripley, Miriam Rothschild, Carolyn Sinclair-Smith, Peter Slater and Derrick Witty.

Finally, it may be worth pointing out that this is perhaps the last time that this subject will be approached both comprehensively and in a single volume. In 15 or 20 years, this area of interest will be too large to tackle in anything like this format. There will be just too many extinct species! Perhaps in another generation someone will find the impetus to produce a supplementary volume.

Figure ii Dodo and exotic birds. Engraving from H.E. Strickland and A.G. Melville's *Dodo and its Kindred* (London, 1848), after a painting attributed to Roelandt Savory (*c.* 1625) in the British Museum (Natural History). On the right of the Dodo could be depicted the extinct Red Hen of Mauritius, *Aphanapteryx bonasia*, in the act of swallowing a frog.

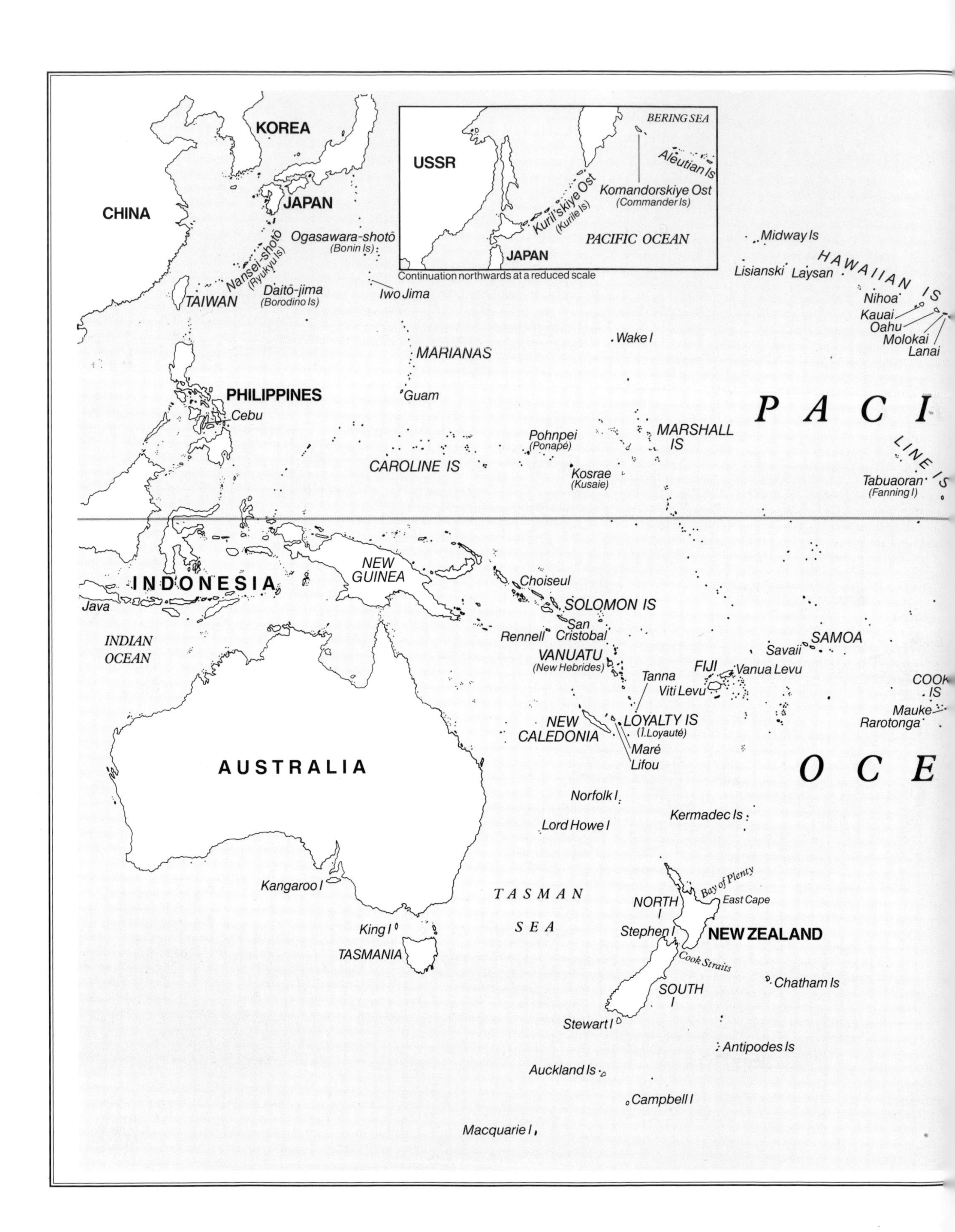
KOREA
CHINA
JAPAN
USSR
BERING SEA
Aleutian Is
Komandorskiye Ost
(Commander Is)
Kuril'skiye Ost
(Kurile Is)
PACIFIC OCEAN
JAPAN
Continuation northwards at a reduced scale
Ogasawara-shotō
(Bonin Is)
Nansei-shotō
(Ryukyu Is)
Daitō-jima
(Borodino Is)
TAIWAN
Iwo Jima
Midway Is
Lisianski
Laysan
HAWAIIAN IS
Nihoa
Kauai
Oahu
Molokai
Lanai
Wake I
MARIANAS
PHILIPPINES
Cebu
Guam
P A C I
Pohnpei
(Ponapé)
MARSHALL IS
CAROLINE IS
Kosrae
(Kusaie)
LINE IS
Tabuaoran
(Fanning I)
NEW GUINEA
INDONESIA
Java
Choiseul
SOLOMON IS
San Cristobal
Rennell
INDIAN OCEAN
SAMOA
Savaii
VANUATU
(New Hebrides)
FIJI
Vanua Levu
Tanna
Viti Levu
COOK IS
Mauke
Rarotonga
NEW CALEDONIA
LOYALTY IS
(Î. Loyauté)
Maré
Lifou
AUSTRALIA
O C E
Norfolk I
Kermadec Is
Lord Howe I
Bay of Plenty
East Cape
Kangaroo I
TASMAN SEA
NORTH I
King I
Stephen I
NEW ZEALAND
TASMANIA
Cook Straits
Chatham Is
SOUTH I
Stewart I
Antipodes Is
Auckland Is
Campbell I
Macquarie I

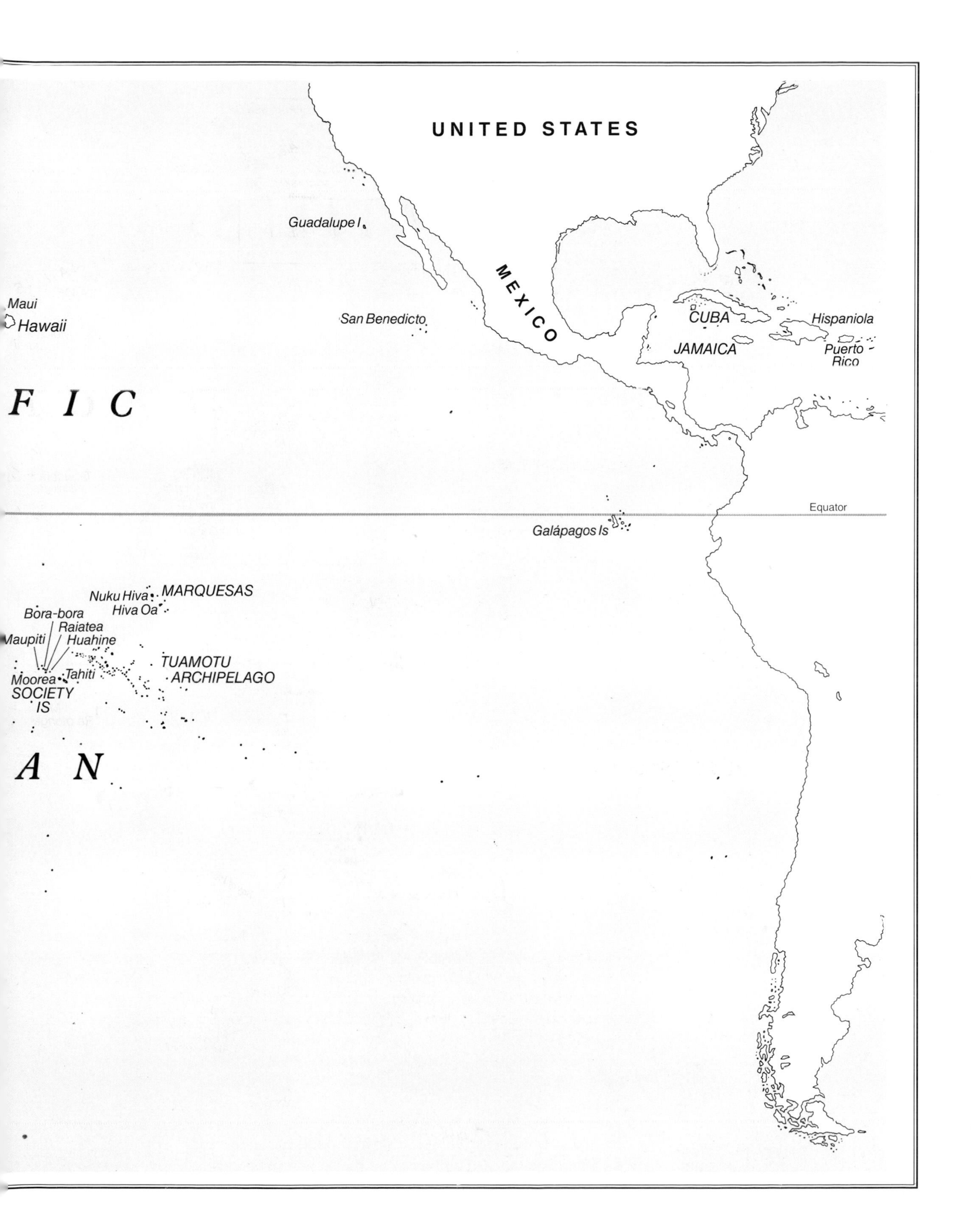

UNITED STATES
Guadalupe I.
MEXICO
Maui
Hawaii
San Benedicto
CUBA
Hispaniola
JAMAICA
Puerto Rico
F I C
Equator
Galápagos Is
Nuku Hiva
MARQUESAS
Hiva Oa
Bora-bora
Raiatea
Maupiti
Huahine
TUAMOTU ARCHIPELAGO
Tahiti
Moorea
SOCIETY IS
A N

RATITES

ORDERS
Aepyornithiformes, Dinornithiformes, Apterygiformes, Casuariiformes, Struthioniformes and Rheiformes

EXTINCT SPECIES

Aepyornis (*Aepyornis maximus*)
Slender Moa (*Dinornis torosus*)
Greater Broad-billed Moa (*Euryapteryx gravis*)
Lesser Megalapteryx (*Megalapteryx didinus*)

The ratite birds are almost always placed at the head of listings of living avian groups. They are the most obviously primitive of birds – so much so that in various respects they appear quite unbird-like. Yet birds they clearly are and to call them 'ostrich-like' conveys a good idea of their general nature. Among the group are the ostriches themselves, the emus and cassowaries and the rheas, all – by bird standards – quite huge and all sharing the characteristic of a keel-less breastbone.

Although all ratites are very different from other birds, the relationships of the various ratite orders one to another are obscure. Two particularly intriguing questions remain unanswered. Did some of the orders evolve separately or did all descend from a common ancestral stock? Was this ancestral stock – or were these stocks – made up of true flying birds that over aeons lost the power of flight, or did these creatures never fully acquire flight? Most ratites have remnants of wings but in moas – the most perfect of bipeds – no vestige of a wing remains (Figure 2).

Among the ratites are most of the largest birds known to us. Assuming they adopted a more-or-less upright stance, the tallest of the moas would have stood easily over 3.5 m (12 ft). By no means as lofty, *Aepyornis maximus* of Madagascar was a much bulkier bird that laid eggs of quite colossal proportions – the largest ever recorded. The smallest ratites – the kiwis – are dwarfs by comparison but are actually quite large for birds, resembling farmyard chickens in size. With their adaptation to the forest floor, strictly nocturnal habits and long, thin beaks carrying nostrils close to the tip, kiwis are in some respects the most aberrant ratites.

Although representatives of all six groups existed after 1600, two of the orders are now completely extinct.

1 (Previous page): Pacific islands where populations of recently extinct birds lived.

The great elephantbirds (order Aepyornithiformes) of Madagascar have all disappeared. About a dozen species are described but in reality there may have been fewer; all are known only from skeletal remains, egg fragments or native tales, and probably only the largest, the Great Elephantbird (*Aepyornis maximus*), survived into recent historical times.

The moas (order Dinornithiformes) of New Zealand are likewise extinct but almost certainly some still survived when Tasman first sighted the islands in 1642. In size, they ranged from the incredibly tall but comparatively slender birds of the genus *Dinornis* down to the smaller, dumpier forms included in *Megalapteryx*; one species of medium height, *Pachyornis elephantopus*, approached the Great Elephantbird of Madagascar in bulk. From skeletal remains a large number of species have been described and the resulting situation is very muddled. Walter Rothschild (1907) listed thirty-seven species but G. Archey (1941) and W.R.B. Oliver (1949) were more conservative, giving twenty (see Table 1) and twenty-eight, respectively. More recently, J. Cracraft (1976) allowed even fewer, believing only around thirteen could be recognized.

In whichever way the remains are interpreted, the sheer number of moa species inhabiting so small a land area is surprising; the whole African continent supports just one type of ratite – the Ostrich (*Struthio camelus*). Two factors presumably account for this anomaly. First, not all described species lived at the same time – forms evolved and died out over millions of years; secondly, numbers and variety were influenced by particular circumstances that applied in New Zealand but in no other land of comparable size and equable climate. Birds adapted to niches normally occupied by mammals simply because there was no competition from warm-blooded, four-legged beasts. Not until human colonists shattered New Zealand's long isolation from the rest of the world did terrestrial mammals arrive. How moas themselves came to inhabit this remote land when mammalian rivals apparently failed to do likewise is a mystery.

2 Typical reconstruction of a moa (with kiwis). Engraving from F. von Hochstetter's *New Zealand, its Physical Geography, Geology and Natural History* (London, 1867).

New Zealand appears to be the remnant of a splinter of land that around 130 million years ago broke away from the ancient southern supercontinent of Gondwanaland; from this time onwards it could be colonized only from the sea or from the air. It seems improbable – although not impossible – that stocks later giving rise to the moas arrived by flight; these ancestral birds were probably already on the fragment of land as it broke free. Assuming this to be the case, what were the circumstances enabling the moas' forerunners to be present in New Zealand while terrestrial mammals were absent?

Primitive mammals did exist at the time of the split so it may be unrealistic to suppose, as some scientists have, that none were trapped with ancestral moas upon the land splinter. If mammals were present, then these birds, due to some special circumstance of which we know nothing, were able to triumph and take possession of the land in the struggle for existence that surely ensued. Although it is generally stated that terrestrial mammals were entirely lacking in New Zealand, there is some evidence that one small creature may have survived in the 'Land of Birds' until quite recently. Early settlers and naturalists spoke of the *waitoreke* but no specimen or living example was ever procured and the references remain enigmatic.

Whether moas were restricted to New Zealand is an open question. A moa from Queensland, Australia was described as *Dinornis queenslandiae* by C.W. De Vis in 1884 but its remains are fragmentary and some researchers exclude them from the order. Oliver (1949), however, assigned these

Table 1

SPECIES OF MOA RECOGNIZED BY ARCHEY (1941)

Species
Anomalopteryx didiformis Owen, 1844
Anomalopteryx antiquus Hutton, 1892
Megalapteryx didinus Owen, 1883*
Megalapteryx benhami Archey, 1941
Pachyornis elephantopus Owen, 1856
Pachyornis pygmaeus Hutton, 1891
Pachyornis mappini Archey, 1941
Pachyornis oweni Haast, 1886
Emeus crassus Owen, 1846
Emeus huttonii Owen, 1879
Euryapteryx gravis Owen, 1870*
Euryapteryx geranoides Owen, 1848
Euryapteryx exilis Hutton, 1897
Euryapteryx curtus Owen, 1846
Dinornis novae-zealandiae Owen, 1843
Dinornis ingens Owen, 1844
Dinornis giganteus Owen, 1844
Dinornis torosus Hutton, 1891*
Dinornis robustus Owen, 1846
Dinornis maximus Haast, 1869

*Probably became extinct after 1600 so description is included here

bones to the genus *Pachyornis*, being sure enough of his ground to call his account of the order, *The Moas of New Zealand and Australia*.

Most moas became extinct well before 1600 but three species – *Dinornis torosus*, *Euryapteryx gravis* and *Megalapteryx didinus* – probably all survived into recent historical times; others may also have lingered on to quite recent dates.

New Zealand is, of course, home to the kiwis (order Apterygiformes) of which three species are recognized – the Brown (*Apteryx australis*), the Great Spotted (*A. haasti*) and the Little Spotted (*A. oweni*) (Figure 3). The Great Spotted Kiwi was for long considered a rare bird but is now acknowledged as far more widespread than previously supposed. The reverse is the case with the Little Spotted; it is believed to inhabit extensive areas of the South Island but recent surveys have failed to locate it there.

3 Kiwis. Engraving from *Cassell's Natural History* (London, 1889).

4 Kangaroo Island Emu? Chromolithograph after a painting by J.G. Keulemans, from W. Rothschild's *Extinct Birds* (London, 1907), Pl.40. Courtesy of The Hon. Miriam Rothschild.

Fortunately, it is found in good numbers on Kapiti Island off the southwestern tip of the North Island to which it was introduced some years ago.

Maoris expressed belief in a fourth species of kiwi, one considerably larger than the others – about the size of a turkey – that they seem to have called *roa-roa*. No naturalist ever found it although there exists a cloak of kiwi feathers made for a Maori chief in which the feathers are very much larger than those of any known kiwi. The *roa-roa* may even have been a small species of moa, perhaps a *Megalapteryx*! In *A History of the Birds of New Zealand* (1887–8), Buller mentioned that the natives claimed a kiwi lived on the Chatham Islands until about 1835 but nothing can be added to their account; there may be truth to it or there may not.

Although no recent extinction of entire species has occurred among the other ratite orders, four races have now vanished: the Arabian Ostrich (*Struthio camelus syriacus*), the Tasmanian Emu (*Dromaius novaehollandiae diemenensis*), the Kangaroo Island Emu (*D.n. demenianus*) and the King Island Emu (*D. n. minor*).

Distinguished by its comparatively small size, the apparently extinct Arabian Ostrich (*Struthio camelus syriacus*) was fairly common until about the time of World War I, but as firearms became more plentiful in Arabia and it became possible to pursue birds in automobiles, numbers rapidly declined. By the outbreak of World War II, this subspecies of the largest of living birds was virtually extinct. A bird shot and eaten in Bahrain during 1941 may constitute the last legitimate record of the race but a report of a dying bird brought down by floodwaters in southwestern Jordan in February of 1966 is generally accepted.

The history of the emus of King Island and Kangaroo Island off Tasmania is rather confused. That populations inhabited both islands is not open to doubt but the precise status of each has never been satisfactorily determined.

The name *minor* was proposed for the King Island birds on the basis of skeletal remains; and, according to J.C. Greenway (1958), this population is known only from these. The Kangaroo Island form, *diemenianus*, is usually considered to be known not just from bones but also from one stuffed specimen. The living bird from which this specimen originates was collected with others in the first years of the nineteenth century by a French naval expedition to Australia under the command of Nicholas Baudin. Some of the captives died during the long journey back to Europe but at least two arrived alive and these were sent to the residence of the Empress Josephine. They survived until 1822 and at the time of their deaths may have been the very last representatives of their race. Another bird, perhaps dead on arrival in France, was sent to the Jardin des Plantes in Paris and provided the only remaining skin, the type of *diemenianus*. Although it has been assumed that this was a Kangaroo Island bird, P. Slater (1978) has pointed out that Baudin took aboard emus on both King and Kangaroo islands and there is actually no great certainty over locality data for individual birds. An illustration of the preserved bird, by J.G. Keulemans, appeared in Rothschild's *Extinct Birds* (1907) (Figure 4).

It seems likely that both these island forms disappeared during the first decades of the nineteenth century and the Tasmanian Emu, *diemenensis*, probably vanished around 50 years later.

Of extant races of ratite birds, only a subspecies of the Lesser Rhea (*Pterocnemia pennata tarapacensis*) is considered endangered.

AEPYORNIS

Aepyornis maximus

PLATE I

Aepyornis maximus (St Hilaire, 1851)

Aepyornis maximus St Hilaire, 1851

Height ~ 315 cm (10 ft)

Description
Appearance in life unknown.

An egg over a foot in length – larger than any known dinosaur egg – big enough to hold the contents of 7 ostrich eggs, 180 chicken eggs or more than 12,000 hummingbird eggs, is enough to stir the imagination of even the most incurious person. The Great Elephantbird of Madagascar, the Aepyornis, laid eggs that reached just such gigantic size – a wonder sometimes thought to have originated the Arabian tales of the monstrous *roc* or *rubk*.

Perhaps these mighty objects did provide the basis for the stories, but the colossal raptor of Arabian legend in no way resembles the bird that laid the eggs, for the Aepyornis was very much a ratite and presumably appeared something like a gigantic and excessively ponderous ostrich. Standing around 3 m (10 ft) high, this great creature was by no means as tall as the tallest of the moas, yet it was much more massive in build than any of these species: D. Amadon (1947) estimated its weight at around 454 kg (1,000 lb); for comparison, an Ostrich's weight can be given at about 136 kg (300 lb). The great egg is quite possibly as large as any egg has ever been; engineers calculate that structurally and functionally it is impossible for an egg to be bigger.

Although *Aepyornis maximus* is the largest and best known of the Madagascan elephantbirds, there were, in fact, several species; these survived until fairly recently and seem to have been distributed quite widely over the island in much the same way as moas were once spread across New Zealand. It seems that their natural enemies were few, perhaps only crocodiles – apart from man – posing any real threat. The fossil record seems confined to Pleistocene and Recent remains (from the past

I Egg of Aepyornis (⅓ natural size). Hand-coloured lithograph by J. Erxleben from G.D. Rowley's *Ornithological Miscellany*, Vol. 3 (London, 1875–8).

two million years only), leaving the early history of the Aepyornithiformes little understood. Presumably, they developed in the isolation of Madagascar but this may not necessarily be the case; fragments of bone and egg shell from elsewhere have been assigned to the group – although upon grounds that might be considered rather insufficient.

It seems certain that several species of elephantbird survived until just a few thousand years ago but probable that by recent historical times the smaller ones had all disappeared leaving only the monstrous *A. maximus* extant. It can be guessed that these birds existed either by cropping the lower branches of trees and shrubs, or by grazing; maybe their livelihood depended on a combination of both feeding methods. As man's presence on the island made itself increasingly felt, the birds must have been pinned back into the loneliest and most inaccessible parts of Madagascar.

Little is known of the settlement of Madagascar before the arrival of Europeans; whether or not Madagascans actively hunted the great birds is a matter for speculation – presumably some did. Perhaps of rather more value than the birds themselves would have been the eggs, which, surely, were prized by the natives for their food content and also for the ornamental and utilitarian value of the shells.

When the French claimed Madagascar as a possession in 1642, the Great Elephantbird probably still survived in isolated places. Under the heading of 'vouroupatra' the first French Governor of Madagascar, Étienne de Flacourt, described in 1658 'a large bird which haunts the Ampatres and lays eggs like the ostriches; so that the people of these places may not take it, it seeks the most lonely places'. Whether de Flacourt actually saw the Aepyornis or whether he relied solely on the testimony of others is not clear. On his journey back to France he was killed by Algerian pirates without further elaborating on his fleeting account.

For the next 200 years the French kept their foothold on the coasts of Madagascar without ever being able thoroughly to explore the interior, and also without gathering further firm word of the Aepyornis. Then, in the early 1830s, a French naval officer, Victor Sganzin, is supposed to have seen a gigantic egg and perhaps even acquired it. It was rumoured that an

egg was sold to the natural history dealers Verreaux but that the ship carrying this treasure back to France ran aground on the rocks of La Rochelle, the prize sinking to the bottom. A few years later another Frenchman, this time by the name of Dumarele, claimed to have been shown by natives the shell of an enormous egg. This shell he had tried without success to buy; the owners regarded it as a very rare item and the bird that laid it even rarer.

In 1851, rumours of giant eggs were finally substantiated. Three were obtained and taken to France by a Captain Abadie, together with fragments of bone. Within a few more years, enough bone had been accumulated for a complete skeleton to be reconstructed and any lingering doubts about the nature of Madagascar's giant bird were resolved.

For how long the Aepyornis survived the arrival of Europeans was, and remains, a mystery. It seems probable that it still lived at the time of de Flacourt but the last of the elephantbirds is likely to have expired well before the beginning of the nineteenth century and the awakening of interest in the amazing fauna of Madagascar.

SLENDER MOA

Dinornis torosus

FIGURES 5–8

Dinornis torosus (Hutton, 1891)

Dinornis torosus Hutton, 1891 (Takaka, South Island, New Zealand)
Palapteryx plenus Hutton, 1891
Dinornis strenuus Hutton, 1893

Height ~ 215 cm (7 ft)

Description
Appearance in life unknown.

So far as my skill in interpreting an osseous fragment may be credited, I am willing to risk the reputation for it on the statement that there has existed, if there does not now exist, in New Zealand, a struthious bird nearly, if not quite, equal in size to the Ostrich, belonging to a heavier and more sluggish species.

These bold words, written during 1839 for the journals of the Zoological Society of London, were used by the distinguished British anatomist Richard (later, Sir Richard) Owen to introduce to the civilized world the moa – a gigantic, snake-necked bird whose final eclipse, now as then, is shrouded in mystery. Owen's concept of a large, flightless bird was built around the most meagre of evidence. Before him, in support of his rash statement, was nothing more than a fragment of bone, a 15-cm (6-in) length of femur brought from New Zealand earlier that year by a Dr John Rule. If Owen's credibility as a man of science suffered following his pronouncement, it was soon vindicated; within months, much more complete remains of giant birds arrived in London from the far-distant dominion. Against what was probably the general expectation, it became undeniable that New Zealand was indeed the home of huge ratite birds.

Although little caution was shown in the assessment of general form, rather more had been attached to the question of whether or not the birds still existed, raising a problem never satisfactorily resolved.

Owen's original remarks were made at a time when the interior of New Zealand remained largely unexplored. Not only was the country many weeks' sailing distance from the great centres of civilization, but this new land was the home of a war-like people then numbered among the fiercest cannibals inhabiting islands of the southern seas. The signing of the Treaty of Waitangi, the charter used by European settlers to intensify their campaign of wresting land from the Maoris, was still a year away as Owen made his pronouncement.

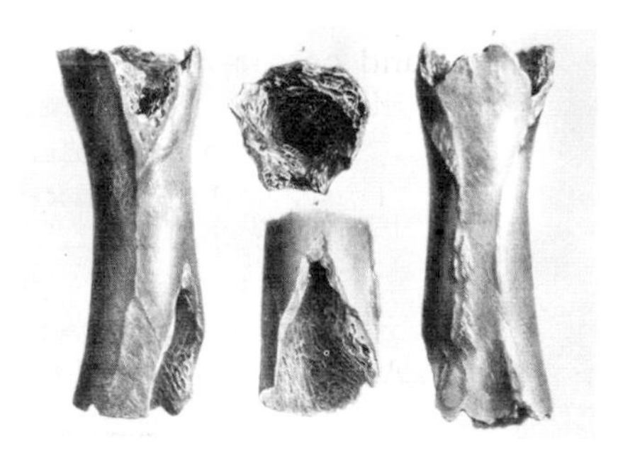

5 Rule's bone – the 15-cm (6-in) long fragment from which the English anatomist Professor Richard Owen deduced that gigantic flightless birds once inhabited New Zealand. Lithograph from the *Transactions of the Zoological Society of London*, Vol. 3 (1842).

6 Skeleton of a moa. Lithograph by James Erxleben from the *Transactions of the Zoological Society of London*, Vol. 11 (1883).

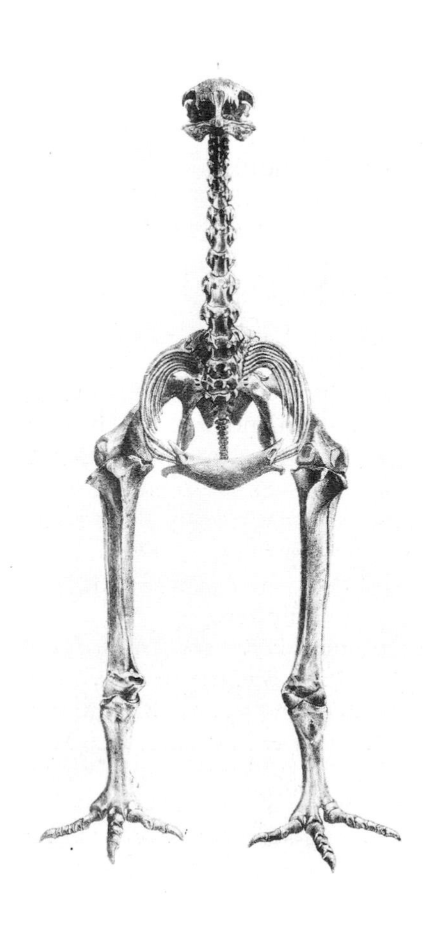

In these circumstances, there seemed little reason to suppose that monstrous creatures could not exist hidden somewhere in the interior, and good grounds for suspecting that some might. But as the country was opened up for settlement, it became ever more apparent that if moas did survive they were at the verge of extinction. Optimists speculated that a few still lived; those with a less romantic inclination maintained that all were recently extinct; and a third school believed that moas died out during a past so remote that humans could never have beheld living examples – an idea now completely discredited, although from the large number of species recognized today can be selected some that did disappear during distant epochs. That the various Polynesian peoples who started invading New Zealand in waves about a thousand years ago were familiar with other species is no longer doubted. Which moas these were and when exactly each line failed cannot be precisely determined. Bones taken from middens show that a number of species were well known to the first Polynesian settlers and their successors, the Maoris.

Whether moas were still extant during the early years of European settlement is a matter of some controversy. Should remains be found in kitchen refuse together with pig bones this would indicate survival of giant birds beyond the beginnings of European colonization; there were no pigs in New Zealand until Captain James Cook left them there in the late eighteenth century. Although there are no moa bones in middens containing the remains of pigs, bones of butchered birds show what appear to be marks of iron knives on them.

In very different circumstances, remains of moas have been found with an amazingly fresh appearance, but this freshness is itself an uncertain indicator, the state of preservation of animal tissue depending so much on prevailing conditions.

Even more difficult to interpret are the various sightings alleged. Owen's announcement in London aroused enormous interest. Not only were moas widely discussed, in the Colony itself their traces were intensively sought. Prestige and celebrity would certainly have fallen to anyone making verifiable sightings of living moas – perhaps a considerable inducement to fantasize.

Whether honest in spirit or not, reports remain. Sealers and whalers recalled seeing monstrous birds running along wilder shores of the South Island. One George Pauley claimed a meeting with a bird 6 m (20 ft) high, close by a lake in the Otago area. The encounter, supposed to have taken place during the 1820s, was brief. When the man saw the bird and the bird saw the man, each turned and ran one from another – a train of events that at least carries some ring of truth! Two birds resembling emus were reported from a South Island hillside during 1850 by members of an expedition under the command of Lieutenant A. Impey. A story published in the *Nelson Examiner* for 12 January 1861 told of the finding of three-toed footprints, 36 cm (14 in) long, on the ranges between Takaka and Riwaka, by members of a surveying party. Although followed, these tracks were lost among scrub and rocks. This story ends with characteristically romantic speculation that a last, solitary moa might even then have been in existence. During 1878, the *Otago Witness* published one of the more detailed accounts:

> The story is current, and generally believed in the Waiau district, and our informant states that he took the trouble to see the runholder on whose station the bird had been seen, who states there can be no doubt but that a

> very large bird – much larger than any emu – exists in the back portion of his run on the west side of the Waiau, and adjoining the large bush which stretches to the west coast. This gentleman has repeatedly seen its tracks and footmarks, and on a recent occasion his shepherd – an intelligent man – started the bird itself out of a patch of manuka scrub, with his sheep-dog. The bird ran from the dog till it reached the brow of a terrace above him, some thirty or forty yards off, when it turned on the dog, which immediately ran into the shepherd's heel. The Moa stood for fully ten minutes on the brow of the terrace, bending its long neck up and down exactly as the black swan does when disturbed. It is described as being very much higher than any emu ever seen in Australia, and standing very much more erect on its legs. The colour is described as a sort of silver grey with greenish streaks through it.

Such tales make tantalizing reading and to them may be attached whatever belief the reader wishes to give. Whether or not genuine sightings were ever made by Europeans remains an open question but from the accumulation of evidence of various kinds it can be assumed that at least a few individuals still survived when Cook arrived in New Zealand. To which species these may have belonged is altogether more uncertain.

The genus *Dinornis* contains those gigantic species, the tallest of all known birds, with which the name of 'moa' is most familiarly associated. Several species have been founded on remains discovered on New Zealand's South Island and several more from material taken on the North, it seeming a general rule that North Island birds were more lightly built than their southern counterparts. They were all truly spectacular birds, the tallest known examples towering to heights in excess of 3.6 m (12 ft). Species of the genus are characterized not just by their great loftiness but also by their long, broad, downcurved beaks, their bones more slender in the legs than those of other members of the order Dinornithiformes, and peculiarities of the sternum.

One species, *Dinornis torosus*, is listed in *The ICBP Bird Red Data Book* (King, 1981) as having disappeared since 1600 but, as with all moas, any actual date suggested for extinction can be no more than very approxi-

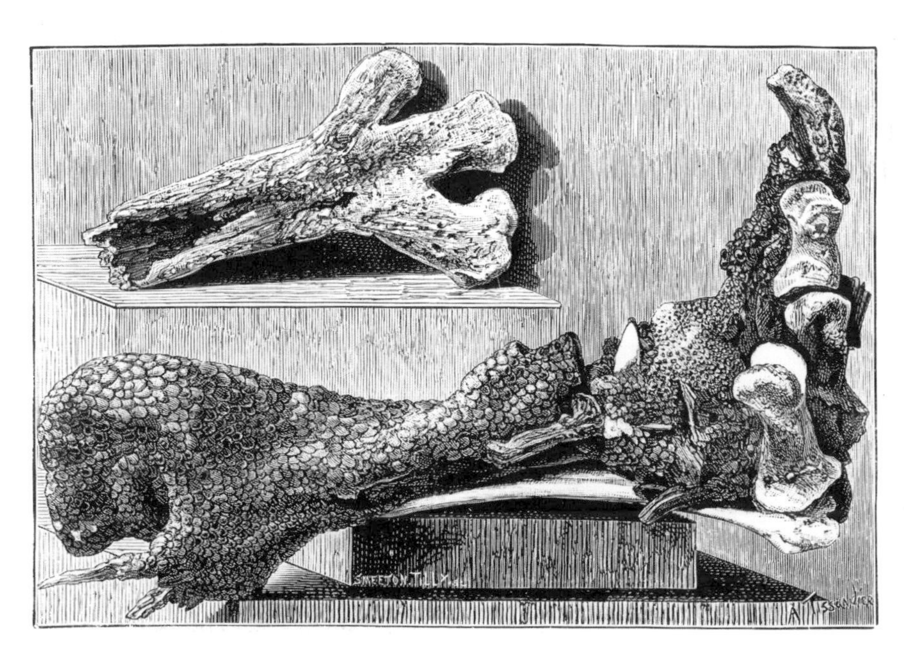

7 Right foot of a moa discovered before 1874 in an unusually fine state of preservation in a fissure among mica-schist rocks, Knobby Ranges, Otago, New Zealand. Engraving from W.L. Buller's *History of the Birds of New Zealand*, Vol. 1 (London 1887–8).

8 Egg of a moa ($\frac{1}{4}$ natural size). Lithograph by James Erxleben from G.D. Rowley's *Ornithological Miscellany*, Vol. 3 (London, 1875–8).

mate. Although a gigantic bird, this was not one of the larger members of its genus. As defined by G. Archey (1941) and W.R.B. Oliver (1949), it was restricted to the South Island, bones having been found at widely scattered localities in the Canterbury, Otago and Nelson areas. Despite this considerable range, remains attributable to this species are not among the most plentiful. In swamp deposits they are scarce and, for this reason, Archey (1941) suggested that these moas preferred hill country to lowland and, perhaps, comparative lightness saved individuals from the death that so favours preservation.

A clue as to the appearance of the Slender Moa may be provided by the deep pits on the tops of some skulls – perhaps an indication that one of the sexes carried a crest.

Archey (1941) designated an almost complete skeleton in the War Memorial Museum, Auckland as the type. Curiously, this skeleton was found quite close to the site at Takaka from which came reports of footprints belonging to a last, solitary giant moa. Although this skeleton, found during the early 1890s, is nominated as the type, Owen's original bone has been attributed to this species also; T.L. Buick (1931) referred the famous fragment to *D. strenuus*, a form now incorporated with *torosus*. Since, however, this relic is supposed to have come from the North Island, its alignment with *torosus* cannot be correct.

GREATER BROAD-BILLED MOA

Euryapteryx gravis

FIGURES 9–11

Euryapteryx gravis (Owen, 1870)

Dinornis gravis Owen, 1870 (Kakanui River, South Island, New Zealand)
Emeus gravipes Lydekker, 1891
Emeus parkeri Rothschild, 1907
Emeus boothi Rothschild, 1907
Emeus haasti Rothschild, 1907
Euryapteryx kuranui Oliver, 1930

Height ~ 190 cm (6 ft)

Description
Appearance in life unknown.

Richard Owen's deduction in 1839, made from just a fragment of bone (see page 20), that gigantic 'struthious' birds had inhabited New Zealand, quickly acquired renown as an outstanding achievement in comparative anatomy. It is not so widely realized that the deduction was almost never made. Without a measure of persistence from Dr John Rule, the bone's owner, the privilege of introducing moas to the scientific world might never have fallen to Owen.

Rule was told that his bone came from a colossal eagle, recently extinct but known to the Maori as a *movie* and the relic was put before Owen with the claim that it came from a bird of flight once inhabiting New Zealand.

On unwrapping the fragment (see Figure 5) in Rule's presence, Owen immediately realized two things. First, although clearly an object of some age, the fossilization process was yet to take place; secondly, the fragment did not come from any bird capable of flight – the great eagle was a myth. Busy that afternoon and anxious to deliver a scheduled lecture, Owen was disinclined to investigate further, venturing the opinion that before him was nothing more interesting than an ordinary beef bone. Presumably, he paid little attention to Rule's argument that structure was not as might be expected in a mammal's bone but was more suggestive of a bird.

As Owen prepared to dismiss the matter, Rule produced something that really caught the great man's eye – a greenstone *mere*, the warclub of the Maori. This characteristic weapon of a formidable people was generally used for quick thrusts but the heel was designed for the downward delivery of a *coup de grâce* to the skull or back of neck as a stumbling opponent fell forwards. To European eyes these short clubs were objects of great curiosity and beauty; perhaps it is not being unfair to Owen to add

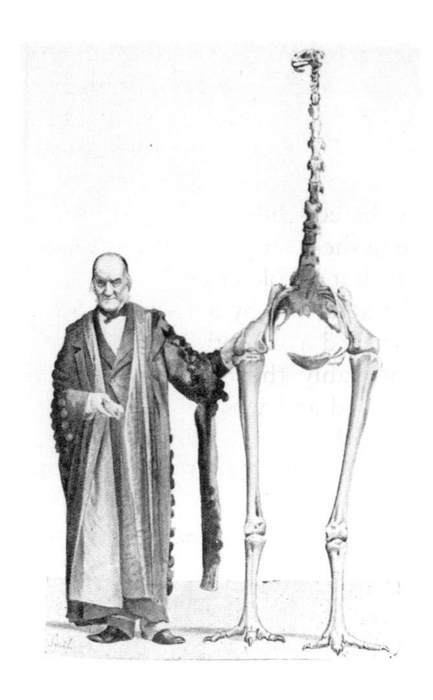

9 Professor Richard Owen and a moa skeleton, *c*. 1850.

that they were also pieces of considerable value, a fact with which he would hardly have been unacquainted. Softened by his interest, Owen now felt more disposed to pay the bone due attention and agreed that, should it be left, he would give it close study after his lecture.

It is easy to imagine Owen, later, in the silence of his study poring over the osseous fragment as the afternoon wore on into evening. One by one bones of other creatures would have been compared with the mysterious relic then rejected, one by one assumptions made then abandoned. At last, Owen was forced to conclude that before him was part of the thigh bone of a bird which he described as,

> As large as, if not larger than, the full-sized male ostrich, with the more striking difference that, whereas the femur of the ostrich, like that of the rhea and the eagle, is 'pneumatic', or contained air, the present huge bird's bone had been filled with marrow, like that of a beast.

Even after this seal of approval had been given, acceptance of the relic as an object of significance was not universal. With the backing of the great scientist, Rule offered it to the Museum of the College of Surgeons in London. The sum required for his historic fragment was ten guineas but the invitation to purchase was declined. Eventually, the bone did find its way into a public institution. In 1873, it was presented to the British Museum by the grandson of a private collector, Benjamin Bright, who had acquired it direct from Dr Rule.

The lifestyle and appearance of moas were, and are, a mystery. Presumably, the various species had adapted to a range of ecological niches. Some, probably, were inhabitants of open, grassy plains, whereas others may have preferred the shelter of the forest. Some undoubtedly lived on the high country, others in the lowlands. None seem to have been fast-running birds of the plains comparable to the Ostrich (*Struthio camelus*) or the Rhea (*Rhea americana*). In the sheltered environment of prehistoric New Zealand there was little reason why they should have been so.

For the most part moas were probably slow-moving, sluggish species and reconstructions of them, most often based broadly on the appearance of the Emu (*Dromaius novaehollandiae*), may be quite misleading. Although feathers have been found and skeletons exist with skin and muscle attached to the bones, attempts to convey the appearance of living moas are largely conjectural. It is possible – indeed likely – that some species did resemble the familiar representations; others may have looked quite different. Maori tradition suggests that a moa perhaps resembling a cassowary existed, with a brightly coloured neck and a comb upon its head.

Towards the end of the nineteenth century an old man recalled in London an incident that he claimed had occurred in New Zealand during his youth, 40 years previously. He alleged that he had seen a giant, long-legged bird with tassels hanging from its neck and a fleshy comb on top of the head. There is osteological evidence that some species carried crests. T.J. Parker (1893) noted that pits in the skulls of various moas, *Dinornis robustus*, *Anomalapteryx didiformis* and *Pachyornis pygmaeus* among them, were consistent with such a feature; as only some skulls show pits it is assumed that crests were sexual characteristics.

Although when hard-pressed they doubtless made formidable adversaries, moas were not flesh-eating (carnivorous) birds; food consisted mainly, if not exclusively, of vegetable matter. In addition to feeding on items that were readily available, moas probably used their powerful legs and feet, equipped with strong claws, for scratching or digging in the ground and turning up roots or tender shoots. When succulent grubs and insects were uncovered, these, too, most likely were eaten. As an aid to trituration, stones of varying sizes were swallowed and these are now often found with moa remains.

Maori tradition asserts that the hen incubated while the male supplied her with food. Pieces of surprisingly thin egg shell – cream or pale-green in colour – have been found in many places but whole eggs are rare.

Birds of the genus *Euryapteryx* are characterized by a broad, round-tipped beak. Several species have been described from both the North and South islands. *Euryapteryx gravis* was probably the most widespread species; it occurred throughout New Zealand and was also present on Stewart Island off the southern tip of the South Island. Although lacking the great height of the larger species of *Dinornis* or the incredible stoutness of *Pachyornis elephantopus*, *E. gravis* was a large and bulky bird – the biggest member of its genus. Of remains found in Polynesian middens, bones of this species are said to be the most common.

While the appearance of moas in life is a subject shrouded in some mystery, so, too, is the disappearance of these birds. Their final extinction was hastened by the arrival of Polynesians about AD 900 but a general

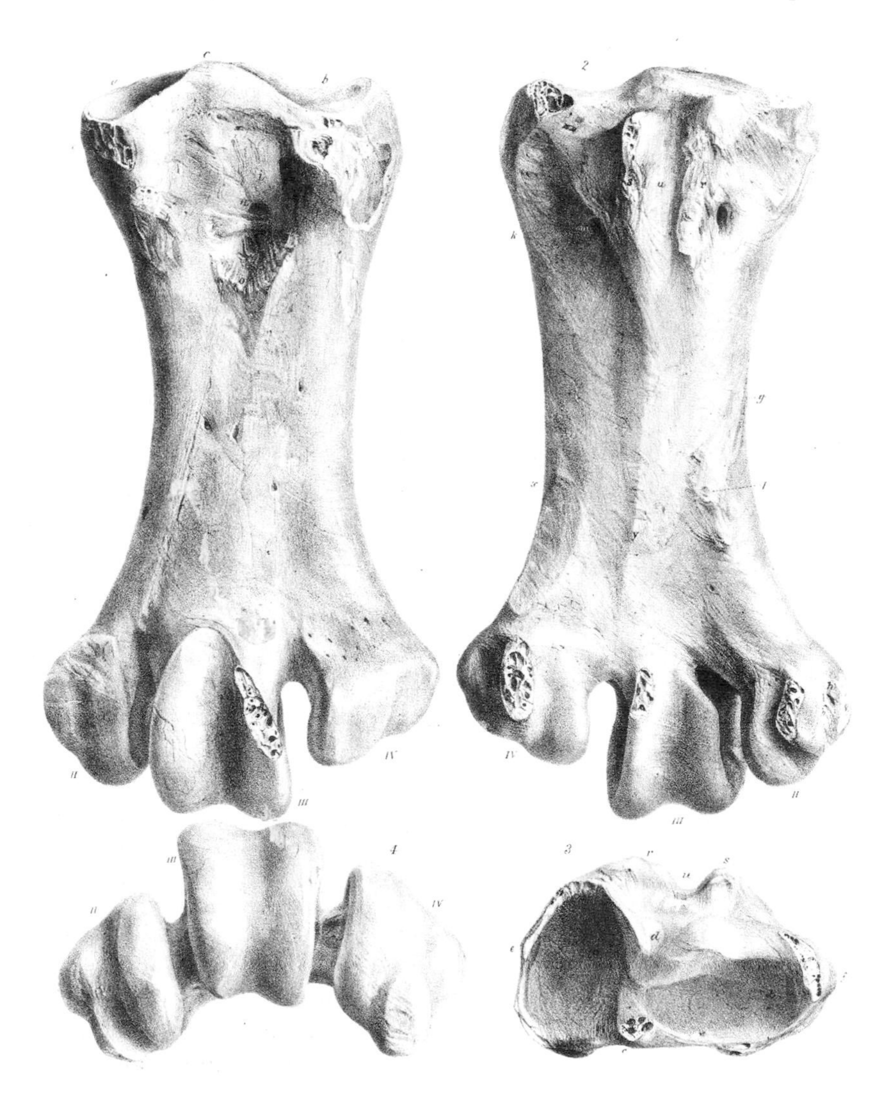

10 Foot bones (metatarsi) of the Greater Broad-billed Moa. Lithograph by James Erxleben from the *Transactions of the Zoological Society of London*, Vol. 8 (1873).

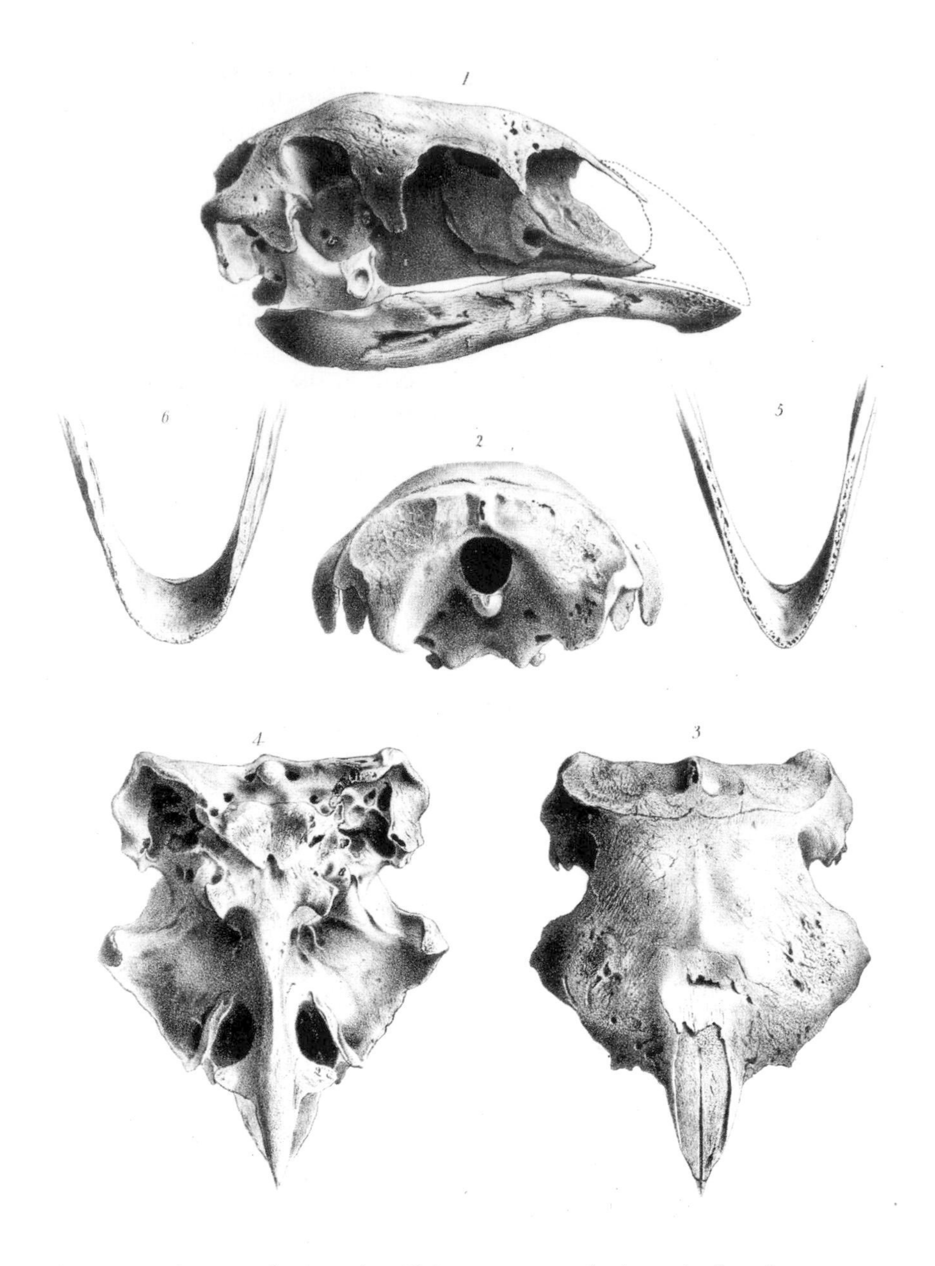

11 Skull of the Greater Broad-billed Moa. Lithograph by Joseph Smit from the *Transactions of the Zoological Society of London*, Vol. 7 (1870), Pl.14.

decline had set in during the Pleistocene epoch, long before humans set foot upon New Zealand shores.

Climatic changes are usually considered an important factor in the moas' extinction. During comparatively recent geological times – since the end of the last ice age – drier periods have been succeeded by periods of much heavier rainfall. There is no way of determining whether these changes in the environment affected moas. Perhaps they proved disastrous. Archey (1941) concluded that the combined effect of a wet climate and extension of the forests may have restricted opportunities for some kinds of moa. Changes in climate, though, have presumably been occurring throughout the entire period of moa existence.

The North Island of New Zealand is subject to violent volcanic activity, a factor occasionally advanced to explain the disappearance of some forms. Boiling mud and hot pools, geysers and thermal areas are characteristic of large areas; New Zealand's largest city, Auckland, is actually built on and around some sixty volcanoes. The most recent serious explosions took place in 1886 when the volcano Tarawera blew apart, destroying human life and also the famous Pink and White Terraces of Rotomahana, which, even at this early date, were attracting visitors from all over the world. During prehistoric times, explosions occurred in the Taupo and Rotorua areas on a scale that dwarfs all recent activity. There can be little doubt that thousands of moas must have perished beneath a rain of pumice, cinders and boiling mud as cataclysmic explosions tore and wrenched the land apart. Vast areas, devastated by the destructive power of the eruptions, would have been rendered unsuitable for occupation by moas for many years. Although this could not have caused extinction for any but very localized species, other kinds may have been reduced to levels making recovery difficult. Similar reductions in population levels may have followed extensive flooding at the end of the last ice age, about 12,000 years ago.

It seems possible that such reductions in numbers occurred many times during past ages and on each occasion moas were able to adapt and diversify according to the differing opportunities offered. Perhaps the ultimate factor in the extinction of the entire order was simply that never before were periods of adversity followed by an intense onslaught from formidable predators – people.

Although the heyday of moas may have been long past when man first chanced upon New Zealand, Polynesian invaders were responsible for the extirpation of the remnant population. In a land supporting no terrestrial mammals, meat was in short supply. It may have been possible to eke out a living from fish, small birds and whatever roots and other vegetable matter were available but two apparently more attractive food sources presented themselves. The new arrivals could eat giant birds or they could feed on each other. They did both.

LESSER MEGALAPTERYX

Megalapteryx didinus

PLATE II; FIGURES 12–13

Megalapteryx didinus
(Owen, 1883)

Dinornis didinus Owen, 1883 (Queenstown, South Island, New Zealand)
Megalapteryx hectori Haast, 1886
Megalapteryx tenuipes Lydekker, 1891
Megalapteryx hamiltoni Rothschild, 1907
Megalapteryx huttoni Rothschild, 1907

Height ~ 110 cm (3 ft 6 in)

Description
Appearance in life unknown.

By any standard, Richard Owen's concept of monstrous birds – built from a single piece of bone – represents a staggering piece of intuition. Yet there is little doubt that he could be a devious, some have said unpleasant, character and the deduction may not be quite so straightforward as is generally supposed. The idea of large, flightless birds occurring in New Zealand might not have been – to Owen – something entirely unfamiliar.

The famous bone fragment (see Figure 5) was examined during the autumn of 1839 but a year earlier a book published in London contained certain remarks which may have come to the great man's attention: 'That a species of emu, or a bird of the genus *Struthio* formerly existed in the island [North] I feel well assured.' These words were written by a Jewish merchant, Joel S. Polack, in a chronicle of travels and adventures in New Zealand between 1831 and 1837. If Owen carried this passage in mind while making his deduction, he cannot be said to have come completely fresh to the subject.

An intriguing feature of the history of the moas' discovery is the intense pride, rivalry and bitterness surrounding various stages of it. Argument raged over who was first to mention gigantic birds in print, or first to use

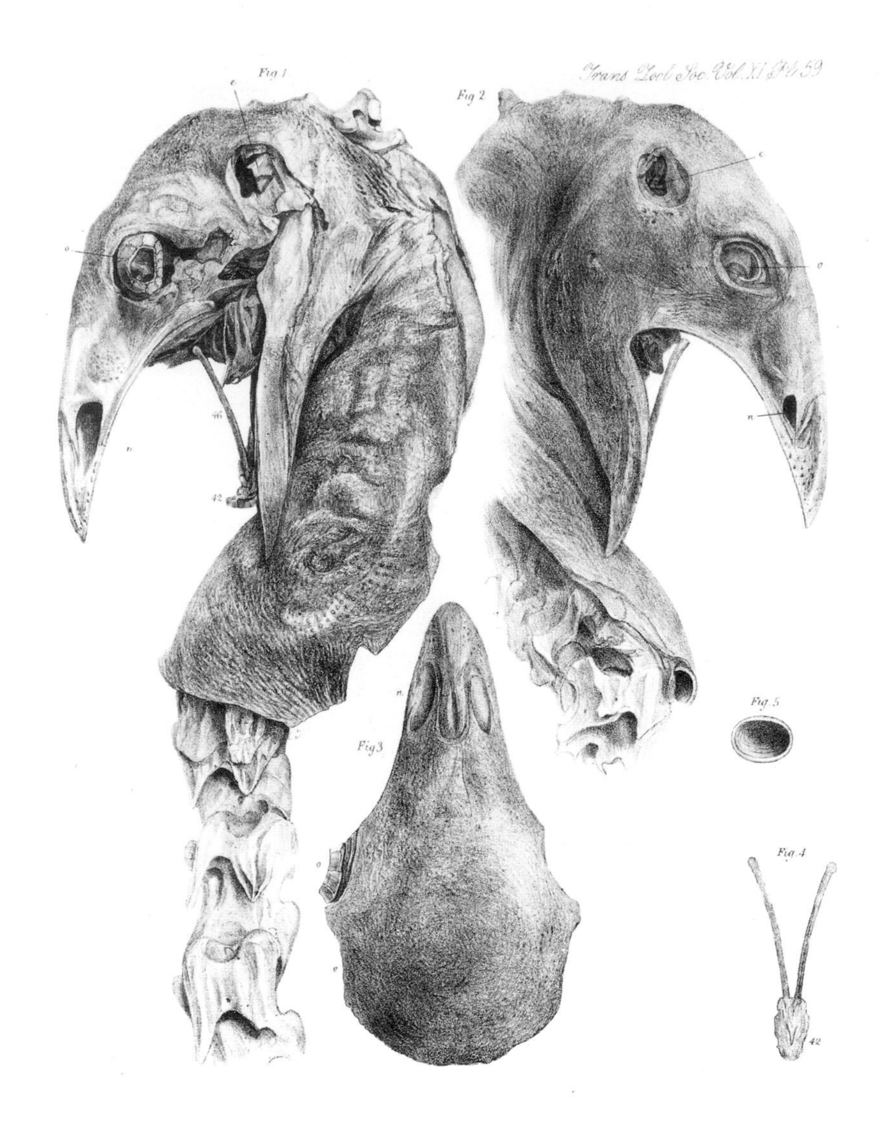

12 Dried head and neck of the Lesser Megalapteryx. Lithograph by James Erxleben from the *Transactions of the Zoological Society of London*, Vol. 11 (1883), Pl.59.

the word 'moa', who was first to find bones or to understand the significance of remains then being found.

No mention of gigantic birds had been noticed earlier than that of Polack, who traded along the New Zealand coasts from a small boat eventually wrecked near to East Cape, North Island. Here he was shown by the natives fossil bones brought from the inland mountain of Hikurangi and told how in times long past great birds had been known to the Maori. Polack (1838) speculated that such might still survive, 'in parts, which, perhaps, have never yet been trodden by man. Traditions are current among the elder natives of Atuas, covered with hair, in the form of birds, having waylaid native travellers, among the forest wilds, vanquishing them with an overpowering strength, killing and devouring.'

Soon after publication of Polack's book and Owen's pronouncement, a missionary printer based in New Zealand, the Revd William Colenso, became possessed by a fanatical determination to enter history as the man who discovered the moas; apparently working in collusion with two missionary friends, the Revd William Williams and the Revd Richard Taylor, he attempted to exclude all others from any part of the honour.

Colenso first heard stories of giant birds soon after Polack but paid little attention until the existence of such creatures became actual news. During

1838, accompanied by Revd Williams, he toured part of the East Cape district and later recounted (Colenso, 1843) that here he heard of:

> A certain monstrous animal; while some said it was a bird, and others a person, all agreed that it was called a Moa; – that in general appearance it somewhat resembled an immense domestic cock, with the difference, however, of its having a face like a man; – that it dwelt in a cavern in the precipitous side of a mountain; – that it lived on air; – and that it was attended, or guarded, by two immense tuataras, who, Argus-like, kept incessant watch while the Moa slept; also that if anyone ventured to approach the dwelling of this wonderful creature, he would be invariably trampled on and killed by it.

Such distorted tales of old, lone moas terrifying districts may have been current in many areas and appear to have impressed themselves upon Colenso as nothing more than myths – at least initially. Only later did he begin to realize something of real significance might lie behind them.

In November of 1841, well after Owen and Polack had had their say, Colenso left his base at the North Island's Bay of Islands to uncover whatever information he could about moas. On his travels he collected bones and left word with natives that additional finds should be forwarded to his colleague William Williams in Poverty Bay; Williams was himself soon amassing an embarrassment of bones as the natives took advantage of whatever trading opportunities presented themselves. In April 1842, Colenso, now back at the Bay of Islands, began to set his collection in order and establish his priority; since both Owen and Polack had already – as it were – registered their claims, desperate and uncharitable methods were resorted to.

The Jewish merchant Polack was clearly open to Christian attack. His book was declared a hoax – a hotchpotch of writing assembled from the work of others; not content with this broadside, Colenso asserted that Polack couldn't write anyway (able to write or not, Polack's book preceded any work of Colenso's upon the subject of moas by several years). With these two claims behind him, neither having any bearing on the question of priority, Colenso still wished to niggle over details. Why, he required to know, if a Jewish merchant had really seen bones, had that man not cheated the natives of them and later traded at a profit? Such reasoning on the merchant's trading habits was actually very badly founded; Polack must have known very well what he could trade and for how much – large ugly bones presumably did not figure very high on the lists. With better connections or greater understanding he may have thought otherwise, but it is plain to see why these particular relics lacked immediate attraction as objects of merchandizable quality.

Although Polack may have been without power or influence to defend himself, Owen certainly was not. Colenso could only complain that Owen's memoir was unknown in the Colony – certainly as far as he and his associates were concerned – but the anatomist's supporters soon made it clear that the famous paper had received coverage in every New Zealand newspaper. Moreover, Owen had set aside a hundred copies for distribution in the Colony, hoping to stimulate further discovery – a communication expressly designed to come to the attention of traders and missionaries. It hardly seems possible that such a procedure failed altogether in its purpose or that Colenso – a printer by trade – did not read the newspapers.

13 Dried foot of the Lesser Megalapteryx. Lithograph by James Erxleben from the *Transactions of the Zoological Society of London*, Vol. 11 (1883), Pl.61.

II Lesser Megalapteryx. Chromolithograph after a painting by an unknown artist from W. Rothschild's *Extinct Birds* (London, 1907), Pl.41. Courtesy of The Hon. Miriam Rothschild.

The implication of the defence made by Owen's faction was obvious. Colenso had not discovered moas, he had come to the truth about giant birds through the agency of Polack's book and Owen's memoir.

On being so regarded, a furious Colenso hit back spitefully but with little new ammunition. The large accumulation of bones was packed in wicker baskets and shipped to England – not to Owen but to William Buckland, Dean of Westminster, and W. J. Hooker, Director of the Royal Botanic Gardens, Kew. It is an amusing irony that upon receipt of the bones, Buckland and Hooker, little interested in the squabbling that raged in the far-distant dominion, immediately called in the man best qualified to assess them – Richard Owen (Figure 14). With these remains – so carefully collected by self-proclaimed rivals – in hand, Owen produced another paper on the giant flightless birds of New Zealand, founded the genus *Dinornis* and was able to dispel once and for all any lingering doubts over his ability to interpret an osseous fragment.

Colenso, self-styled 'Father of the Moa', has, as his only legitimate claim on posterity, the fact that he was first to use the word; his role in securing bones and generating interest was also, of course, not inconsiderable.

Surprisingly perhaps, even among Colenso's Christian brotherhood there was not total unity, for one of his own associates was to lay claim to the title Colenso so desperately desired. More than 30 years after the events described, the Revd Richard Taylor wrote that he and no other could, 'Justly claim to have been the first discoverer of the Moa.'

The facts, according to Taylor, were that a year after Colenso's initial tour of East Cape he – Taylor – participated in a similar excursion. While Colenso did not this time take part, the Revd Williams again went along and Taylor recalled the relevant details of their journey together. *He* (Taylor) and no one else, had first noticed a piece of bone lodged in a native roof at Waiapu near Poverty Bay. *He*, upon his own initiative had removed it, immediately on doing so noticing the 'cancellated' structure – a peculiarity quickly brought to the attention of his companion. *He*, with this structure in mind, asked his fellow traveller there and then whether this object was not the bone of a bird. The Revd Williams, so explained Taylor, knew nothing of moas – a surprising deficiency in view of Colenso's claim for the trip a year earlier. Unlike Jewish trader Polack, the Christian missionary did deprive the natives of their treasure, making an exchange for a small quantity of tobacco!

Maori traditions relating to moas were widespread and probably several species were known to the Polynesian invaders of *Aotearoa* – Land of the Long White Cloud. What is strange is that Maori tales and legends took so long to come to attention.

In 1642, when the Dutch navigator Abel Janszoon Tasman sailed east in the *Heemskirk* from the island bearing his name, he crossed uncharted seas for more than a thousand miles before sighting the 'great, high and bold land', which he was to call, rather inappropriately, *Nieuw Zeeland*. Dismayed by the hostility of the natives prancing upon the shore, Tasman sailed on without making a proper landing.

Setting aside vague rumours to the contrary, no European ship is known to have visited these shores again until in November 1769 Captain Cook's *Endeavour* sailed into New Zealand waters. Cook, not so easily deterred as Tasman, landed and made contact with the natives. A story was at one time current telling how a huge bird stood on the shore as the *Endeavour* dropped anchor – a creature that disappeared into the bush as a small boat put out. This appears to be a piece of fiction since no mention is

14 Richard Owen with his granddaughter, *c.* 1880.

made of such an incident in official records; nor was any mention made of giant birds by the Maoris with whom Cook and his men communicated. Not until Polack – almost 70 years later – did Europeans record Maori tales of these creatures, but then the stories flooded out.

In 1844, Robert Fitzroy (once captain of the *Beagle*, now Governor of New Zealand) interviewed an ancient Maori named Haumatangi who claimed to have seen, when only a boy, Captain Cook and – 2 years earlier than this – the trapping of a moa. Another aged Maori, Kawane Paipai, recalled taking part in moa hunts on the plains of Waimate during the final years of the eighteenth century. He remembered the birds being hounded, encircled and then speared to death, sometimes with weapons designed to snap easily once the body was stuck. Trapped moas defended themselves vigorously with terrible blows from their feet but while administering these, the monstrous bipeds were forced temporarily to support their weight upon one leg. A party of hunters would launch a frontal attack – a feint – while another crept behind waiting for the moment when the Moa raised a leg; then the party attacking from the rear would strike, knocking away the supporting leg. Once down, the victim was either despatched immediately or such grievous wounds were inflicted that the final outcome was no more in doubt.

One Maori hunting story, if true, may concern individuals belonging to the species *Megalapteryx didinus*. Sir George Grey, once Prime Minister of New Zealand, was told at Preservation Inlet in 1868 of the recent capture and killing of a small moa, taken from a drove of six or seven.

Remains of *Megalapteryx didinus* have been found in a surprisingly fresh condition, especially in the caves of Otago; three skeletons were listed by W.R.B. Oliver (1949) as of particular interest. A dried head, neck and legs with skin, ligaments and feathers attached were found in a cave near Lake Wakatipu in 1878. Purchased by the British Museum, these remains were described by Owen as his type of *didinus*. Although the feathers remaining are few in number, it is apparent that in living birds they were present from the base of the beak right down to the toes. A dried left leg, once in the Otago Museum and presumably still there, was found during 1895 in a cave at the head of the Waikaia River. Flesh, skin and feathers are attached, again indicating feathering to the toes. An almost complete skeleton from Cromwell, now in the Dominion Museum, Wellington, lacks feathers but has skin and flesh on the head and neck with the right eye preserved; ligaments are still joined to the leg bones. Feather pits in the skin show that the top of the head was covered with small feathers and the neck with larger ones. Feathers show an open structure, lacking barbicles and giving a very hair-like appearance; some are greyish brown in colour, some have a rufous tinge and others are tipped with white.

Oliver (1949) tentatively restricted this species to the South Island where it appears to have been limited in range and in numbers. Archey (1941) suggested that its strongholds were on the Takaka tableland and in western Otago, and inferred that this was a bird of the high country.

Hopes are occasionally expressed that the remote areas of the South Island may harbour a few surviving moas; and truly in parts this is a formidable wilderness. If the birds dwelt only in the depths of the forest, if they were strictly nocturnal and, above all, if they were by nature silent, secretive creatures, then perhaps a few could exist unnoticed, but such hopes are hardly realistic.

'We are lost as the Moa is lost' run the words of a Maori lament – *Ka ngaro i te ngaro a te Moa.*

TINAMOUS, PENGUINS, DIVERS AND GREBES

ORDERS Tinamiformes, Sphenisciformes, Gaviiformes and Podicipediformes

Avian classifiers usually place the orders Tinamiformes (tinamous), Sphenisciformes (penguins), Gaviiformes (divers) and Podicipediformes (grebes) after the ratites. These orders contain no known species or subspecies that appear recently extinct.

Classed among the tinamous, however, is at least one very vulnerable species. The Magdalena Tinamou (*Crypturellus saltuarius*) is known only from the type specimen, taken on the northwestern foothills of the Sierra de Ocaña, Colombia in 1943. No further individuals have been collected and, with the area now largely deforested, fears concerning the species' survival cannot be thought groundless.

Although penguin species all seem relatively safe, the still-substantial population of Jackass Penguins (*Spheniscus demersus*) has plunged dramatically over the past 150 years and a spiral of decline may have set in that will see the final eclipse of this bird before the demise of other, presently much rarer, species.

The status of the four species of diver seems secure but several grebe species are endangered (see Table 2). One of the most vulnerable is the Madagascan Red-necked Grebe (*Tachybaptus rufolavatus*), a bird threatened in a rather unusual way. The closely related Little Grebe (*T. ruficollis*) has recently become widespread in Madagascar and wherever its spreading range brings it into contact with *T. rufolavatus*, hybrids are produced. Madagascan Red-necked Grebes may very well be bred out of existence entirely in future years.

Table 2

RARE OR ENDANGERED GREBES
Colombian Grebe *Podiceps andinus*
Hooded Grebe *Podiceps gallardoi*
Junin Grebe *Podiceps taczanowskii*
Giant Pied-billed Grebe *Podilymbus gigas*
Madagascan Red-necked Grebe *Tachybaptus rufolavatus*
Madagascar Little Grebe *Tachybaptus pelzelni*

ALBATROSSES AND PETRELS

ORDER
Procellariiformes

EXTINCT SPECIES

Guadalupe Storm Petrel (*Oceanodroma macrodactyla*)

The order Procellariiformes consists of albatrosses, shearwaters, storm and diving petrels or, as they are sometimes all picturesquely named, the tube-nosed swimmers. There are about a hundred species, all with a particularly noticeable physical feature in common – the peculiar tubing that extends onto the bill and encases the nostrils.

One full species and one subspecies appear extinct; the Guadalupe Storm Petrel (*Oceanodroma macrodactyla*) was lost during the early years of this century and the Jamaican Diablotin (*Pterodroma hasitata caribbea*) vanished rather earlier – around 1880. This extinct, dark form of the Diablotin was known to Jamaicans as the Blue Mountain Duck and was virtually exterminated by hungry humans. The introduction of mongooses (*Herpestes auropunctatus*) to the island in 1872 probably signalled its final destruction. The remaining subspecies, *P. h. hasitata*, is endangered but still breeds in some numbers on Hispaniola.

Among rare species in this order (see Table 3) is the legendary Cahow (*Pterodroma cahow*). This bird provides one of the most celebrated instances of the rediscovery of a supposedly extinct species. The Cahow was known to colonists in Bermuda during the early part of the seventeenth century, these people relying quite heavily on the birds as food items. Probably as a direct result of their persecutions the Cahow disappeared on the island long before the century's end. Then in 1951, after not very much short of 300 years' presumed extinction, Cahows were found again breeding on five islets off the Bermuda coast. The species is very rare even so, with perhaps no more than thirty pairs existing.

Rediscoveries in this group are by no means unique as is shown by the recent finding of two more 'lost' birds – Macgillivray's Petrel (*Pterodroma macgillivrayi*) and the Chatham Island Taiko (*P. magentae*).

Pterodroma macgillivrayi was described from a single specimen – a fledgling collected on Ngau Island, Fiji in October 1855 – and this remains

Table 3

RARE OR ENDANGERED ALBATROSSES AND PETRELS
Short-tailed Albatross *Diomedea albatrus*
Black Petrel *Procellaria parkinsoni*
Westland Black Petrel *Procellaria westlandica*
Bird of Providence *Pterodroma solandri*
Réunion Petrel *Pterodroma aterrima*
Barau's Petrel *Pterodroma baraui*
Cahow *Pterodroma cahow*
Black-capped Petrel (Diablotin) *Pterodroma hasitata*
Macgillivray's Petrel *Pterodroma macgillivrayi*
Chatham Island Petrel *Pterodroma axillaris*
Chatham Island Taiko *Pterodroma magentae*
Dark-rumped Petrel *Pterodroma phaeopygia*
Heinroth's Shearwater *Puffinus heinrothi*

the only skin in scientific collections. The colony from which it came has been searched for several times but without success. In 1984, the British naturalist Dick Watling, resident in Fiji and author of a book on Fijian birds, claimed to have evidence of the species' continuing existence. According to his account in a local newspaper, and following a year-long search, a petrel belonging to the species crashlanded on his head after being lured in from sea by night lights. After recovery and examination the bird was released.

The Taiko, a gadfly petrel from the Chathams in the South Pacific, was well known to islanders during the nineteenth century, being used by them as a food resource. To science, however, it was with certainty known only by skeletal remains uncovered in recent deposits. Apparently absent as a skin from museum collections, this species seemed gone also from its island home by the beginning of the twentieth century, although reports of occasional sightings persisted. In 1978, two individuals were captured on the Chathams, measured, photographed and released.

This rediscovery seems to have helped clear up another ornithological mystery. An enigmatic petrel form was described from a single specimen taken south of Pitcairn Island in the South Pacific in 1867. It was never again encountered in this area or for that matter anywhere else. This particular form went into ornithological literature as the Magenta Petrel (*Pterodroma magentae*) but always remained something of a puzzle. Even though the type locality for it is several thousand miles from the Chathams, a suspicion was long voiced that the Magenta Petrel and the Chatham Island Taiko are one and the same. Measurements taken during the recent capture of taikos appear to confirm this.

III Guadalupe Storm Petrel. Hand-coloured lithograph by J.G. Keulemans from F. du C. Godman's *Monograph of the Petrels* (London, 1907–10), Pl.5A.

GUADALUPE STORM PETREL

Oceanodroma macrodactyla

PLATE III

Oceanodroma macrodactyla
(Bryant, 1887)

Oceanodroma leucorhoa macrodactyla Bryant, 1887 (Guadalupe Island off the coast of Baja California)

Length 21 cm ($8\frac{1}{4}$ in)

Description
Adult: upperparts mostly sooty black with glossing of plumbeous grey; scapulas and lesser wing coverts black; median and greater series drab brown; primary coverts black; upper tail coverts white with broad black tips; tail feathers black with narrow white edging at base of outermost; undersurface of body sooty chocolate brown but showing a patch of white at sides of vent; bill and feet black. Sexes alike.

Measurements
Wing 157 mm; tail 92 mm; culmen 16 mm; tarsus 23 mm.

A little-known species of storm petrel – not very different from the more familiar kinds – once lived on the small Pacific island of Guadalupe, an insignificant area of land lying off the coast of Baja California some 320 km (200 miles) to the south-west of San Diego.

It was first discovered by Walter E. Bryant, whose name is also associated with the extinct caracara of Guadalupe (*Polyborus lutosus*), from specimens collected in January of 1885 but seems to have become extinct by the outbreak of World War I. Definite records of the species' existence later than 1911 are lacking so it may be assumed that soon after this date the birds were gone.

As the known nesting grounds were confined to Guadalupe, a very small island – some 32 km (20 miles) long by 9.6 km (6 miles) wide – the continuing presence of this bird is unlikely to have remained overlooked for long. Also, there are good reasons for supposing that man's interference with the island brought about changes that were detrimental to birdlife in general and likely to bring about the extinction of this species in particular. Goats were released onto Guadalupe more than 150 years ago (Lever, 1985) and these animals caused a fairly rapid deterioration in the environment. More particularly, cats were introduced to the island and it is this agency that probably brought about the end for the Guadalupe Storm Petrel. Cats are known to have combed the nesting areas; visitors to the island witnessed them roaming over the high places where the birds had their strongholds and at these times many small, mangled bodies were found.

Against these rather gloomy indications, it can fairly be said that the absence of sight records might not be of absolute significance because an almost identical, albeit slightly smaller, species – Leach's Storm Petrel (*Oceanodroma leucorhoa*) – occurs on Guadalupe and on the adjacent shores of California; if any true Guadalupe petrels survive they might easily be mistaken for birds of the more widespread species. The most important distinguishing mark between the two species is the paler underside to the wing shown in *Oceanodroma macrodactyla* but this would, of course, not often be detectable in the field.

Although confusion is certainly possible – even likely – the burrows of the Guadalupe Storm Petrel have remained undisturbed by their one-time owners for many decades now and this in itself seems an indication that the birds are gone. These burrows were carved out of soft soil under pines and oaks mostly along the steep northeastern ridge of the island. Although so small in size, Guadalupe climbs out of the ocean to a height of some 1,220 m (4,000 ft); the birds characteristically nested at altitudes of around 760 m (2,500 ft).

The species' population was probably never very large but a thriving group seems to have been well established. Breeding began in March, usually coming to an end during May. A single white egg, marked at its larger end with minute spots of faint reddish brown or pale lavender, was laid in the burrow on a bed of pine needles and leaves.

And little more is known of this bird. Bryant described the call as, 'here's a letter, here's a letter', with an answering refrain of, 'For you, for you', which seems as picturesque a way of finishing an account of one of the lost birds of Guadalupe as any other.

PELICANS
AND RELATED BIRDS

ORDER
Pelecaniformes

EXTINCT SPECIES

Spectacled Cormorant (*Phalacrocorax perspicillatus*)

The order Pelecaniformes is based, as might be expected, around the pelicans (Pelecanidae); but it contains several other bird families that might seem unconnected, although all share anatomical and behavioural characteristics. In the order are such diverse-seeming groups as the tropicbirds (Phaethontidae), gannets and boobies (Sulidae), cormorants and shags (Phalacrocoracidae), darters (Anhingidae) and frigatebirds (Fregatidae).

Just one species in the order is extinct – the Spectacled Cormorant (*Phalacrocorax perspicillatus*); there are no losses at subspecific level.

Several species within the order are threatened (see Table 4). One of the most interesting of these is the Galápagos Flightless Cormorant (*Nannopterum harrisi*), a very large cormorant that, in the isolation of the Galápagos Islands, has lost power of flight. It nests on only two islands where perhaps 2,000 individuals survive all told.

At present, the King Shag (*Phalacrocorax carunculatus carunculatus*) is the most seriously endangered subspecies in the order, but this is simply because it breeds over such a restricted area of the New Zealand coast. There is no evidence of any recent serious decline in numbers, which seem always to have been small. The total number of birds at the various colonies may amount to no more than 300.

Table 4

RARE OR ENDANGERED PELICANS AND RELATED FORMS

Dalmatian Pelican *Pelecanus crispus*	Christmas Frigatebird *Fregata andrewsi*
Abbott's Booby *Sula abbotti*	Ascension Frigatebird *Fregata aquila*
Galápagos Flightless Cormorant *Nannopterum harrisi*	

SPECTACLED CORMORANT

Phalacrocorax perspicillatus

PLATE IV

Phalacrocorax perspicillatus (Pallas, 1811)

Phalacrocorax perspicillatus Pallas, 1811 (Bering Island)
Graculus perspicillatus Elliot, 1869
Pallasicarbo perspicillatus Coues, 1869
Carbo perspicillatus Rothschild, 1907

Length 97 cm (38 in)

Description
Male: colour of naked area around base of bill varied as in the Turkey with vermilion, blue and white; thick skin forming 'spectacles' around the eyes white; double crest projecting from occiput, greenish blue; rest of head dark greenish blue but decorated with long, hair-like pale-yellow feathers that extend onto the upper part of the neck; body and wings deep bronze green showing steel-blue reflections on neck and with a large whitish patch on each flank; tail black.
Female: similar to male but smaller; apparently also lacking crest and spectacles.

Measurements
Wing 342 mm; tail 180 mm; culmen 74 mm; tarsus 72 mm.

IV Spectacled Cormorant. Hand-coloured lithograph by Joseph Wolf from D.G. Elliot's *New and Heretofore Unfigured Species of the Birds of North America*, Vol. 2 (New York, 1869), Pl.50.

One of the saddest yet most fruitful voyages of zoological discovery was undertaken by the young German naturalist Georg Wilhelm Steller aboard Russian vessels commanded by the Dane Vitus Bering. The main purpose of this multinational expedition to the far north was to determine whether Asia was joined by a land bridge to North America, but among its lasting contributions to the record of human endeavour is the written account of the voyage left behind by the youthful ship's doctor who doubled as naturalist.

Following an edict of Tsar Peter the Great, Bering sailed from Petropavlorsk in Kamchatka, through the sea that now bears his name and eventually made the Alaskan coast; Steller became the first naturalist ever to set foot upon Alaskan soil. On his return to the vessel, Bering brewed a cup of chocolate – a rare and choice commodity – in recognition of the event. But almost immediately, and rather against Steller's wishes, the Danish commander ordered a return to Russian soil. The weather had deteriorated and scurvy broke out almost as soon as they were under sail. Bad water, taken on board from the Shumagin Islands, added to the expedition's misfortunes and the debilitated crew were hardly able to work the ship.

On 5 November 1741, after running through the Aleutian Archipelago before the storm, the ship *St Peter* encountered an unknown piece of land that came to be known as Bering Island (Ostrov Beringa). In desperate straits, the men aboard elected to abandon ship and go ashore. A week later a northeasterly gale piled the stricken ship on the beach and it broke up leaving the crew and scientific staff stranded on this inhospitable and hitherto uninhabited island.

Just over a month later, Bering died of cold, exhaustion and scurvy complicated by other diseases and in the following weeks many crew members also succumbed; the total death toll for the entire expedition was to reach thirty. Steller managed to rally the remaining men, organize the building of a new boat and, after enduring months of privation, the survivors regained Kamchatka.

For Steller, however, the story had no happy ending. He failed to get back to Europe and for 4 years wandered through Siberia enduring, among other hardships, a period of wrongful imprisonment, until, worn out by his exertions, he died of fever at Tyumen on 12 November 1746.

Steller's record is noteworthy not just for its documentation of spirit and determination in the face of great adversity. It is also an amazing account of the observation of little-known species and a remarkable number of creatures are catalogued, then new to science.

The most celebrated of his 'finds' was encountered in the waters around Bering Island itself – the enormous sirenian that came to be known as Steller's Sea Cow (*Hydrodamalis* (*Rhytina*) *stelleri*) (Figure 15). Steller was the only naturalist ever to observe this extraordinarily docile and touching 915-cm (30-ft) long creature; within 25 years or so of the return to Kamchatka, fur trappers and sealers had wiped out the defenceless mammals in their home waters leaving only a remnant of displaced individuals to wander the seas of the far north. Scattered and aimless, these lost sea cows are thought to have been entirely gone by the early 1800s.

Among the remainder of Steller's remarkable discoveries is one that

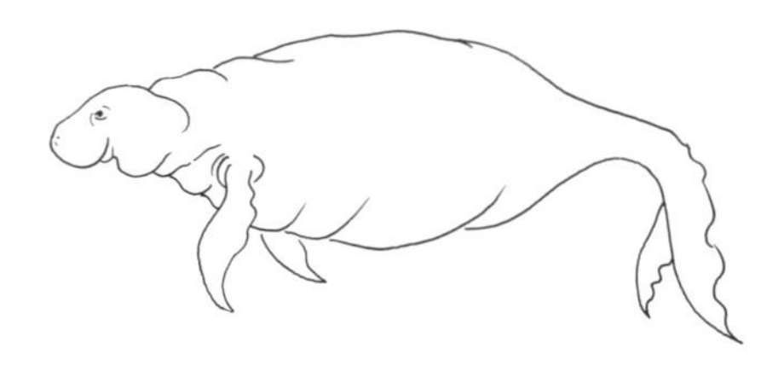

15 Steller's Sea Cow (*Rhytina stelleri*), an extinct sirenian.

parallels the story of the sea cow to an extraordinary degree. On Bering Island not only did Steller find giant sirenians, he also found a very large cormorant with white markings around the eyes reminiscent of spectacles.

> In size and plumpness they exceed the allied species and since the flesh of one would easily satisfy three hungry men [the birds weighed between 5.4 and 7.3 km (12–14 lb)] they were a great comfort . . . From the ring around the eyes, and the clown-like twistings of the neck and head, it appears quite a ludicrous bird.

Although he found them very numerous, nowhere but on Bering did Steller encounter these birds. Later, they were found also to occur on nearby Commander Island (Komandorskiye Ostrova) and they probably lived on other islands in the same general area.

As is the case with the sea cow, Steller's is the only description given by a naturalist of the creatures in life. During the 100 years after Bering's death, the Spectacled Cormorant was systematically exterminated; by the middle of the nineteenth century it was probably extinct. Since it appears to have been thriving when Steller met with it, the loss can only be attributed to human agency. Sealers and other hunters who visited the area (often in pursuit of yet more of Steller's discoveries) caught and ate them, and Aleuts, imported to the islands by commercial interests during the 1820s, made cormorant flesh their principal diet when other meat was difficult to obtain. Apparently, the birds sometimes straggled to Kamchatka and the natives had a method of cooking them whole – feathers and all – in clay.

Its inexperience of man, together with its slowness of motion on land made the Spectacled Cormorant easy prey; added to this, its short wings rendered it almost flightless. In 1882, the ornithologist L. Stejneger, then engaged in an unsuccessful search for the cormorant, was told by natives that the last stronghold of the birds was a small island by the name of Aij Kamen where they had been seen 30 years previously.

Surviving specimens of the Spectacled Cormorant are very rare; according to J.C. Greenway (1958), they are to be seen only in the British Museum (Natural History) and the museums of Leningrad, Leiden, Helsinki and Dresden. The British Museum has two skins, one perfect, the other lacking the tail. One of these was given to Captain Edward Belcher in 1939 by the Russian Governor of the Sitka district – Kuprianof; presumably this is the bird figured by John Gould as plate 32 of *The Zoology of the Voyage of H.M.S. Sulpher Under the Command of Captain Sir Edward Belcher, 1836–42* (1843–4). All the known examples seem to have come from the same source – Governor Kuprianof – and were apparently all traded for or received as gifts during the 1830s and 1840s.

In many of the published illustrations of this species, the white 'spectacles' giving the species its name are coloured wrongly, which may cause confusion to anyone delving into the old literature.

OTHER PUBLISHED ILLUSTRATIONS OF THE *SPECTACLED CORMORANT*

Hinds, R.B., 1843–4. *The Zoology of the Voyage of H.M.S. Sulphur Under the Command of Sir Edward Belcher, 1836–42*, pl. 32. Artist, J. Gould.

Rothschild, W. 1907. *Extinct Birds*, pl. 39. Artist, J.G. Keulemans.

HERONS, STORKS AND RELATED BIRDS

ORDER
Ciconiiformes

EXTINCT SPECIES

Rodrigues Night Heron (*Nycticorax megacephalus*)

The order Ciconiiformes is a collection of bird families consisting of herons (Ardeidae), storks (Ciconiidae), ibises and spoonbills (Threskiornithidae) and flamingos (Phaenicopteridae). Also included within the order are two distinct species, each placed in families of their own – the Hammerkop (*Scopus umbretta*) in Scopidae and the Whale-headed Stork (*Balaeniceps rex*) in Balaenicipitidae.

Partly because of their specialized feeding and habitat requirements and partly, perhaps, because of their generally large size, many species are subject to increasing pressure as human population and technological ability grow. Despite threats to existing birds, however, only one full species and three subspecies have become extinct in recent times. The extinct species is the Rodrigues Night Heron (*Nycticorax megacephalus*), the extinct subspecies the Bonin Nankeen Night Heron (*N. caledonicus crassirostris*), the New Zealand Little Bittern (*Ixobrychus minutus novaezelandiae*) and the Princípe Olive Ibis (*Lampribis olivacea rothschildi*).

The Rodrigues Night Heron was an early casualty of the Mascarene avifauna and particularly interesting for a heron in that it may have been flightless. Another species in the same genus, the widespread Nankeen Night Heron (*N. caledonicus*), has lost the race *crassirostris* formerly an inhabitant of the Bonin (Ogasawara) Islands to the south of Japan.

Like the Mascarenes, the Bonin Islands have lost a comparatively large number of forms. Bonin night herons, distinguished from others of their species by thicker, straighter bills, were discovered on Peel Island (Chichi-jima) during the voyage of HMS *Blossom* in the 1820s and last collected in 1889. The rather mysterious Little Bittern of New Zealand (*Ixobrychus minutus novaezelandiae*), known only from a few specimens taken during the nineteenth century, is in all probability extinct. The individuals associated with this form have sometimes been considered as a species in their own right, sometimes as a well-marked race of the widespread Little Bittern

Table 5

RARE OR RARELY SEEN HERONS AND RELATED FORMS		
New Guinea Tiger Heron *Zonerodius heliosylus*	Swinhoe's Egret *Egretta eulophotes*	Bald Iris *Geronticus calvus*
African Tiger Heron *Tigriornis leucolophus*	Slaty Egret *Egretta vinaceigula*	Waldrapp *Geronticus eremita*
Zigzag Heron *Zebrilus undulatus*	Madagascan Pond Heron *Ardeola idae*	Madagascan Crested Ibis *Lophotibis cristata*
White-eared Night Heron *Gorsachius magnificus*	Imperial Heron *Ardea imperialis*	Japanese Crested Ibis *Nipponia nippon*
White-backed Night Heron *Gorsachius leuconotus*	Madagascan Heron *Ardea humbloti*	White-shouldered Ibis *Pseudibis davisoni*
	Milky Stork *Mycteria cinerea*	Giant Ibis *Thaumatibis gigantea*

and sometimes merely as wind-blown vagrants belonging to the subspecies that occurs in Australia. Because of the extreme scarcity of specimens it is difficult to assign this form meaningfully but it seems clear that New Zealand birds were distinct enough from those generally found in Australia to justify subspecific ranking at least. Recently, however, individuals have been found in Australia that are said to be very similar to those known to have occurred in New Zealand a century ago.

An ibis, *Lampribis olivacea rothschildi*, vanished from Principe in the Gulf of Guinea during the first half of this century. The species to which it belonged, the Olive Ibis, is still widely distributed in Africa.

Of threatened species within the order (see Table 5), the Japanese Crested Ibis (*Nipponia nippon*) can be singled out as one of the rarest birds in the world and a recent survey (1976) on Sado Island off the west coast of Japan, which may be the very last stronghold, numbered the population at only eight individuals. Perhaps it also survives in very small numbers in eastern Manchuria and Siberia.

RODRIGUES NIGHT HERON

Nycticorax megacephalus

Nycticorax megacephalus (Milne-Edwards, 1874)

Ardea megacephala Milne-Edwards, 1874 (Rodrigues)
Nycticorax megacephala Günther and Newton, 1879
Megaphoyx megacephala Hachisuka, 1937

Length ~ 60 cm (24 in)

Description
Appearance in life unknown.

It is possible to be quite definite about a species of heron that once inhabited the Mascarene island of Rodrigues in the Indian Ocean (Figure 16). Two early sources mention it, the first account being given by the Huguenot refugee François Leguat in his journal published in English during 1708. He mentioned 'Bitterns as big and good as capons'.

According to the journal these were remarkably bold and confiding birds. Leguat tells how he and his companions formed a sentimental attachment to some quite tame lizards, allowing them sometimes to feed from their table. For entertainment, the Huguenots often shook these reptiles down from the trees but they then had difficulty in offering them protection from the herons, so aggressive and determined were the birds in pursuing dainty morsels.

An anonymous author, in the document known as the *Relation de l'Île Rodrigue*, also mentioned herons on the island. These he described as the size of an egret with only very poor powers of flight but having the ability to run well.

The population almost certainly disappeared during the eighteenth century as no later traveller mentions similar birds and the anonymous *Relation* is thought to have been written around 1730. Since this time, however, skeletal remains of herons have been found in Rodrigues, which, presumably, can be linked with the earlier reports. A. Milne-Edwards (1874), in reviewing bone fragments, felt able to assign the Rodrigues species to the genus *Ardea*.

Five years later, A. Günther and E. Newton, working with more complete material, expressed themselves in no doubt that this was a night heron and assigned it to *Nycticorax*. These authors pointed out that the bill of the Rodrigues birds was stronger than that of the familiar European Night Heron (*Nycticorax nycticorax*) and the metatarsus more strongly developed although in overall size the birds were not dissimilar. The length of the wing was considerably reduced, which, together with the conformation of the sternum and the stoutness of the legs, indicates that this was a species losing, or maybe already without, the power of flight. It is assumed that prey was chased on land rather than hunted in more typically heron-like fashion.

Nothing more is known of this species but about other extinct Mascarene herons there is even more uncertainty. Although populations that no longer exist seem to have inhabited all three Mascarene islands (Rodrigues, Réunion and Mauritius) during the not-too-distant past, those of Mauritius and Réunion are very difficult to classify.

An account of herons living in Réunion was given by a travelling gentleman, a M. Dubois, in a journal of his stay in the island from 1669 to 1672. In size, these birds were as big as chickens, their colour was grey – each feather spotted with white – and their legs were green. To this description Walter Rothschild (1907) appended the name *Ardea duboisi*, but the value of this action is quite uncertain. Dubois' account is probably truthful enough but whether he saw an otherwise unknown bird is not clear. J.C. Greenway (1958) pointed out that the description could conceivably apply to *Egretta gularis dimorpha*, the Madagascan race of the Western Reef Heron. Nothing can be added to the Dubois account.

Although bones referable to an extinct species of heron have never been found in Réunion, from Mauritius have come skeletal remains that cannot be aligned with any living species of heron. In the Mare aux Songes, Mauritius, bones were discovered on the basis of which the species *Butorides mauritianus* was founded. These remains are not sufficiently extensive for any particular genus to be realistically proposed, although both *Ardea* and *Nycticorax* have been. Whatever the population was, no early accounts have been found relating to it and it may – although not necessarily – have vanished at a comparatively distant date.

16 The Indian Ocean island of Rodrigues. Foldout papyrograph from H.E. Strickland and A.G. Melville's *The Dodo and its Kindred* (London, 1848).

WATERFOWL

ORDER
Anseriformes

EXTINCT SPECIES

Labrador Duck (*Camptorhynchus labradorius*)
Auckland Islands Merganser (*Mergus australis*)
Pink-headed Duck (*Rhodonessa caryophyllacea*)
Korean Crested Shelduck (*Tadorna cristata*)

The waterfowl, the collective name for swans, geese and ducks, have always caught people's attention. Partly perhaps because of the tastiness of their flesh, but also because their mode of life makes them easy prey, waterfowl have long been sought after by those who enjoy shooting birds. Probably only the raptors and the gamebirds have attracted similar interest. Nevertheless, many kinds of waterfowl remain relatively unscathed by this persecution; it is drainage of wetlands that contributes more seriously to the decline of many species.

Waterfowl can often be successfully bred in captivity and this is certainly one reason why out of almost 150 species extant in the nineteenth century only four now seem extinct – the Korean Crested Shelduck (*Tadorna cristata*), the Pink-headed Duck (*Rhodonessa caryophyllacea*), the Labrador Duck (*Camptorhynchus labradorius*) and the Auckland Islands Merganser (*Mergus australis*). Furthermore, at least four otherwise plentiful species appear to have lost races: the Bering Canada Goose (*Branta canadensis asiatica*), Coues' Gadwall (*Anas strepera couesi*) the Rennell Island Grey Teal (*A. gibberifrons remissa*) and the Niceforo Brown Pintail (*A. georgica niceforoi*).

At one time it seemed as if two subspecies of the widespread Canada Goose had become extinct. *Branta canadensis asiatica*, the Bering Canada Goose, once bred upon the Kurile (Kuril'skiye) and Commander (Komandorskiye) islands off northeastern USSR but only five specimens now exist and these are divided between the American Museum of Natural History, New York and the US National Museum (Smithsonian Institution), Washington, DC. Because of the small number of known examples it has proved difficult to assess the validity of this form. R.S. Palmer (1976) and W.B. King (1981) regarded it as inseparable from *B. c. leucopareia*, the Aleutian Canada Goose, which is itself endangered. J. Delacour and P. Scott (1954–64) remarked that the series of specimens of *asiatica* are

Table 6

RARE WATERFOWL OR SPECIES WITH UNSTABLE POPULATIONS

Cuban Whistling Duck *Dendrocygna arborea*
Trumpeter Swan *Cygnus buccinator*
Swan Goose *Anser cygnoides*
Snow Goose *Anser caerulescens*
Ross's Goose *Anser rossi*
Hawaiian Goose (Nene) *Branta sandvicensis*
Red-breasted Goose *Branta ruficollis*
Cape Barren Goose *Cereopsis novaehollandiae*
Freckled Duck *Stictonetta naevosa*
Blue-winged Goose *Cyanochen cyanopterus*
Ruddy-headed Shelduck *Chloephaga rubidiceps*
Paradise Shelduck *Tadorna variegata*
Radjah Shelduck *Tadorna radjah*
White-winged Wood Duck *Cairina scutulata*
Blue Duck *Hymenolaimus malacorhynchos*
Madagascan Teal *Anas bernieri*
Chestnut Teal *Anas castanea*
New Zealand Brown Teal *Anas aucklandica*
Madagascan Mallard *Anas melleri*
Philippine Duck *Anas luzonica*
Marianas Mallard *Anas oustaleti*
Pink-eared Duck *Malacorhynchus membranaceous*
Canvasback *Aythya valisineria*
Redhead *Aythya americana*
Madagascan White-eyed Duck *Aythya innotata*
New Zealand Scaup *Aythya novaeseelandiae*
Brazilian Merganser *Mergus octosetaceus*
Chinese Merganser *Mergus squamatus*
White-headed Duck *Oxyura leucocephala*
Australian Blue-billed Duck *Oxyura australis*

distinctly lighter in colour, especially on the underparts, but F.S. Todd (1979), too, now believes the race invalid. If the population from the Kurile and Commander islands was distinct then this subspecies became extinct around 1914. A giant form of the Canada Goose (*B. c. maxima*) was once listed extinct but subsequently rediscovered in Minnesota during 1962 and now exists in considerable numbers.

Coues' Gadwall (*Anas strepera couesi*) was a dwarf form of the Common Gadwall that once inhabited the Fanning (now the Tabuaeran) group of islands in the mid-Pacific. The race is known from just a pair of birds collected on the lake and in the peat bogs of Washington Island during January 1874. These are now in the US National Museum and no similar birds have since been recorded.

Another insular form from the central Pacific, the Rennel Island Grey Teal (*Anas gibberifrons remissa*), seems to have vanished after the introduction of a certain species of fish to Rennell's only lake. This subspecies of the Australasian Grey Teal, first described by S.D. Ripley in 1942, has not been reliably recorded since 1959.

A pintail from central Colombia, *Anas georgica niceforoi*, may be extinct. First described in 1946, this subspecies of the Yellow-billed Pintail has not been reliably recorded since 1952. Despite the fact that it appears to have been well known to local shooters only a small number of naturalists ever observed it.

Several extant species of waterfowl are rare and others are endangered or potentially threatened (see Table 6) even though large populations may still exist. Population levels among many species can be very unstable and these levels are liable to speedy reduction if prevailing conditions alter. Wetland drainage, habitat deterioration, subtle climatic alteration or excessive hunting can quickly effect a drastic fall in numbers. Although very few waterfowl appear to be facing imminent extinction, this situation could change rapidly. It is interesting that both the Labrador Duck and the Pink-headed Duck had reached the point of extinction almost before their decline was widely appreciated. The Canvasback (*Aythya valisineria*) and its close relation the Redhead (*A. americana*), each with a population of probably more than half a million individuals in North America, can hardly be called rare birds. Yet, according to P.A. Johnsgard (1978), the long-term security of both is in question. Vulnerability to hunting and sensitivity to pollution combined with human interference in their favoured habitats make their prospects bleak.

Similarly, two rather aberrant Australian species, the Pink-eared Duck (*Malacorhynchus membranaceous*) and the Freckled Duck (*Stictonetta naevosa*), might easily become endangered species in the near future. The highly nomadic Pink-eared Duck is likely to be found over a vast area but depends for its living upon floodwaters, which are, of course, ephemeral. As the inland waterways of Australia become increasingly under human control, such flooding is much less extensive and the consequences for this species could be disastrous. The chief stronghold of the Freckled Duck is the Murray Darling Basin of southeastern Australia. Although the area in which it occurs is very extensive, fewer and fewer of the permanent swamps that provide its breeding grounds remain available and a marked drop in numbers is more than possible.

The Trumpeter Swan (*Cygnus buccinator*) is a regularly cited example of an endangered bird. Certainly, it did at one time appear seriously threatened and only comparatively recently has it been realized that breeding populations in Alaska are actually much larger than previously thought. Numbers are now considered to be in the low thousands.

The Hawaiian Goose (*Branta sandvicensis*), or Nene, is perhaps chiefly celebrated for the great success with which it has been reared in captivity. Once common on Hawaii, by the early 1950s numbers had dropped to around fifty individuals and may even have fallen as low as thirty. Clearly, the species was at this time faced with extinction but a breeding programme for captive birds was initiated and great success achieved, both at Pohakuloa and at the Wildfowl Trust in England. Birds raised in captivity were first released back into the wild during 1960, since which time well over a thousand more have been reintroduced to their natural habitat on the island of Hawaii and also to Maui. The wild population probably now numbers about 800 individuals; a rather higher number still live in captivity and colonies can be seen in several zoos and waterfowl collections.

Another species that seems to have been saved from immediate danger is the curious but handsome Cape Barren Goose (*Cereopsis novaehollandiae*), which survives on islands off the coast of southern Australia. The precise affinities of this strange bird to other kinds of waterfowl have not been determined and certainly no surviving member of the order appears closely related. However, two species showing considerable similarity to it occurred in New Zealand in the not-too-distant past. These have been described from skeletal material found in geological deposits at several localities but although the remains are of comparatively recent origin, it is not known at what date either species died out. One, described under the name of *Cnemiornis calcitrans*, stood 91 cm (3 ft) high – very large for a goose; but a smaller form has been listed under the names of *C. gracilis* or *C. minor*.

An apparently valid species even closer to the Cape Barren Goose was described by W.R.B. Oliver in 1955 from a skull found in New Zealand. This was given the name of *Cereopsis novaezealandiae* but as a distinct species it is no longer recognized. The skull upon which it was founded proved attributable to the still-extant Australian species, a population of which was introduced to New Zealand in 1869 with further introductions being made in later years. It seems that a small number of these introduced birds still survive. The main strongholds of the Cape Barren Goose are the windswept and rather lonely islands of the Furneaux Group, although the birds also inhabit other islands off the southern coast of Australia and during the non-breeding season may be seen on the mainland. Recent estimates at numbers have varied between 5,000 and 16,000. The overall ash-grey colouring marked with a subtle spotting of darker greyish brown on the wing coverts gives this species a very distinctive appearance. Still more unusual in colour is the short cere-covered bill: pale yellow apart from the tip and cutting edges, which are black.

The Blue Duck (*Hymenolaimus malacorhynchos*) of New Zealand is another distinctive and rare species. Originally, this bird occurred from sea level to tree line but has retreated before the advance of civilization and is now found only along remote and fast-flowing mountain streams. Since there is no shortage of such habitat in parts of New Zealand, the future of this species might be assured but it is not known to what extent these ducks are vulnerable to introduced mammalian predators. In the past, they have shown themselves to be sometimes tame and rather trusting birds. Trout, which have been introduced to New Zealand waterways, may also pose a threat to the Blue Duck since they almost certainly compete with it for the aquatic insects upon which the birds depend. The subtle blue-grey colouring of the upperparts contrasting with the heavy reddish-brown spotting on the breast make this a very beautiful duck. One curious feature is that the colour of the pinkish-white bill can vary greatly in intensity depending upon the emotional state of the bird. No reliable figures for the total population exist but it is guessed to number around 5,000.

An apparently endangered form about which there has been considerable confusion is the Marianas Mallard (*Anas oustaleti*), recorded from the Pacific islands of Guam, Tinian and Saipan. Two colour morphs occur, one similar to the Mallard (*A. platyrhynchos*) and the other resembling the Spotbill Duck (*A. poecilorhyncha*). Because of these resemblances, and taking into account the form's isolation, it is believed that the population resulted from hybridization. Presumably, a few individuals of both parent species chanced upon the islands, interbred and a viable hybrid colony has been maintained ever since. This colony might almost be said to represent the evolution of a species through hybridization. Alternatively, it is possible that these ducks do actually constitute a valid species. Whichever is the truth, due to hunting pressure and draining of wetlands, the colony is now close to extinction and seems to have disappeared altogether from Guam.

Another rather mysterious form was described by a British resident in Taiwan, a Mr Swinhoe, during the nineteenth century under the name *Cygnus davidii*, Père David's Swan. The description was based upon a smallish, white swan with a red bill and orange yellow feet that had been found in China. No similar bird came to attention until the 1960s when comparable birds were seen in Britain. Individuals showing this particular combination of colouring and size are now considered to be abnormally coloured Bewick's Swans (*Cygnus columbianus bewickii*).

Among threatened races two perhaps merit special mention. Both the Laysan and Hawaiian ducks were originally described as full species but each is usually now regarded as a race of the mallard being named *Anas platyrhynchos laysanensis* and *A. p. wyvilliana*, respectively. The Laysan Duck, or Teal as it is sometimes called, is an inhabitant of Laysan Island, one of the Hawaiian group, and has several times been at the point of extinction. Considering the appalling deterioration of the environment during early years of this century and the attitudes of sealing parties and

V Labrador Ducks by L.A. Fuertes from J.C. Phillip's *Natural History of the Ducks*, Vol. 1 (Boston, 1922), Pl.78.

Japanese plume hunters who on occasion invaded the island, it is remarkable that this duck survived at all. During the years 1911 to 1912 the total population, reportedly, had dropped to just seven individuals and it has been alleged that at one time during the early 1930s only a single female existed in the small colony then surviving. Following these low ebbs a remarkable comeback was staged, but recently the population of wild birds again appears on the decline although numbers do seem to fluctuate greatly quite naturally. Fortunately, captive breeding began in 1958, since when many birds have been maintained in captivity and Laysan Duck are now common in waterfowl collections (Todd, 1979).

The Hawaiian Duck or Koloa has never been reduced to anything like the pitifully small remnant that once represented the Laysan Duck. Excessive hunting, habitat destruction and introduction of the Small Indian Mongoose (*Herpestes auropunctatus*) are the prime causes of this bird's decline. It formerly inhabited several islands in the Hawaiian group but is presently restricted to Kauai. Although captive birds are kept in collections around the world, most of these, according to Todd (1979), do not seem perfectly pure.

A large number of species known only from fossils have been attributed to the order Anseriformes. A few now-extinct forms have been found as subfossils (that is, not fully fossilized) and it is not possible to be precise as to when these birds died out. Some of these species may have survived into fairly recent historical times. King (1981), for instance, puts the date of extinction for *Cygnus sumnerensis*, a swan from the Chatham Islands, between 1590 and 1690, but this appears to be rather speculative. Geese seen by seventeenth-century travellers in Réunion and Mauritius may correspond to bones subsequently found in Mauritius and described under the name of *Sarcidiornis mauritiana*.

LABRADOR DUCK

Camptorhynchus labradorius

PLATE V

Camptorhynchus labradorius (Gmelin, 1789)

Anas labradoria Gmelin, 1789 (Arctic America)
Anas labradora Latham, 1790
Rhynchaspis labradora Stephens, 1824
Fuligula labradora Bonaparte, 1826
Somateria labradora Boie, 1828
Camptorhynchus labradorus Bonaparte, 1838
Kamptorhynchus labradora Eyton, 1838
Fuligula grisea Leib, 1840
Camptolaimus labradorus Gray, 1841

Length 54 cm (22 in)

For a bird that frequented such a well-populated area as the northeastern seaboard of the United States, surprisingly little is known of the Labrador Duck. Much of what is said must be qualified by 'probably' or 'perhaps'. It is not at all clear, for instance, quite why this fast-flying species became extinct, since it was shot at no more regularly than any other kind of waterfowl. The likelihood is that this beautiful little sea-duck was persecuted rather less than most, for it was not common and always shy and wary. Although it is undeniable that these birds were occasionally available in the meat markets of New York and Baltimore during the first half of the nineteenth century, they did not make popular eating. Their flesh tasted 'fishy' and often carcasses rotted before they could be sold.

Several reasons for extinction have been put forward but all are neither more nor less than speculative. It has often been suggested that breeding grounds may have been localized and therefore vulnerable to attack, but T. Halliday's (1978) claim that Labrador Ducks bred on islands in the Gulf of St Lawrence and that eggs were harvested by collectors appears conjectural upon both counts.

Perhaps the most interesting theory relating to the disappearance, and one at least partially supported by anatomical peculiarities, springs from

Camptorhynchus labradorius (cont'd)

Description

Male: head, neck and scapulars white, except for stiff yellowish feathers on cheeks, a black stripe extending from crown to nape and a black collar dividing neck from white breast; wings white apart from primaries, which are black; back, rump upper tail coverts, tail and underparts black; iris reddish hazel to yellow; bill blackish brown for most of length, this colour separated from greyish-blue basal area by a band of yellow orange or flesh tone; legs and feet probably greyish blue but possibly yellow.
Female: generally light brownish grey, mantle, scapulars and wing coverts with a bluish-slate tone; tail dark, underparts lighter; greater secondary coverts and secondaries white; primaries blackish brown; colours of soft parts probably similar to male's.
Immature male: similar to adult female during the first year with white feathering on head, throat and upper breast probably increasingly evident as the year ends.

Measurements

Male: wing 215 mm; tail 78 mm; culmen 44 mm; tarsus 40 mm.
Female: wing 210 mm; tail 75 mm; culmen 42 mm; tarsus 38 mm.

the hypothesis that these birds were very specialized feeders; the bill in some ways resembled that of the Australian Pink-eared Duck (*Malacorhynchus membranaceous*). There was a curious softness around the edges with the cere rather swollen and a surprisingly large number of lamellae contained inside. If feeding requirements were indeed highly specialized, the species could obviously have been vulnerable to even the slightest of alterations in the environment. Following the rapid increase in human population, a number of subtle changes are thought to have taken place among the molluscan fauna of the Atlantic seaboard.

P.A. Johnsgard (1978) speculated that feeding may have involved both dabbling, shoveller-like, at the surface and also diving. Labrador Ducks certainly ate shellfish and old records tell of birds captured by fishermen using trotlines baited with mussels.

According to John James Audubon, the species ranged from as far north as Labrador, where the summer months were spent, south to Chesapeake Bay. It seems always to have been restricted to the coast and the only inland record – of a bird in Michigan – is generally discounted. Populations appeared regularly in autumn and winter on Long Island and the coasts of New Jersey and New England, where they frequented sandy bays and inlets.

The nesting grounds were never unequivocally located; they may have been in Labrador or even further north, or, as has sometimes been suggested, they may have been on islands in the Gulf of St Lawrence. Although Audubon's son John was shown nests in Labrador supposedly belonging to this species, there is some doubt as to whether they really did. Two pale-olive eggs measuring 62 mm x 42 mm, the property of the Dresden Museum, may be attributable to the species but this, too, is doubtful.

What is known with certainty is that from the time of their initial description in 1789, Labrador Ducks were uncommon. Whether or not numbers had been dwindling over a long period is not clear but the species' decline was rapid between 1840 and 1870 and during these last years mature males were said to be particularly scarce. The last specimen to be taken seems to be one obtained on Long Island waters during the autumn of 1875 – a male example now in the US National Museum (registered number 77126). Although there is some doubt concerning the accuracy of this date, even more doubt exists over an individual alleged to have been shot on 2 December 1878 in an overflow of the Chemung River near Elmira, New York, a specimen which in any case has now been lost.

Taxonomically, the Labrador Duck seems to provide a link between the scoters (genus *Melanitta*) and the Harlequin Duck (*Histrionicus histrionicus*).

AUCKLAND ISLANDS MERGANSER

Mergus australis

PLATE VI

Mergus australis (Hombron and Jacquinot, 1841)

Mergus australis Hombron and Jacquinot, 1841 (Auckland Islands) (cont'd)

The most curious feature of the Auckland Islands Merganser was its range, which at the time of discovery was apparently very limited. In itself such a restriction would not be of any special peculiarity; what makes it so remarkable is the distribution of other mergansers. Five different kinds are found north of the Equator, among them species enjoying very wide distribution across the Palaearctic and Nearctic. A sixth species,

Mergus australis (cont'd)
Nesonetta aucklandica Gray, 1844
Merganser australis Salvadori, 1895
Promergus australis Mathews and Iredale, 1913

Length 58 cm (23 in)

Description
Male: head, crest and neck dark reddish brown; chin and throat lighter; back, scapulars, upper tail coverts and tail dark bluish black; wings dark slate grey except for black primaries, middle secondaries, which are white on outer vanes, and greater tipped white secondary coverts; breast dull grey with lighter crescent-shaped markings; rest of underparts mottled grey and white; iris dark brown; legs and feet reddish brown perhaps tending to orange; most of upper and tip of lower mandible black, cutting edge of upper and remainder of lower yellowish orange.
Female: similar to male although slightly smaller with crest perhaps shorter and a less red, rather greyer colouring of the crown; one white wing bar.
Immature: similar to adults but with crest shorter, light crescents lacking on breast, and lower breast and abdomen white with fewer greyish markings.
Chick: dark olive brown above, showing faint traces of lighter colouring on wing, scapular and rump patches; chin, throat and upper breast rusty red; rest of underparts light yellowish grey; bill dark olive; legs and feet dull olive; toothed character of mandible well developed.

Measurements
Male: wing 192 mm; tail 87 mm; culmen 60 mm; tarsus 42 mm.
Female: wing 180 mm; tail 82 mm; culmen 54 mm; tarsus 40 mm.

---

rather poorly known, occurs in Brazil but across the whole of the Southern Hemisphere no other now exists.

During March 1840, two French corvettes, *L'Astrolabe* and *La Zelée*, in the process of exploring southern seas, came to the seldom-visited Auckland Islands (Figure 17). Here, Lieutenant Charles Jacquinot obtained a merganser, which he later described jointly with the ship's surgeon Jacques Hombron. It must have been with great surprise that this description was received for the Auckland Islands are separated by over 320 km (200 miles) of open sea from the southern coasts of New Zealand and are thousands of miles from any land that is now occupied by mergansers.

For such a small part of the earth's surface, so far from even the outposts of civilization, the Auckland Islands are associated with a surprising number of bird mysteries. While the French were surveying New Zealand waters in their two corvettes, the British were exploring southern oceans in the *Erebus* and *Terror*. This expedition also arrived at the Aucklands in 1840 and one of the curiosities collected was a bird that has come to be known as the Auckland Island Shore Plover (*Thinornis rossi*). This creature very much resembled the New Zealand Shore Plover (*T. novaeseelandiae*) and clearly bears some connection to it. What that connection is no one knows for no identical individual has since been seen.

Another mystery surrounds a rail from the Auckland Islands, named *Rallus muelleri* by Walter Rothschild on the basis of a single specimen. No others were with certainty obtained from this locality, the form was presumed extinct and eventually the type in Stuttgart was lost during World War II. Then, quite against expectation, an individual was found near to a dump on Adams Island, one of the Auckland group, in 1966. On the basis of this specimen, S.D. Ripley (1977) assigned any surviving population to *Rallus pectoralis*, relegating the form to subspecific status. Some New Zealand specialists, however, still list *muelleri* as a full species.

After Jacquinot's record of mergansers on the Aucklands, for a period of 30 years no further examples were obtained. Something of the collector's lust for the rare and the curious, and the excitement at finding it can be felt in the words of Baron von Hugel who, during the 1870s, was on the right spot at the right time:

> I procured a pair of Mergansers with a few other skins in Invercargill, from a man who had just returned from a surveying trip to the Auckland Islands. He had not even turned the skin after taking it off the body; but as soon as I saw the back through the opening and felt the beak through the skin of the neck, I knew what I had.

Without doubt, interested parties were visiting the islands from time to time and the New Zealand Government steamboat occasionally called in, so that during the next quarter of a century additional examples were procured. At least twenty-five skins are distributed among museums around the world, a collection made up of twelve males, nine females and four ducklings; and also in existence are one or two skeletons and a pickled individual. The last birds to be seen were probably a pair shot on 9 January 1902 and subsequently presented to the British Museum by the Earl of Ranfurly. According to J. Delacour and P. Scott (1954–64), the species was last seen and collected by E.R. Waite in 1909, but there is no mention of this in Waite's report (in Chilton, 1909) on observations and collections carried out by an exploratory team in the steamer *Hinemoa*. Waite simply

VI Auckland Islands Mergansers. Hand-coloured lithograph by J.G. Keulemans from W.L. Buller's *Supplement to the 'Birds of New Zealand'*, Vol. 2 (London, 1905), Pl.6.

17 The French corvettes *L'Astrolabe* and *La Zelée* among the icebergs of the Antarctic, from A. Mangin's *Mysteries of the Ocean* (London, 1874).

says that although a sharp look-out was kept along the shores, the species was not recorded. It has been searched for several times since but with no success and almost certainly is now extinct.

Little is known about the Merganser's habits. According to J.G. Myers (in Phillips, 1922–6) and J. Kear and R.J. Scarlett (1970), the species favoured inland waterways, being found at the coast only around sheltered inlets and creeks. Although most of the known specimens were taken at the coast, this may reflect the habits of those searching for the species rather than the preference of the birds themselves. Because they were often taken in pairs during the southern summer it has been assumed that these birds were monogamous. A brood of four ducklings, all of which were collected, was being tended by both parents until disturbed. These chicks showed none of the patterning that might normally be expected in a merganser but this tendency is by no means without parallel among species developing on isolated islands free from mammalian predators. Similarly, the tendency of the sexes to appear almost alike, resembling immature stages of the species' more spectacular relatives, is fairly typical of the effect of the evolutionary process in such an environment. So, too, is the reduction of the wings although this was not a bird completely without the ability to fly.

The exact limits on the range of this species at the time of its discovery are unknown. From the Aucklands it has been recorded on the main island and on Adams. A record of the bird's occurrence on Campbell Island, many miles to the south-east, is regarded as dubious. Although living birds were never recorded by Europeans in any other locality, it has become apparent that during quite recent historical times the species occurred in New Zealand itself. Skeletal remains associated with the middens of Polynesian settlers have been found in widely separated districts. Bones have also been uncovered in sand dunes and in deposits beneath the volcanic ash of Rangitoto – the volcano dominating the northern harbour of Auckland, New Zealand's largest city (the Auckland Islands are many hundreds of miles distant from the city).

Why this merganser became extinct on the Aucklands and also in New Zealand is something of a mystery. Presumably, it could not cope with one or more of the many changes to the environment that man brought about. Various mammalian predators have been introduced to New Zealand by both Polynesians and Europeans, and pigs, cats, rats and mice were taken to the Auckland Islands during early years of the nineteenth century. These mammals may have had a disastrous effect on populations. Maybe a clue to the extinction lies in an observation made by the Austrian collector Andreas Reischek who visited the Merganser's last stronghold in 1888. When pursued, he stated (1930), these birds conceal themselves among rocks and do not dive like their European counterparts.

At what date the Merganser disappeared from the main islands of New Zealand is not certain. Perhaps a few individuals still survived even as the species was being discovered hundreds of miles to the south.

PINK-HEADED DUCK

Rhodonessa caryophyllacea

PLATE VII

Among the crocodile-infested swamps and reed beds, which a century ago studded the plains surrounding the lower reaches of the Ganges and Brahmaputra, Pink-headed Ducks made their homes. The lowlands extended across a wide area, stretching for many miles to either side of the mighty rivers. Here, tall grasses, once the haunt of countless numbers of tigers, made journeys difficult and laborious; the traveller venturing into this vast tract of country ran many risks, not the least of which was exposure to malaria. A sparse scattering of poor villages was strung out over this wilderness, separated from one another by virtually unexplored territory – land divided by deep, slow-moving streams whose sluggish waters frequently overflowed creating huge areas of marsh. Waterfowl abounded, providing additional attractions for those sportsmen who, in the heyday of the British Empire, braved the rigour of the terrain in search of tigers. No other local water bird acquired quite the mystique of the Pink-headed Duck whose comparative rarity and most peculiar appearance made it highly sought after.

During the twentieth century the home grounds of this bird have undergone drastic changes. The waterways have increasingly been brought under control, wetlands have been drained and claimed, and cultivation and settlement have repeatedly pushed back the borders of the wilderness that once was Bengal. Today, parts of the area are among the most densely populated places on earth. In the midst of change and development, the Pink-headed Duck disappeared, caught between the contraction of suitable habitat and the guns of sportsmen who, in India, often failed to observe a close season. One irony of this is that Pink-headed Ducks were never really classic sporting birds. Shy, wary and difficult to

Rhodonessa caryophyllacea
(Latham, 1790)

Anas caryophyllacea Latham, 1790 (India)
Rhodonessa caryophyllacea Reichenbach, 1852

Length 60 cm (24 in)

Description
Male: slightly tufted head bright pink, this colour extending onto hindneck, with a black line on the forehead and a blackish band from chin to breast; upperparts brownish black with some pinkish indications on mantle, but speculum bright pink; under tail coverts dark chocolate brown; under wing coverts pale pink; breast brownish black showing some pink or whitish vermiculation, rest of underparts brownish black; narrow bill bright pink; iris brownish orange; legs and feet brownish black.
Female: similar to male but pink of head less bright, sometimes rather whitish; rest of plumage duller.
Immature: similar to adults but paler with neck and head rosy white. Birds apparently became darker brown as they aged.

Measurements
Adult: wing 250–270 mm; tail 106–115 mm; culmen 50–55 mm; tarsus 38–40 mm.

---

flush, they were not even particularly attractive on the table. F.B. Simson remarked in 1884 that he preferred to eat any other kind of duck with the single exception of the Shoveller (*Anas clypeata*), but it is doubtful whether such distinctions held much meaning for the hungry and swelling population of what is now Bangladesh.

The species has been called beautiful but in honesty this is hardly justified. It was certainly both unusual and eye-catching, the pink of the head, matched by a similar shade on the speculum, extraordinary in its depth of hue. The lines and proportions were perhaps rather lacking in grace and on the water a peculiar stiff-necked posture seems to have been adopted. Much of the hot day was, in fact, spent in the water, the birds usually resting or foraging for food at the surface. One account suggests that they fed by night but this, perhaps, is inaccurate. Presumably, they were fairly omnivorous feeders – a gizzard was found to contain waterweeds and small shells. Although their inclination was for the surface, these ducks may also have been very able divers. F. Finn (1915) watched them performing very deftly and remarked that they could remain beneath the water for as long as a pochard. J. Delacour and P. Scott (1954–64), on the other hand, said that they neither perched nor dived, and Delacour did keep captive individuals during the 1920s.

Pink-headed Ducks were usually found on enclosed waters surrounded by heavy vegetation. Sometimes, particularly in winter, small groups would gather, usually numbering between six and eight but occasionally incorporating as many as forty individuals. In these colder months birds were now and then encountered upon open rivers but they were rarely seen moving from one place to another and comparatively few accounts of their flight exist. One report states that it was rapid and powerful while another describes it as light and easy.

Breeding began during the months of April or May when well-formed circular nests, for which J.C. Phillips (1922–6) gives measurements of 227 mm in diameter and 102–127 mm in depth, were constructed from dry grasses and feathers. These were carefully hidden within tufts of tall grass, sometimes well away from the water's edge although rarely more than 460 m (500 yds) distant. Size of clutch varied between five and ten; the eggs, found chiefly in June or July, were peculiar for a duck's in that they were almost spherical in shape. White in colour or faintly yellowish, they measured about 46 mm × 43 mm.

Delacour and Scott (1954–64) described how drakes displayed in company, puffing out short head feathers while the neck, at first withdrawn to rest upon the back, suddenly stretched upwards as the call was delivered. The voice of the male has been described as a sort of wheezy whistle but, according to Phillips (1922–6), a mellow two-syllable cry with a metallic ring was given. Females offered a loud quack, a more powerful tone perhaps resulting from considerable differences in the tracheas of the two sexes. On the basis of the trachea of the male, P.A. Johnsgard (1961) associated the species with pochards and in general body shape it does resemble these birds. G.E. Woolfenden (1961) confirmed the affinity on osteological grounds while more recently A.H. Brush (1976) provided additional evidence for the connection in a work on feather proteins.

As far as can be determined, the decline of the Pink-headed Duck took place over a period of about 50 years. Its comparatively inaccessible haunts made the species always something of a rarity but during the last quarter of the nineteenth century it was encountered regularly and often offered for sale in the markets of Calcutta. By the early 1900s, Pink-headed Ducks

VII Pink-headed-Ducks by H. Gronvold from E.C.S. Baker's *Indian Ducks and their Allies* (London, 1908), Pl.4.

were no longer easily obtainable and the decline began to be commented upon. T. Halliday (1978) mentioned that as late as 1922 Phillips was suggesting that the rarity might be more apparent than real, but this is hardly fair on the author of the magnificent four-volume monograph of the ducks. Phillips made it very clear that he was quoting the opinion of E.C.S. Baker, a renowned authority on Indian birds, and elsewhere mentioned that the species was fast approaching extinction. In fact, the last fully substantiated records of Pink-headed Ducks in the wild occurred quite soon after the completion in 1926 of *A Natural History of the Ducks*. The date of the last sighting of a wild member of the species is a matter of some controversy but 1936 is sometimes suggested.

As is often the case with creatures thought to be extinct, rumours persist that some individuals linger on. During the 1960s, a small group was reported near the Burmese-Tibetan border and other similar claims have been made. There may be some truth to these, perhaps the species does still survive. It should be remembered, though, that unless a good view is obtained confusion with the Red-crested Pochard (*Netta rufina*) could arise, so reports have to be received with a degree of scepticism.

Over the years a few individuals have been kept in captivity. Surprisingly, as late as 1925 a consignment of three pairs was received by Alfred Ezra who kept a wonderful collection of waterfowl at Foxwarren Park, Surrey. Even more surprisingly, another ten pairs, or perhaps just ten individuals (records do not define this clearly), were obtained 4 years later. These birds thrived but unfortunately never bred nor, apparently, made any attempt to do so. Gradually they died. The date for the death of the last is, curiously, something of a mystery. The year 1936 is given by some authors; by others, 1939 is proposed. A rumour indicates that the last bird, a male, may not have died until 1945.

It seems rather strange that the last Pink-headed Duck of all may have expired in England, so far from its native Bengal.

KOREAN CRESTED SHELDUCK

Tadorna cristata

PLATE VIII

Tadorna cristata (Kuroda, 1917)

Pseudotadorna cristata Kuroda, 1917 (Naktung River, near Fusan, Korea)

Tadorna casarca × *Querquedula falcata*? Sclater, 1890

Length 63–71 cm (25–28 in)

Description

Male: entire top of head dark metallic green, separated from lower parts of face by a black line extending from forehead to middle hindneck but passing just

Conflicting opinions have been expressed in regard to this rather mysterious bird of which the whereabouts of just three specimens are known. The earliest surviving example, a female (some authors erroneously list this specimen as a male) now in Copenhagen, was taken near to Vladivostock in April 1877 but not described until 13 years later when P.L. Sclater suggested that it might be a hybrid of the Ruddy Shelduck (*Tadorna ferruginea*) and the Falcated Teal (*Anas falcata*).

This designation is discounted by more recent commentators (Kuroda, 1917; Phillips, 1922–6; Delacour and Scott, 1954–64) on a variety of grounds. Firstly, the plumage of the Korean Crested Shelduck is not unequivocally intermediate between the suggested parent species and what marks it has cannot be considered definite enough to support Sclater's hypothesis. Secondly, a pair was taken in 1913 or 1914 by a resident of Seoul at the mouth of the Kun-Kiang River, near Kunsan, western Korea. The male, a unique example, was given to N. Kuroda some 10 years later and is now in the Kuroda collection at Tokyo, but the female, presented to a friend of the collector, has since vanished. Thirdly,

Tadorna cristata (cont'd)
below the eye; long nuchal feathers greyish at base but tipped with greenish black; almost all of the rest of face and upper parts of neck smoke grey with indistinct bars of chocolate brown, but just under the eye is a small whitish area and the chin shows a conspicuous blackish-green patch; lower part of hindneck to upper part of mantle deep blackish green extending forwards to form a wide band across the breast; back, lower breast and abdomen dark grey with very thin, wavering white lines; greater wing coverts white; median wing coverts grey marked white shading to brown; inner secondaries chestnut; outer secondaries green; primaries and tail greenish black; shorter under tail coverts ochraceous orange; longer under tail coverts white; bill red or pinkish with small knob-like appendage at base of culmen.
Female: crown and nape black, this colour looping below each eye to encircle an area of white; forehead, rest of face, throat and upper neck, white; back, flanks and underparts greyish brown marked with thin, wavering whitish lines; greater wing coverts white; median wing coverts greyish brown; secondaries green; primaries and tail greenish black; under tail coverts pale brown shading to white; bill and feet light coloured.

Measurements
Male: wing 320 mm; tail 117 mm; culmen 45 mm; tarsus 49 mm.
Female: wing 310 mm; culmen 41 mm; tarsus 47 mm.

there is a body of historical evidence suggesting that birds identical to those known as specimens were at one time quite familiar to the Japanese who called them *chosen-oshi.*

Both sexes are, apparently, described in the old avicultural work called *Kanbun-Kinpu* and illustrations have been traced in Japan (reproduced in Kuroda, 1924) and also found on tapestries in China. S. Uchida (1918) claimed that birds referable to this species were imported alive from Korea into Japan around 200 years ago. This suggests that during historical times the Korean Crested Shelduck was a reasonably common bird but had been in decline for a considerable period before its scientific description. Although the limits of its former range are, of course, unknown, J.C. Greenway (1958) speculated that the species may have bred in eastern Siberia perhaps migrating to Korea and Japan.

The third known museum specimen, along with the unique male, is in the Kuroda collection, Tokyo, and although taken after the two previously mentioned, is, in fact, the type. Like the Copenhagen bird it is a female example but was taken near to Fusan, Korea in December of 1916. Another specimen is alleged to have been seen around the turn of the century by a Mr G.D. Wilder of Peking University in the possession of a Chinese hunter. Having, at this time, no idea of the bird's rarity, Mr Wilder made no attempt to obtain it. Writing during the 1920s, Kuroda claimed to have heard that Japanese hunters still regularly took individuals in Korea and it is recorded that in 1917 three birds were dropped from a flock of six, but none were preserved.

The species may not be quite extinct but this seems only a remote chance. An alleged sighting is on record for late March 1943 at Chushinhokudo and, still more recently, in 1964, two Russian students claimed to have seen three birds, two females and a male, among a flock of Harlequin Ducks (*Histrionicus histrionicus*) close to Vladivostock.

ANOTHER PUBLISHED ILLUSTRATION OF THE *KOREAN CRESTED SHELDUCK*

Proceedings of the Zoological Society of London, 1980, pl. 1. Artist, J. Smit.

VIII Korean Crested Shelducks by N. Kobayashi from J.C. Phillips's *Natural History of the Ducks*, Vol. 1 (Boston, 1922), Pl.101.

DIURNAL RAPTORS

ORDER
Falconiformes

EXTINCT SPECIES

Guadalupe Caracara (*Polyborus lutosus*)

Diurnal birds of prey are found, with the exception of Antarctica, throughout the world and are characterized by sharply hooked beaks, powerful feet armed with long curved claws, a generally slow reproductive rate and a need for animal food either taken alive or found dead. From the owls, which share most of these characteristics and with which the diurnal raptors were once classified, they differ in a number of important respects. Owls are, of course, largely nocturnal and this is certainly the most obvious mark of separation. More fundamental, however, are significant skeletal dissimilarities and the notably soft feathering of the night birds.

Although many species within the order Falconiformes are rare and a large number have been relentlessly persecuted, only one, the Guadalupe Caracara (*Polyborus lutosus*), can with certainty be listed as extinct.

Of the several extremely rare species (see Table 7), the Madagascar Serpent Eagle (*Eutriorchis astur*) is perhaps the closest to extinction, if not extinct already. This little-known bird was found in the dense, humid forest of eastern parts of the island but has not been recorded since 1930 when two individuals were seen and collected. The forests of Madagascar have, of course, been seriously depleted.

Another very rare and little-known bird, Gundlach's Hawk (*Accipiter gundlachi*) from Cuba, was feared extinct but has recently been rediscovered. *Accipiter princeps*, the New Britain Grey-headed Goshawk, is known from just three specimens all of them adult. According to Brown and Amadon (1968) no naturalist has ever seen this bird and its status is quite unknown. The magnificent Monkey-eating Eagle (*Pithecophaga jefferyi*) is seriously threatened with extinction and it is doubtful whether the small population can withstand the destruction of forested country in the Philippines.

Another very large yet vulnerable member of the order is the California

Condor (*Gymnogyps californianus*). Despite the introduction of a variety of measures aimed at conserving the species, the population of these long-lived birds appears to be dwindling and may now have passed beyond the point at which recovery is possible.

Islands in the Indian Ocean are the home of two species of kestrel with very small populations. *Falco araea*, the smallest species in the genus, is restricted to three or four small islands of the Seychelles group. This species appears to have adjusted to human colonization of the Seychelles and a seemingly stable population of around one hundred birds exists. The Mauritius Kestrel (*Falco punctatus*), on the other hand, is a seriously endangered bird and in 1973 only six or seven individuals could be located. By 1976, numbers had doubled but such a tiny population is obviously very vulnerable and destruction of habitat together with predation of nests by the Crab-eating Macaque (*Macaca fascicularis*) and other introduced mammals (Lever, 1985) remain major causes for concern.

The problematical form known as Kleinschmidt's Falcon (*Falco kreyenborgi*) often occurs in lists of very rare birds. Described as a new species in 1929, it is represented in museums by just five specimens. A number of sightings have been made, photographs taken, and living birds kept at the Münster Zoo. Some have considered this form to be merely a colour phase of the South American Peregrine (*Falco peregrinus cassini*), while others have thought of it as a separate and valid Chilean race of the Peregrine. In December 1980, a pair of typical South American Peregrines was found with a brood of four recently fledged young, one of which appeared identical to the pale-coloured Kleinschmidt's Falcon. This bird was trapped, examined in the hand and photographed. Its seems most probable, therefore, that birds listed as *kreyenborgi* are, as previously suspected, rare colour morphs of the race *cassini*.

Table 7

RARE OR LITTLE-KNOWN DIURNAL RAPTORS

California Condor *Gymnogyps californianus*	Madagascar Sparrowhawk *Accipiter madagascariensis*	Harpy Eagle *Harpia harpyja*
Cape vulture *Gyps coprotheres*	Gundlach's Hawk *Accipiter gundlachi*	Crested Eagle *Morphnus guianensis*
Madagascar Serpent Eagle *Eutriorchis astur*	Celebes Little Sparrowhawk *Accipiter nanus*	Monkey-eating Eagle *Pithecophaga jefferyi*
Madagascar Sea Eagle *Haliaeetus vociferoides*	American Collared Sparrowhawk *Accipiter collaris*	Cassin's Hawk Eagle *Spizaetus africanus*
White-tailed Sea Eagle (Erne) *Haliaeetus albicilla*	Plumbeous Hawk *Leucopternis plumbea*	Javan Hawk Eagle *Spizaetus bartelsi*
Steller's Sea Eagle *Haliaeetus pelagicus*	Barred Hawk *Leucopternis princeps*	Philippine Hawk Eagle *Spizaetus philippensis*
Black Harrier *Circus maurus*	Mantled Hawk *Leucopternis polionota*	Sclater's Forest Falcon *Micrastur plumbeus*
Red Goshawk *Erythrotriorchis radiatus*	White-necked Hawk *Leucopternis lacernulata*	Slaty-backed Forest Falcon *Micrastur mirandollei*
Grey-bellied Goshawk *Accipiter poliogaster*	Grey-backed Hawk *Leucopternis occidentalis*	Traylor's Forest Falcon *Micrastur buckleyi*
New Britain Grey-headed Goshawk *Accipiter princeps*	Galápagos Hawk *Buteo galapagoensis*	Seychelles Kestrel *Falco araea*
Ovampo Sparrowhawk *Accipiter ovampensis*	Hawaiian Hawk (Io) *Buteo solitarius*	Mauritius Kestrel *Falco punctatus*
		Taita Falcon *Falco fasciinucha*

IX Guadalupe Caracara. Oil painting by Errol Fuller.

GUADALUPE CARACARA

Polyborus lutosus

PLATE IX

> The 'Calalie' is abundant on every part of the island; and no bird could be a more persistent or more cruel enemy of the poultry and domestic animals. It is continually on the watch, and in spite of every precaution often snatches its prey from the very doors of the houses. The destruction of the wild goats is not so great, as these animals are better able to protect themselves than the tame ones. No sooner is one kid born – while the mother is in labor with the second – than the birds pounce upon it; and should the old one be able to interfere, she is also assaulted. No kid is safe from their attacks. Should a number be together, the birds unite their forces, and, with great noise and flapping of their wings, generally manage to separate the weakest one and dispatch it. They sometimes fasten upon the tongue when the poor creature opens its mouth to bleat, and have been known to tear it out, leaving the animal to perish, if not otherwise destroyed. Sometimes the anus is the point of first attack. The birds are cruel in the extreme, and the torture sometimes inflicted upon the defenceless animals is painful to witness ... Even when food is plenty, they often attack living animals instead of contenting themselves with the carcasses of those already dead, seeming to delight in killing. Should one of their number be disabled or wounded, it is instantly dispatched by the rest.

Thus did Edward Palmer (in Ridgway, 1876b) describe the Caracara of Guadalupe. From this description of a bird possessed by an almost demoniac drive, and also from material brought by Palmer from Guadalupe, Robert Ridgway named the species *Polyborus lutosus* in 1876. Within 25 years this bird, which so tyrannically held sway over the island, was gone, exterminated by a predator far more ruthless and cruelly efficient.

As every man's hand turned against the 'Calalie' or 'Queleli', the very boldness of the bird turned to its undoing. W.E. Bryant (1887) described

Polyborus lutosus (Ridgway, 1876)

Polyborus lutosus Ridgway, 1876 (Guadalupe Island off the coast of Baja California)

Length 54 cm (22 in)

Description
Adult: entire top of head, under wing coverts, primaries and terminal tail band blackish brown; neck, back, middle wing coverts and most of tail buffish brown irregularly barred with blackish brown and dull white; sides of face buff; throat whitish; breast, abdomen and thighs buffish brown barred with blackish brown; iris brown; cere reddish pink; bill bluish grey; legs and feet yellow. Sexes alike. *Immature:* plumage of upperparts darker brown and more uniformly coloured than in adult; underparts duller brown streaked with buffy white.

Measurements
Adult: wing 381–418 mm; tail 260–286 mm; culmen 40 mm; tarsus 83–92 mm.

how easily these creatures could be picked off with a rifle as they came to drink at shallow pools. If missed, they would take not the slightest notice of the shot and would wait quite unaware of their peril. Even by 1885 the population had been drastically reduced and Bryant recorded that during a two-day excursion on the central part of the island he saw only four birds. When Palmer again visited Guadalupe in 1889 he did not find a single individual. The great irony of the species' disappearance lies in the fact that the island was left uninhabited shortly before the last Caracaras were seen; there must, therefore, have been some hope that numbers, although desperately depleted, might recover. During the afternoon of 1 December 1900, the field collector Rollo Beck, then visiting Guadalupe, watched eleven Caracaras fly over him. Nine were brought down. Beck explained afterwards in a letter written to Clinton G. Abbott of the San Diego Society of Natural History, 'Judging by their tameness and the short time that I was on the island I assumed . . . that they must be abundant.' No one ever recorded with certainty a living Guadalupe Caracara again.

Guadalupe, some 32 km (20 miles) long by 9.6 km (6 miles) wide, lies about 224 km (140 miles) off the coast of Baja California and should not be confused with Guadeloupe in the Lesser Antilles. Goats were introduced to the island more than 150 years ago (Lever, 1985), destroying much of the original vegetation. Although this may not have much affected the Caracara, it resulted in the loss of several other avian inhabitants.

The introduction of goats to the island must, in the first instance, have been beneficial to the Caracara. Despite the fearsome reputation, these birds were probably very dependent upon carrion. In addition to this source of food, they almost certainly consumed shellfish, rodents and little birds, and were seen scratching in ploughed areas, like chickens, to disturb worms, caterpillars and other insects. Habits generally were not unlike those of a close relative on the mainland, the Common Caracara (*Polyborus plancus*), and in appearance too these species resembled one another. They clearly form a superspecies, the more generalized *P. lutosus* presumably being closest to the original ancestral stock.

The nest, a very large untidy structure made from sticks, was usually built on cliffs and difficult to reach. One investigated in 1897 was, however, situated upon a pile of rubbish and cacti. An egg alleged to have been taken from this nest was acquired by H. Kirke Swann and described by him in his *Monograph of the Birds of Prey* (1925–36). Not unlike the egg of *P. plancus auduboni*, this example is smaller, measuring 55 mm × 43 mm, with a whitish ground obscured by reddish-brown blotching and spotting. Another egg supposed to belong to this species is considerably larger with recorded dimensions of 67 mm × 50 mm. Eggs seem to have been laid in April in clutches of three.

For the most part, these birds seem to have been rather silent except when excited or about to attack, under which circumstances they sometimes made a peculiar gabbling noise. Only if surprised or wounded would they emit a harsh and prolonged shriek. That these imposing birds might retain dignity even in the direst circumstances is shown by Bryant (1887):

> One which was badly wounded attempted to escape by running, with the assistance of his wings. Being overtaken and brought to bay, instead of throwing himself on his back in an attitude of defence, or uttering a cry for quarter, he raised his crest and with an air of defiance, calmly awaited death as became the Eagle of Guadalupe.

GALLINACEOUS BIRDS

ORDER
Galliformes

EXTINCT SPECIES

New Zealand Quail (*Coturnix novae-zelandiae*)
Himalayan Mountain Quail (*Ophrysia superciliosa*)

The order of gallinaceous or fowl-like birds is large as it includes the megapodes (Megapodiidae), curassows, guans and chachalacas (Cracidae), grouse, ptarmigans, quails, partridges, pheasants, turkeys and guineafowl (Phasianidae) and the strange, aberrant Hoatzin placed alone in the family Opisthocomidae.

Many species within the order are beautiful and some are truly spectacular. Others, less showy and more subtly marked, are equally attractive when viewed closely. The gallinaceous birds also have other rather obvious attractions. Many have proved exceptionally easy to domesticate and, of all birds, their flesh is in the greatest demand.

Within the order are a considerable number of threatened species, most particularly among the curassows, guans and chacalacas and among the pheasants, yet only two species seem extinct – the New Zealand Quail (*Coturnix novae-zelandiae*) and the Himalayan Mountain Quail (*Ophrysia superciliosa*), neither of which has been reliably reported since the last century. Similarly, the losses at subspecific level seem comparatively slight, the Heath Hen (*Tympanichus cupido cupido*) being the only casualty.

Probably because the Heath Hen came from North America, it is one of the most celebrated of extinct birds. It was a race of the Prairie Chicken, a species originally widely distributed across the plains of North America with a range extending eastwards as far as the Atlantic seaboard (Figure 18). It was the most easterly form that became known to colonists as the Heath Hen. This subspecies disappeared form the mainland at an unknown date some time during the nineteenth century but it clung to existence on Martha's Vineyard, an island in Buzzard's Bay off the Massachusetts coast where numbers may have reached as many as 2,000 in 1916. Then a fire destroyed much of the environment, a disaster from which the population only partially recovered before spiralling downwards to oblivion. After December 1928 only a solitary individual seems to have survived, a bird that was probably last seen alive on 11 March 1932. As a species, the Prairie Chicken is in general decline but it cannot yet be regarded as facing extinction.

Of rare species in the order (see Table 8), the most spectacularly beautiful are to be found among the pheasants. Populations of a number of pheasant species have suffered serious depletion and several kinds will probably become extinct in the next few decades. Perhaps the best hope for some lies in the building up of captive stocks; pheasants have long aroused interest among those who keep birds and impressive collections exist in many parts of the world.

The beautiful tragopans and monals are among the hardest hit of pheasants with three species of tragopan and two monals listed by W.B. King (1981) as rare or endangered. Similarly, several species in the genus *Lophura* may be seriously threatened. It is not known to what extent the

18 Prairie Chicken (*Tympanuchus cupido*). Engraving from *Cassell's Book of Birds* (London, 1889).

Table 8

RARE OR ENDANGERED GALLINACEOUS BIRDS

Maleo *Macrocephalon maleo*	Edward's Pheasant *Lophura edwardsi*
White-winged Guan *Penelope albipennis*	Swinhoe's Pheasant *Lophura swinhoei*
Cauca Guan *Penelope perspicax*	Bulwer's Wattled Pheasant *Lophura bulweri*
Horned Guan *Oreophasis derbianus*	White-eared Pheasant *Crossoptilon crossoptilon*
Blue-billed Curassow *Crax alberti*	Brown eared Pheasant *Crossoptilon mantchuricum*
Red-billed Curassow *Crax blumenbachii*	Cheer Pheasant *Catreus wallichi*
Gorgeted Wood Quail *Odontophorus strophium*	Elliot's Pheasant *Syrmaticus ellioti*
Tadjoura Francolin *Francolinus ochropectus*	Hume's Bar-tailed Pheasant *Syrmaticus humiae*
Swierstra's Francolin *Francolinus swierstrai*	Mikado Pheasant *Syrmaticus mikado*
Western Tragopan *Tragopan melanocephalus*	Malay Peacock Pheasant *Polyplectron malacense*
Blyth's Tragopan *Tragopan blythii*	Palawan Peacock Pheasant *Polyplectron emphanum*
Cabot's Tragopan *Tragopan caboti*	Crested Argus *Rheinardia ocellata*
Sclater's Monal *Lophophorus sclateri*	Green Peafowl *Pavo muticus*
Chinese Monal *Lophophorus lhuysi*	
Imperial Pheasant *Lophura imperialis*	

wars and continuing disruption in Indo-China have affected populations of at least two species, *L. edwardsi* and *L. imperialis*.

On the other side of the world, in South and Central America, many curassow species have declined; as with some of the pheasants, their current status can only be guessed at.

The White-winged Guan (*Penelope albipennis*) for many years was believed extinct. Apparently endemic to northern Peru, the species was described from two skins – one now in Warsaw, the other in Lima (a third skin exists in the British Museum (Natural History) collection). For almost exactly 100 years after the original description, the White-winged Guan was not again located. Then, on 13 September 1977, four birds were found and since this time additional populations have been seen. Ironically, because the species was thought extinct, it was not included on the Peruvian list of protected birds.

Another very poorly known species in the order is the Gorgeted Wood Quail (*Odontophorus strophium*) of Colombia. Known from just two skins taken in the nineteenth century and two more obtained around 1915, this species has apparently been rediscovered fairly recently, a specimen allegedly belonging to it having been collected in 1972.

There is at least one 'mystery' bird in the order – the Double-banded Argus, a form that was given the scientific name of *Argusianus bipunctatus* by J. Wood in 1871. Housed in the British Museum (Natural History) is part of a primary feather clearly belonging to an Argus pheasant; it cannot, however, be referred with confidence to either of the two known species of *Argus*. The relic is of unknown origin and perhaps belongs to an otherwise undiscovered species that could, of course, now be extinct. On the other hand, it may simply represent an aberrant development.

X New Zealand Quail. Chromolithograph after a painting by J.G. Keulemans from W.L. Buller's *History of the Birds of New Zealand*, Vol. 2 (London, 1888), Pl 23.

NEW ZEALAND QUAIL

Coturnix novae-zelandiae

PLATE X

Coturnix novae-zelandiae
(Quoy and Gaimard, 1830)

Coturnix novae-zelandiae Quoy and Gaimard, 1830 (Baie Chouraki = Hauraki Gulf, North Island, New Zealand)

Length 19 cm ($7\frac{1}{2}$ in)

Description
Male: lores, line over eyes, sides of head and throat rufous; crown and nape dark brown edged with paler feathers running down the centre, marked in their middles with yellowish white; shoulders, mantle and all upper surfaces rufous varied with black and decorated with many lanceolate stripes of white; primaries and outer secondaries dark brown; inner secondaries and wing coverts as well as tail feathers greyish brown varied with pale rufous and shafted with white; lower part of neck mottled or obscurely spotted with black and white; sides and long plumage overlapping the thighs rufous brown, each feather margined and marked down the centre with white and streaked on the webs with blackish brown; abdomen fulvous white but under tail coverts barred with black; iris light hazel; bill black, paler at tip; legs and feet pale flesh brown.
Female: similar to male but no rufous on face and throat, which is whitish varied with brown.

Measurements
Male: wing 110 mm; tail 40 mm; culmen 12 mm; tarsus 25 mm.
Female: wing 112 mm; tail 42 mm; culmen 12 mm; tarsus 26 mm.

Like other quail, those of the genus *Coturnix* are subtly and beautifully marked. Although there are only five extant species, these are spread across most parts of the Old World. The Australian representative of the genus is the Stubble Quail (*Coturnix pectoralis*), widely distributed in its homeland and in many places quite common. A very closely related bird, some consider it conspecific, was plentiful in New Zealand until the middle of the nineteenth century. Then, during the space of a decade or two, the population level plunged and the species disappeared without trace.

Quite why this disaster should overtake birds so similar in appearance to their still flourishing Australian counterparts is something of a mystery, but the fact that it did may indicate that differences between the two forms had become quite fundamental. Whatever the vital factor that led to extinction, it is unlikely to be determined now; all that can be noted today are the slight variations in plumage. New Zealand Quail were more intensely coloured, with black and rufous shades particularly conspicuous, even though the bright rufous of the throat was actually less extensive. This smaller bright area of the throat was separated from the rest by a blackish border rather more definite than that shown by the Australian species.

Quail were first noticed in New Zealand in 1769–70 by the first Europeans recorded as landing there. The English naturalist Joseph Banks reported seeing them during the great circumnavigation of the world he made with James Cook and those who rapidly followed the trail blazed by the *Endeavour* noted quail to be common. Presumably because of their passing similarity to the already-familiar European Quail (*Coturnix coturnix*), these New Zealand birds seem not to have aroused any great interest among the early explorers. It was not until the visit of Dumont D'Urville in 1827 – nearly 60 years after Cook and Banks – that specimens were taken; from these, Quoy and Gaimard (1830) described a new species.

Writing just over a hundred years after Cook's first New Zealand adventures, Walter Buller (1872) remarked that the indigenous quail was on the verge of extinction but the decline had been so abrupt that even he could easily recall the years of their abundance. On the open country and especially the grass-covered downs of the South Island, they provided good hunting for the early colonists – reminiscent, presumably, of the sport they had once enjoyed at home. Sixty brace are alleged to have been shot one morning on the site of Christchurch Cathedral and, as late as 1848, 43 brace were shot within a day close to the site of the city of Nelson.

Within 20 years from this date, the quails were virtually all gone but no completely satisfying reasons for this disappearance have been advanced. Mammalian predators – including man – are sometimes blamed and doubtless these did take a high toll; a more powerful suspicion hangs over the widespread procedure of 'burning off' used to prepare land for agriculture. It is often assumed that an unidentified avian disease spelt disaster for the species; any or all of these factors may have played a decisive role but none convincingly account for the extinction of a bird that might fairly be expected to be robust.

The last record of North Island birds is probably an observation of several individuals by the Hon. J.C. Richmond in Taranaki during December 1869. On the South Island quail appear to have been seen in the mid-1870s, although the last specimens were taken at Blueskin Bay in 1867 or 1868 (and at some time were in the collection of Walter Buller).

For a brief period in the 1880s it seemed as if there had been a rediscovery on the largest island of the Three Kings group – one of New Zealand's many small offshore islands. The birds eventually proved to be Australian Brown Quail (*Synoicus ypsilophorus*).

Potts (1871) made notes on some aspects of the birds in life but apart from his record little is known. In wet or damp weather the quail were noisiest, giving voice to their characteristic call – twit, twit, twit, twee, twit – rapidly repeated. In addition, Potts remarked on 'a low purring sound that one might suppose to proceed from an insect rather than a bird'.

The nest was just a depression in the ground lined with grass. Eight to ten eggs were laid, yellowish brown or buff spotted and blotched with a darker brown; they measured about 31 mm x 24 mm. Incubation took 21 days; after hatching, the young grew rapidly and by 4 months were virtually indistinguishable from their parents. The diet consisted of seeds and grasses and, in captivity at least, small invertebrates were consumed.

OTHER PUBLISHED ILLUSTRATIONS OF THE *NEW ZEALAND QUAIL*

Buller, W.L. 1872. *A History of the Birds of New Zealand* (first edn), pl. 18. Artist, J.G. Keulemans.

Fleming, C.A. and Lodge, G.E. 1983. *George Edward Lodge, Unpublished Bird Paintings*, pl. 43.

Richardson, J. and Gray, G.R. 1844–75. *The Zoology of the Voyage of H.M.S. Erebus and Terror, under command of Captain Sir James Clark Ross, R.N.*, pl. 8.

HIMALAYAN MOUNTAIN QUAIL

Ophrysia superciliosa

PLATE XI

Ophrysia superciliosa (Gray, 1846)

Rollulus superciliosus Gray, 1846 (India)
Ophrysia superciliosa Bonaparte, 1856
Malacortyx superciliaris Blyth, 1867

Length 25 cm (10 in)

Description
Male: forehead, broad superciliary stripe, spot in front of eye and another behind white; sides of head, chin, throat (cont'd)

Comparatively few recently extinct bird species come from the great continental land masses. Africa has none nor, in all probability, has South America – although this situation is unlikely to remain so for long. Europe has only the Great Auk (*Alca impennis*), a species it shares with North America but one that might more properly be considered a marine bird. In North America itself, extinctions have been more drastic, most noticeably with the demise of the Passenger Pigeon (*Ectopistes migratorius*) and the Carolina Parakeet (*Conuropsis carolinensis*), but the number of actual species lost is still comparatively small. Asia has four extinct species but these losses are by no means evenly distributed over the continent. Three of these vanished birds come – perhaps surprisingly – from the sub-continent of India. Perhaps the most enigmatic is the Himalayan Mountain Quail.

As a living entity, the Himalayan Mountain Quail was known for only a very short space of time. The description of the type, a bird in the private menagerie at Knowsley Hall near Liverpool, was published in 1846. Its place of origin was unknown apart from a vague belief that the bird came

Ophrysia superciliosa (cont'd) and line above each superciliary stripe black; a whitish, rather broken line, sometimes indistinct, runs back from beneath eye; crown pale brownish grey showing black shaft stripes on each feather; rest of plumage dark brownish grey with black margins to feathers (wings distinctly more brownish) except for feathers of under tail coverts, which are black tipped and spotted with white, and tail, which is uniform brown, culmen coral red; legs dull red; eyelids black with small white spot at corner.
Female: overall cinnamon brown but sides of head show a pinkish greyish tinge; small white spot before, and a larger one behind the eye; some feathers of crown and all those of nape have black shafts that pass into triangular black spots bordered with buff on the back, scapulars, rump and upper tail coverts; wing coverts, lower back, rump and upper tail coverts mottled with buff; tail black becoming mottled with buff towards the edges; breast, abdomen and under tail coverts paler than upperparts; culmen dusky red; legs dull red.
Immature: plumage of young male shows some buff mottling on wing.

Measurements
Male: wing 85–93 mm; tail 77–78 mm; culmen 11 mm; tarsus 29 mm.
Female: wing 87–91 mm; tail 65 mm; culmen 11 mm; tarsus 28 mm.

from India. A few additional individuals were then shot during the years 1865, 1867 and 1868 at altitudes between 1,540 and 1,840 m (5,000 and 6,000 ft) in the foothills of the western Himalayas near to Mussooree; a final sighting occurred in 1876 at Sher-ka-danda, near Naini Tal. Since then, there have been no reliable records.

Although the species is generally considered extinct – and almost certainly correctly – there is at least some small chance it may survive. According to S. Ali and S.D. Ripley (1969), known habitats together with similar localities have been thoroughly searched, but in the vast and remote mountainous areas of northwestern India it is conceivable that a bird difficult to flush (as this was) might escape detection. Also, it is by no means clear whether the species' regular home grounds were ever discovered. Although these birds were only reported from parts of the western Himalayas, there may be significance in the fact that all recorded examples were seen during a comparatively short period of time; perhaps some unknown condition disturbed these creatures at their remote but regular haunts, causing them to disperse to other, better scrutinized, areas and thence bringing about their discovery. This is no more than speculation, of course; W.T. Blanford (1898) believed the long, soft plumage to be an indication of an inhabitant of cold climates.

Very little is recorded of the habits of the mysterious bird. Most information is derived from the notes of Captain J. Hutton (in Hume and Marshall, 1879–81). He mentions visitors to Mussooree as occurring in small parties numbering between six and ten. They kept to the high grass and scrub on the steep hillsides and could be flushed only when heavily pressed or almost trodden upon. Their rise into the air was accompanied by a shrill whistling note, but the flight itself was heavy and slow with a pitch back into grass at the earliest opportunity. The birds fed upon grass seeds and, possibly, also on insects and berries; while feeding their contact note was low, short and quail-like. They appeared to arrive in November and, in one instance at least, stayed until the following June. Additional information about the species is lacking; there are no details at all about breeding behaviour. According to Ali and Ripley (1969), just ten specimens remain in the world's museums – half of them in London.

XI Himalayan Mountain Quail. Hand-coloured lithograph by J.W. Moore from J.E. Gray's *Gleanings from the Menagerie and Aviary at Knowsley Hall*, Vol. 1 (1846), Pl.16.

RAILS
AND RELATED BIRDS

ORDER
Gruiformes

EXTINCT SPECIES

Chatham Islands Rail (*Rallus modestus*)
Wake Island Rail (*Rallus wakensis*)
Tahitian Red-billed Rail (*Rallus pacificus*)
Ascension Island Rail (*Atlantisia elpenor*)
Kusaie Island Crake (*Porzana monasa*)
Hawaiian Rail (*Porzana sandwichensis*)
Laysan Rail (*Porzana palmeri*)
Samoan Wood Rail (*Gallinula pacifica*)
Lord Howe Swamphen (*Porphyrio albus*)
Mauritius Red Hen (*Aphanapteryx bonasia*)
Leguat's Gelinote (*Aphanapteryx leguati*)

Four of the twelve living families making up the order Gruiformes contain just a single species – the Pedionomidae (Plains Wanderer, *Pedionomus torquatus*), Aramidae (Limpkin, *Aramus guarauna*), Eurypygidae (Sun-bittern, *Eurypyga helias*) and Rhynochetidae (Kagu, *Rhynochetos jubatus*). The other eight families are the Mesitornithidae (mesites), Turnicidae (button quails), Gruidae (cranes), Psophiidae (trumpeters), Rallidae (rails), Otidae (bustards), Cariamidae (seriemas) and Heliornithidae (sungrebes). This seemingly disparate collection of birds are linked together by skeletal and muscular similarities.

The order is comparatively well represented in the fossil record and additional families are known only from such remains. Taken as a whole, it seems to be in decline. Although eleven of the twelve living families have no recently extinct species among their ranks, they contain an alarming number of seriously endangered birds (see Table 9). For instance, the Madagascan mesites, a little-known family comprising just three species, have, like much else of the indigenous fauna, suffered as the native forests have been depleted. Two species, the Brown Mesite (*Mesoenas unicolor*) and the White-breasted Mesite (*M. variegata*), have been located only on rare occasions during this century and the third, Bensch's Mesite (*Monias benschi*), although probably more plentiful, is recorded from an area very restricted in size. They fly little, if at all, and inhabit what remains of the native forests and bush.

Another family represented in the most fragile manner is the Rhynochetidae: the Kagu (Figure 19) is the only living member. This species survives on the South Pacific island of New Caledonia but is perilously close to extinction, its existence threatened by feral dogs and other introduced predators, and destruction of the forest. Once common, it seems now to have retreated into the most inaccessible mountain valleys. Hopes for survival come as much from an apparent ability to breed in captivity as from anything else. The Kagu's precise affinities with other bird species are uncertain but it may bear some relationship to two forms, *Aptornis otidiformis* and *A. defossor*, known only from material uncovered in recent geological deposits of New Zealand. Although these are usually classed with the rails, S.L. Olson (1977) considered their closest association to lie with the Kagu. Approximately the size of the Great Bustard (*Otis tarda*), both became extinct at an unknown date.

Many of the bustards are themselves becoming rarer and populations of some of these have reduced drastically. The range of the Great Bustard, for instance, is now much smaller than it was, say, a century and a half ago; yet the species has not reached the point where it could be considered seriously endangered. Of all the bustards perhaps only the Great Indian (*Choriotis nigriceps*) belongs, at present, rightly within this category. According to *The ICBP Bird Red Data Book* (King, 1981), it occupies today only about 1.7 per cent of its former range.

Of the cranes – certainly the most spectacular members of this order – none have become extinct in recent times but populations of several kinds have reached alarmingly low levels. Almost half of the fourteen species are facing extinction.

Table 9

RARE OR LITTLE-KNOWN RAILS AND RELATED BIRDS

Brown Mesite *Mesoenas unicolor*
White-breasted Mesite *Mesoenas variegata*
Bensch's Mesite *Monias benschi*
Kagu *Rhynochetos jubatus*
Great Indian Bustard *Choriotis nigriceps*
Whooping Crane *Grus americana*
Japanese Crane *Grus japonensis*
Siberian White Crane *Grus leucogeranus*
Hooded Crane *Grus monacha*
Black-necked Crane *Grus nigricollis*
White-naped Crane *Grus vipio*
Red-winged Wood Rail *Eulabeornis calopterus*
Uniform Crake *Eulabeornis concolor*
Platen's Celebes Rail *Rallus plateni*
Wallace's Rail *Rallus wallacii*
New Caledonian Wood Rail *Rallus (Tricholimnas) lafresnayanus*
Lord Howe Island Wood Rail *Rallus sylvestris*
Barred-wing Rail *Rallus (Nesoclopeus) poecilopterus*
Guam Rail *Rallus owstoni*
Bogota Rail *Rallus semiplumbeus*
Wetmore's Rail *Rallus wetmorei*
Zapata Rail *Cyanolimnas cerverai*
White-striped Chestnut Rail *Rallina leucospila*
Chestnut-headed Crake *Rallina castaneiceps*
Streaky-breasted Crake *Coturnicops boehmi*
Waters' Crake *Coturnicops watersi*
Ocellated Crake *Coturnicops schomburgkii*
Darwin's Rail *Coturnicops notata*
Horqueta Crake *Laterallus xenopterus*
Olivier's Crake *Porzana olivieri*
Dot-winged Crake *Porzana spiloptera*
Columbian Crake *Porzana columbiana*
Henderson Island Crake *Porzana atra*
Isabelline Waterhen *Amaurornis isabellinus*
San Cristobel Gallinule *Gallinula silvestris*
Takahe *Notornis mantelli*
Giant Coot *Fulica gigantea*
Horned Coot *Fulica cornuta*
Caribbean Coot *Fulica caribaea*

19 Kagu (*Rhynochetos jubatus*). Monochrome painting by George Edward Lodge. Courtesy of Dr Pat Morris.

Although recent extinctions are not recorded among eleven of the twelve living families of Gruiformes, one group, the Rallidae, has lost more species during historical times than any other family of birds. One reason for this is not difficult to identify. Ancestral stocks have shown a remarkable tendency to colonize oceanic islands, diversify and lose, or almost lose, the power of flight. As S.D. Ripley (1977) commented: 'What a paradox; to fly poorly, to occur so widely, and to evolve flightlessness so easily.' There is no reason to suppose that this process is a thing of the past.

Random dispersal of birds seems to continue today amoung several common species. The sudden appearance of the Corncrake (*Crex crex*) in unexpected localities provides a good example. Ripley (1977) expressed the belief that wide dispersal of rails is actually facilitated by their apparent dislike of flying. A skulking disposition and natural reluctance in taking to the air are both quite familiar traits but most well-known species are physically perfectly equipped for flight. Ripley's argument is that uncertainty in taking to the air may be balanced by an equal hesitation over alighting. Once in the air, without any particular motivation to stop, the rail may just keep going and with substantial reserves of subcutaneous fat it should be able to continue for very considerable distances. Perhaps the very lack of airborne experience may of itself result in poorly developed powers of orientation and, hence, random dispersal. However the tendency to wander came about, the fact is that it now exists and has in the past existed. When birds colonize islands free from pressures imposed by mammalian predators, natural selection appears sometimes to favour a reduction in the powers of flight. Possibly this is because energy is conserved as flight muscles are reduced.

Many rails that evolved on isolated islands have become extinct – some in historical times, some in the more remote past. There is no doubt that most of the more recent extinctions can be directly related to the arrival of human colonists on such island refuges together with the mammalian predators so often subsequently introduced.

Rails that have become extinct since 1600 fall into one of three categories. The first group consists of rails known from written descriptions and early travellers' illustrations, matched up with skeletal remains found in deposits of comparatively recent geological age. Species within this category are the Mauritius Red Hen (*Aphanapteryx bonasia*), Leguat's Gelinote (*Aphanapteryx leguati*) and the Ascension Island Rail (*Atlantisia elpenor*).

The second group is made up of those birds known from skins and written descriptions: the Chatham Islands Rail (*Rallus modestus*), the Wake Island Rail (*R. wakensis*), the Kusaie Island Rail (*Porzana monasa*), the Laysan Rail (*P. palmeri*), the Hawaiian Rail (*P. sandwichensis*), the Samoan Wood Rail (*Gallinula pacifica*) and the Lord Howe Swamphen (*Porphyrio albus*).

In the third category is a single species that, under normal circumstances, might be considered 'hypothetical' as no specimen of any kind exists. Although examples may have passed through the hands of early naturalists, the bird is known today only by an illustration by George Forster in the collection of the British Museum (Natural History). It cannot be doubted that Forster actually saw the bird and certainly the identity of the species is so clearly marked that it need not be confused with any other. This is the Tahitian Red-billed Rail (*Rallus pacificus*).

Many more extinct species are known from skeletal material discovered in deposits of recent geological age. In several instances, it is possible that such species may have survived into historical times. Perhaps one or two do properly come within the scope of this book although it seems likely that most do not. Olson (1977) points to the probability that remains of more species of extinct rail await discovery on oceanic islands.

Particular mention may be made of *Nesotrochis debooyi* since there is evidence, albeit rather flimsy, to suggest that this bird perhaps lingered on even into the twentieth century. Remains of the species were first discovered on several Caribbean islands among the debris of kitchen middens dating from before the first European contact in the late fifteenth century. In some places it appears to have been eaten frequently and must, therefore, have been well known to the local inhabitants. An account given by James Bond in Nichols (1943) has come to be associated with this species. Bond had been told of a 'flightless waterhen' inhabiting Virgin Gorda which, although once common, had become very rare. Whether or not such a bird actually existed and whether it was, in fact, *N. debooyi* must be left an open question.

Fulica newtonii, a large flightless coot described from remains found in deposits of recent geological age in Mauritius, may have still been extant when European travellers first landed upon the Mascarenes. Even as late as 1693, the Huguenot François Leguat included 'Poules d'eau' in his list of Mauritius birds, remarking that they were very rare. Since his record does not really match any other bird known to have existed on the island it may refer to *newtonii*. The same species, or a closely related one, perhaps occurred on Réunion although the only evidence for this is in the form of a written description by 'Le Sieur' Dubois who visited this island between 1669 and 1672. He too mentions 'Poules d'eau', as large as chickens, black in colour with a white crest upon the head.

An even larger flightless coot, *Fulica (Palaeolimnas) chathamensis*, once existed on the Chathams and New Zealand's North and South islands. Also living on the Chathams during recent geological times was a flightless rail known as *Diaphorapteryx hawkinsi*. Bones of this large and aberrant member of the family are not uncommon but, like the coot, it became extinct at an unknown date.

All of these birds undoubtedly existed but two additional forms have been described about which there is some controversy and these are often included in lists of extinct rails. No specimens of either *Apterornis caerulescens* or *Leguatia gigantea* exist and the validity of their description is discussed in the section entitled 'Problematical and Hypothetical Birds' later in this book.

Several races belonging to otherwise extant species have been lost in recent times (see Table 10). The White-throated Rail (*Canirallus cuvieri*) occurs on Madagascar and islands of the Aldabra group in the Indian Ocean. Although two races still exist a third, *abbotti*, once an inhabitant of Assumption Island, became extinct some time before 1937. Three races of the Uniform Crake (*Eulabeornis concolor*) are generally recognized. Of these, two (*guatemalensis* and *castaneus*) are rare even though they occur over a wide area of South America. The nominate subspecies from Jamaica is now extinct having disappeared during the last years of the nineteenth century.

A large number of subspecies of the widely distributed Banded Rail (*Rallus philippensis*) have been described and the status of many of these is not known, although at least two seem definitely extinct. *Rallus p. macquariensis* from Macquarie Island, south-west of New Zealand apparently vanished around 1880. The extinct form *dieffenbachii* from the

Table 10

EXTINCT SUBSPECIES OF RAIL AND RELATED BIRDS

Assumption Island White-throated Rail *Canirallus cuvieri abbotti*
Jamaican Uniform Crake *Eulabeornis (Amaurolimnas) concolor concolor*
Macquarie Island Banded Rail *Rallus philippensis macquariensis*
Chatham Island Banded Rail *Rallus philippensis dieffenbachii*
Iwo Jima White-browed Crake *Porzana (Poliolimnas) cinerea brevipes*
Tristan Island Cock *Gallinula nesiotis nesiotis*
North Island Takahe *Notornis mantelli mantelli*

20 Takahe (*Notornis mantelli*). Oil painting by Raymond Ching.

Chathams is sometimes regarded as a separate species but is treated by Ripley (1977) as a race that is just rather clearly differentiated. It is well represented by skeletal material, but the form is also known from a single skin collected in the 1840s and now in the British Museum (Natural History).

The White browed Crake (*Porzana cinerea*) is distributed across Indonesia, the Philippines, New Guinea, northern Australia and many small islands in the South Pacific. A race (*brevipes*) from Iwo Jima south-east of Japan was last recorded in 1925 and is presumed extinct, probably as a result of the introduction of cats and rats.

The Tristan da Cunha race of the Island Cock (*Gallinula nesiotis nesiotis*) is usually listed as extinct. The last specimens taken were birds sent to London in 1861. Three of these were alive and at least one was kept in the Gardens of the Zoological Society of London. A claim that the subspecies was rediscovered in 1973 is not recognized by Ripley (1977).

In addition to rails that have already vanished many extant species are facing extinction (Table 9). Of these, one of the most interesting is the Takahe (*Notornis mantelli*) (Figure 20) whose story very much reflects the romance of ornithological discovery in the nineteenth century.

On a freezing day in 1849 a party of sealers pitched camp with their dogs in the hostile country that makes up the ragged southwestern tip of New Zealand's South Island. The exact location was Duck Cove, Resolution Island and while attempting to make themselves comfortable in these bleak but extraordinarily beautiful surroundings the men noticed in the snow the trail of a large heavy bird. Following for some distance, the sailors eventually glimpsed the maker of the tracks and loosed their dogs. After a fierce pursuit the quarry was cornered in a gully where it was taken alive – squawking, screaming and struggling violently. For 3 or 4 days this mysterious bird was allowed to live but then met the usual fate of such captives. The body was cooked and eaten but the sealers preserved the skin, perhaps attracted by its bright plumage.

A few miles to the north of Resolution is the Thompson Sound and here on Secretary Island a Maori trapper took a similar individual 2 years later. Again the skin was not destroyed. In December of 1879, close to a lonely homestead 14 km (9 miles) south-east of Lake Te Anau, a station manager making routine rounds passed the camp of a rabbiter. Although the tent had just been struck, hanging dead from the ridge pole was a large brightly coloured bird. After a short exchange, the rabbiter quite happily told his manager to take it – an offer immediately accepted.

On an August evening towards the close of the nineteenth century, one Donald Ross strolled with his dog by the shore of Lake Te Anau when the animal took suddenly into the bush and reappeared carrying a large, colourful and not quite dead bird.

Through just these four instances living Takahes became known to the world of science – only to vanish again, just as quickly and seemingly forever. From 1898 for a space of almost 50 years the species was lost and presumed extinct, the evidence of its passing divided between four institutions – museums in London, Dresden, Wellington and Dunedin. It was in November 1948 that Dr G.B. Orbell returned from an expedition to a glacial valley high above Lake Te Anau carrying photographs of living Takahes. The ornithological world was surprised and fascinated and this intriguing bird once again found itself at the centre of scientific and popular attention. Today, however, only a small and dwindling population survives in one or two remote valleys of the Southern Alps and despite the most strenuous efforts of the people of New Zealand to keep this population intact it seems likely that the living Takahe, in the not-too-distant future, will finally be lost to the world.

The Takahes that still survive belong to the subspecies *hochstetteri* as the nominate race was described in 1848 from fossil remains unearthed at Waingongoro, North Island, New Zealand. This was, of course, a year before the taking of the first living bird on the South Island. Although there are no authenticated records of living Takahes being seen by Europeans north of the Cook Straits, it was known on the North Island to the Maoris who called it *moho*. Tantalizingly, there is some evidence to suggest that a moho was taken at the remarkably late date of 1894.

Most rails have hardly been studied at all. The skulking and secretive nature of rails being what it is, it seems hardly surprising to find that many species are incompletely known. Two species are known from just the type specimen and their status is difficult to determine. One of these, the Horqueta Crake (*Laterallus xenopterus*), was collected for the only time in November of 1933. According to Ripley (1977), this is a very distinct Paraguayan species but apart from its appearance absolutely nothing is known of it. The other unique specimen is of a bird that has come to be called the San Cristobel Gallinule (*Gallinula silvestris*). The Whitney South Sea Expedition of 1929 acquired it from a native hunter during a stay in the Solomon Islands. Obviously very rare, the species has never again been located.

Several more forms are known only from a very few examples – for instance, the Columbian Crake (*Porzana columbiana*), the Dot-winged Crake (*P. spiloptera*) and Olivier's Crake (*P. olivieri*). Among these poorly known birds is Platen's Celebes Rail (*Rallus plateni*). Only ten museum skins exist; the species was last seen in 1930 but it is probable that this rail still survives as suitable habitat is fairly abundant. Another rare bird in museum collections is the New Caledonian Wood Rail (*Rallus lafresnayanus*). No specimen has been taken for more than 50 years and the species is often listed as extinct but it is thought to survive in the more inaccessible jungle-covered parts of its island home. The Lord Howe Island Wood Rail (*Rallus sylvestris*) is perilously close to extinction and the Barred-wing Rail (*R. poecilopterus*) from the Solomon and Fijian islands may be similarly endangered. Also seriously threatened is a Madagascan species, Water's Crake (*Coturnicops watersi*), and a related bird, Darwin's Rail (*C. notata*). The latter is rather enigmatically known from just a handful of specimens taken, remarkably, in various widely scattered spots spread over a vast area of South America.

Of threatened subspecies perhaps the one with the most intriguing history is a race of the Slate-breasted Rail – *Rallus pectoralis muelleri*, a form described by Rothschild in 1893 as a distinct species. The description was based upon a single specimen supposed to have been collected on Auckland Island, New Zealand but unfortunately this type, formerly in the Stuttgart Museum, was lost or destroyed during World War II. A picture of it, by J.G. Keulemans, forms plate 7 of the *Proceedings of the Fourth International Ornithological Congress 1905* (1907) and illustrates Walter Rothschild's article on extinct and vanishing birds (pp. 191–217), a paper later expanded into Rothschild's book *Extinct Birds* (1907). A specimen alleged erroneously by Buller (1905) to be similar is now in the American Museum of Natural History but the provenance of this bird is in doubt and it is generally considered to be attributable to a more common race *R. p. pectoralis*. As no additional material came to light and especially since the Stuttgart bird was no longer available for examination, doubts about the value of Rothschild's description grew and some regarded it as altogether invalid. In 1966, however, a living individual was captured close to a rubbish dump on Adams Island, one of the Auckland Islands group. On the basis of this find, Ripley (1977) lists the form *muelleri* as a rather distinct subspecies of the Slate-breasted Rail.

CHATHAM ISLANDS RAIL

Rallus modestus

PLATE XII

Rallus modestus (Hutton, 1872)

Rallus modestus Hutton, 1872 (Mangare, Chatham Islands)
Rallus dieffenbachii (young), Buller, 1873
Cabalus modestus Hutton, 1874
Ocydromus pygmaeus Forbes, 1892

Length 21 cm (8½ in)

Description
Male: upperparts chocolate brown becoming more olivaceous on mantle and upper back; cheeks and throat greyish each feather tipped with brown; lower throat, breast, sides of body and remainder of underparts greyish brown barred with buff, outer primaries and tail similarly but more faintly banded; iris, bill, legs and feet reddish brown.
Female: similar to male but smaller in size with finer barring of the underparts.

Measurements
Male: wing 88 mm; tail 40 mm; culmen 37 mm; tarsus 28 mm.
Female: wing 76 mm; tail 37 mm; culmen 29 mm; tarsus 24 mm.

From the rather bleak and windswept Chatham Islands, which lie several hundred miles to the east of New Zealand, a number of bird species and races have vanished. At an unknown but, in geological terms at least, comparatively recent date a large and aberrant rail (*Diaphorapteryx hawkinsi*) disappeared. Living on to a time perhaps closer to our own were a very sizeable flightless coot (*Fulica chathamensis*) and a huge swan (*Cygnus sumnerensis*). At the beginning of the twentieth century the Chathams lost distinct races of two species still to be found on New Zealand's main islands – a bellbird (*Anthornis melanura melanocephala*) and a fernbird (*Bowdleria punctata rufescens*).

Until the middle of the nineteenth century this archipelago was also the home of two distinct but closely related rails, the Chatham Island Banded Rail (*Rallus philippensis dieffenbachii*) and the Chatham Islands Rail (*R. modestus*). Both are thought to have originated from a similar stock that invaded the islands in two separate waves. *Rallus p. dieffenbachii*, representing much the later of these, does not seem, by the time of its extinction, to have actually passed the point at which it could be regarded as specifically distinct from the widespread Banded Rail (*R. philippensis*). Known from skeletal remains and a single skin brought from the Chathams during the early 1840s, the form appears to constitute a very distinct subspecies with a downcurved beak in marked contrast to the straighter bill of other races.

Its smaller relative – known as Hutton's or the Chatham Islands Rail – long ago became separated as a species, its ancestors presumably invading at a comparatively early date. This bird showed all the traits commonly associated with a rail adapting to conditions on an island free from mammalian predators. The plumage, soft and slightly hair-like in texture, had assumed what S.D. Ripley (1977) called 'neotonic characters' and what Walter Buller (1888) described as 'the indeterminate character peculiar to young rails'. Indeed, upon first seeing specimens of the Chatham Islands Rail, Buller thought them to be the young of the Chatham Banded Rail. *Rallus modestus* was flightless and in the *Supplement to the 'Birds of New Zealand'* Buller mentions a minute spur at the flexure of each wing. An additional peculiarity seems to have been that the birds were nocturnal, or partly so.

The species is known to have occurred on at least three islands of the Chatham group. From Chatham Island itself it seems to have disappeared early in the nineteenth century, the only tangible evidence of its former existence being skeletal remains. A downy chick in the Dominion Museum, Wellington and an adult specimen in the museum at Christchurch are alleged to have been collected on Pitt Island, but all other

XII Chatham Islands Rails. Hand-coloured lithograph by J.G. Keulemans from W.L. Buller's *Supplement to the 'Birds of New Zealand'*, Vol. 1 (London, 1905), Pl.3.

recorded skins were taken from Mangare, which seems to have been the species' last refuge. *Rallus modestus* was described from a specimen obtained on this island at the start of the 1870s and during the next 25 years more examples were taken for museums and private collections. Both Buller and Walter Rothschild owned a series of skins and others were sent to institutions in Britain, France, North America and New Zealand. By 1900, the Chatham Islands Rail appeared to have vanished and it has never been seen since. It is to be doubted whether collecting added significantly to the decline of the species for the original vegetation of Mangare had been burnt off and the ground replanted to provide pasture for sheep. Cats had overrun the island. Almost certainly the birds would have been unable to withstand this combination of circumstances.

Little is known of the birds' habits. They were reported by a Mr Hawkins to nest in burrows in the ground or hollow trees. Their food consisted chiefly of insects and, most particularly, of sand-hoppers (amphipod crustaceans). An egg, measuring 40 mm x 21 mm, is creamy white with faint spotting of purplish grey and pale rufous.

WAKE ISLAND RAIL

Rallus wakensis

PLATE XIII

Rallus wakensis (Rothschild, 1903)

Hypotaenidia wakensis Rothschild, 1903 (Wake Island, Pacific Ocean)

Length 22 cm (9 in)

Description
Adult: lores, cheeks, top of head and rest of upperparts dark greyish brown; a pale-grey superciliary line joins with grey chin, throat and breast and on breast a band of pale rufous sometimes shows; sides of lower breast, abdomen and under tail coverts are darker than upperparts but marked with narrow white bars and centre of abdomen is whitish; bill, feet and legs brown. Sexes apparently similar.

Measurements
Male: wing 85–100 mm; tail 45 mm; culmen 25–29 mm; tarsus 33–37 mm.
Female: as male, perhaps slightly smaller.

The extinction of some species has been accomplished in quite extraordinary circumstances and in several instances these have acquired a celebrity outweighing the intrinsic fascination of the bird itself. The Stephen Island Wren (*Xenicus lyalli*) is known as the bird that was exterminated single-handed by a lighthouse keeper's pet cat (see page 192). The vast array of Passenger Pigeons (*Ectopistes migratorius*), whose seemingly endless ranks darkened North American skies, was thinned within just a few decades to a point where only a single individual survived, caged in the Cincinnati Zoo (page 112). The Wake Island Rail is the bird rumoured to have been eaten out of existence by Japanese forces occupying its island home during World War II. These men are known to have been living on what amounted to virtually a starvation diet for at least part of their stay and birds belonging to the only endemic land species must have provided a very welcome supply of food.

Wake, situated in the middle of the Pacific, hundreds of miles from even the farthest-flung member of the Hawaiian group, is one of the most remote of islands. It is very low and scrub-covered, spreading over an area of just 23 km² (9 sq. miles). In such an environment, flightless rails – for the flight feathers in this species were no more than rudimentary – would have been helpless before the concerted attacks of humans. Visitors to Wake before the outbreak of hostilities recorded that the rail was quite plentiful, but when the island was repossessed in 1945 not one individual could be found.

The Wake Island Rail seems to have been a member of a superspecies that includes the widespread and generally successful Banded Rail (*Rallus philippensis*). In fact, *R. wakensis* seems to have only barely passed the point at which separate status as a species can be recognized. In size, it was smaller, the plumage coloration paler and less bright and lacking the rather spectacular spotting of its more successful counterpart. The loss of the power of flight is, of course, not unusual for a rail developing upon an island far from the major land masses.

First described by Walter Rothschild in 1903, little has been published

XIII Wake Island Rail. Oil painting by Errol Fuller.

regarding the Wake Island Rail's habits. Breeding took place between July and August but perhaps also at other times. A low chattering and clattering, fairly typical for a smallish rail, betrayed the bird's presence and close at hand a gentle cluck could be heard. Also characteristic of a rail were the rather deliberate and cautious movements together with an occasional twitch of the tail. As might be expected, when disturbed in the open these birds would scamper back into cover very quickly. Their desperate attempts to avoid capture as one after another they were hunted down by the starving soldiers, must have made a pitiful sight. In an environment such as that on Wake, land birds could not afford to be too choosy in matters of feeding and presumably this species took advantage of most food items that came its way. Individuals were seen digging in the earth with their bills, searching for molluscs, insects and worms.

TAHITIAN RED-BILLED RAIL

Rallus pacificus

PLATE XV

Rallus pacificus (Gmelin, 1789)

Rallus pacificus Gmelin, 1789
Hypotaenidia (?) *pacificus* Rothschild, 1907

Length 23 cm (9 in)

Description
Adult: top of head and sides of face black; superciliary line running from bill to occiput white; nape rust coloured; back and wings black with white spots or bars; tail coverts and tail black; chin, throat, breast and abdomen white, a black collar dividing throat and breast; bill blood red; iris blood red; legs and feet fleshy pink.

An assortment of allegedly distinct kinds of bird are known only from descriptions given by early travellers or naturalists, rather than from actual specimen material, but as a rule it is impractical to regard these as valid species. The inherent difficulties that these early accounts present are so considerable and the interpretation of whatever information exists in them so subjective that when forms described are unsupported by material evidence they are best considered as 'hypothetical' (see last chapter). In cases where only a description upholds the validity of a species, there exists always the possibility that the original observer mistook what he saw or confused it with a well-known species with which he was unfamiliar. In some puzzling instances it is possible that interesting creatures were actually invented to fill out the pages of a narrative. Attempting to assay the truth from a flimsy account perhaps 200 or 300 years after it was penned is hardly worthwhile.

About the validity of *Rallus pacificus* as a distinct species there need be little doubt, however, even though no kind of specimen exists. The precise identity of this rail is clearly defined and the bird can be described with confidence because of the special circumstances surrounding the species' discovery. It was found by naturalists who took part in James Cook's second voyage around the world and is known now solely from an illustration by George Forster kept in the British Museum (Natural History) (pl. 128 in the collection). The picture shows a black-and-white rail of small size with a red beak and pink legs and although rather crude is fairly explicit – serving as the basis for all subsequent illustrations, including subtler works by John Gerrard Keulemans (in Rothschild, 1907) and Fenwick Lansdowne (in Ripley, 1977).

Although there need be no confusion over the separate identity of this species, there has, in fact, been more than a little – its name becoming mixed up in the synonymy of a completely different form, a race of the widespread Banded Rail, *Rallus philippensis ecaudata*. The two forms do not look alike and have, in reality, no connection other than that they belong to the same genus. Confusion apparently arose over locality records for *ecaudata* (see Lysaght, 1953), and solely because the name *pacificus* has occasionally been included in the synonymy of *R. p. ecaudata*, *ecaudata* itself sometimes occurs in lists of extinct birds (e.g. King, 1981) but the race survives still, an inhabitant of Tonga.

As far as *Rallus pacificus* is concerned, virtually nothing is known. That it once lived on Tahiti, where it was called *tevea*, *oomnaa* or *eboonaa*, seems certain; and it may also have occurred on the nearby island of Mehetia until much more recently. It has been suggested that on Mehetia the species was still extant during the early years of this century although it had long vanished from Tahiti. Presumably the extinction was associated with the predations of introduced mammals – cats and rats.

ASCENSION ISLAND RAIL

Atlantisia elpenor

FIGURE 21

Atlantisia elpenor (Olson, 1973)

Atlantisia elpenor Olson, 1973

Length unknown; probably about 22 cm (9 in)

Description
Appearance in life unknown.

During an age of the world long passed, islands of the South Atlantic were invaded by birds probably much resembling the ancestors of today's long-billed, modestly sized rails included in the genus *Rallus*.

The competitive pressure applied to relatives left behind by these ancestral stocks on the larger land masses was something not experienced by the colonizing individuals and their descendants. Neither did these antediluvian wanderers undergo the correspondingly potent surge of evolutionary activity as they perpetuated themselves for generation after generation in their quiet, if barren, little backwaters. Conditions upon one island might over many, many years have led to the development of a giant breed while on another the emergence of more delicately proportioned rails may have been favoured. Here, a population may have clung tenaciously to existence for long ages; there, the line perhaps failed within a very few lifespans. Sometimes, maybe, an island was recolonized over and over again by successive waves of ralline invaders, each population doomed in its turn to eventual extinction. On how many islands such ancestral populations settled, flourished for a while but ultimately dwindled, will never be known but at least three have supported colonies of seemingly related rails in the recent past.

Birds attributed to the genus *Atlantisia* have been described from Inaccessible Island (one of the Tristan da Cunha group), from Ascension and from St Helena. Of these, one still survives, one was extant until at least 1656 and the third became extinct at an undetermined date.

The rail of Inaccessible Island, *Atlantisea rogersi*, is small – around 13 cm (5 in) in length – blackish in colour with rather hair-like plumage. The species is common in the tussock grass that covers much of the island, and as long as its refuge remains free of mammalian predators this situation seems likely to remain unchanged.

On St Helena there once lived a gigantic form, *A. podarces*. This bird was about the size of a coot and its bones have been recognized in rocks ranging in age from several thousand to hundreds of thousands of years old. It is not known when or why it became extinct. Perhaps the species still survived in 1502 when the island was discovered but it may have already gone by this date. S.L. Olson (1977) drew attention to the very large claws and the rather large wings that (for a flightless bird) this creature possessed, speculating that these were used for climbing and fluttering up the slopes and cliffs of its home.

On Ascension lived a bird much closer in size to the still-living rail of

21 The rail of Ascension Island as figured by Peter Mundy.

Inaccessible Island. This bird is known to have survived into comparatively recent historical times. What it fed upon is something of a mystery for Ascension is notably barren; presumably a living was eked out by seeking the eggs and regurgitated prey of seabirds. The date at which it died out and the reasons for extinction are unknown.

Bones of this rail were first found in 1958 but in 1970 superior material was discovered from which it proved possible to describe a new species. This was not the first record of *Atlantisia elpenor*. The much travelled and very engaging English writer Peter Mundy, who visited Mauritius in 1638 and made observations there of the extinct rail *Aphanapteryx bonasia*, had arrived at Ascension 18 years later, during the summer of 1656. For the month of June he wrote:

> Alsoe halfe a dozen of a strange kind of fowle, much bigger then our sterlings ore stares: colour gray or dappled, white and blacke feathers intermixed, eies red like rubies, wings very imperfitt, such as wherewith they cannott raise themselves from the ground. They were taken running, in which they are exceeding swift, helping themselves a little with their wings . . . It was more than ordinary dainety meat, relishing like a roasting pigge.

The obsessive preoccupation shown by early mariners in regard to the gastronomic mysteries afforded on lonely islands is, perhaps, something of a curiosity to modern readers; remarks like Mundy's occur over and over again in seventeenth-century journals and diaries. The reasons for it are really not hard to uncover.

In those distant days the lot of a seaman was usually far from easy. Man's mastery of the environment was not such that he could construct ocean-going ships with any view to comfort. Not only were a seaman's surroundings cramped and spartan, provisions available were likely to be exceptionally unappetizing. Cutting costs was an important consideration for ships' provisioners but they had also to take into account the fact that methods of food preservation were crude and that their vessels, once upon the high seas, were largely at the mercy of the elements – in a dead calm a ship might float aimlessly for weeks on end, when the wind whipped up it might be blown many miles off course. Hardtack – a bread dried so thoroughly that mould could not ruin it – was the staple diet, supplemented with heavily oversalted meat of the very lowest grade. Anything fresh needed to be used up quickly before it either rotted or was subjected to infestation by insects.

Little wonder, then, that when men were set loose for a short time on deserted islands – where their attention was not distracted by other pleasures – they asked two questions concerning the living creatures they saw. The first was, 'Could you eat it?'; the second, 'Could you catch it?' Usually the answer to both questions was yes!

Mundy, in fact, asked a further question:

> I have heretofore asked the question concerning Mauritius henns and dodos, thatt seeing those could neither fly nor swymme, being cloven footed and withoutt wings on an iland far from any other land, and none to be seence elce where, how they shold come thither.

It may not be generally known that Darwin had so early a predecessor! Mundy continued:

The question is, how they shold bee generated, whither created there from the beginning, or thatt the earth produceth them of its owne accord, as mice, serpentts, flies, wormes, etts. insects, or whither the nature of the earth and climate have altered the spape [shape] and nature of some other foule into this, I leave it to the learned to dispute of.

KUSAIE ISLAND CRAKE

Porzana monasa

PLATE XIV

Porzana monasa (Kittlitz, 1858)

Rallus monasa Kittlitz, 1858 (Kusaie, Caroline Islands)
Ortygometra tabuensis Finsch, 1880
Aphanolimnas monasa Sharpe, 1892

Length 18 cm (7 in)

Description
Adult: generally black showing bluish-grey reflections with quills slightly browner and chin and middle of throat paler; upper surface of tail brownish black, under tail coverts spotted with white; inner wing coverts brownish with white spotting; outer edge of first primary dull brown; iris, feet and legs red; bill blackish.

XIV Kusaie Island Crake. Oil painting by Carolyn Sinclair-Smith.

Kusaie (or Kosrae) Island lying to the east of the main Caroline group in the mid-Pacific was a favourite resting place for whalers during the early years of the nineteenth century. Here they would beach their ships and strip them down for cleaning or repair. Through this agency, probably, rats were released onto the island and two species endemic to it appear to have become extinct as a result – the Kusaie Island Starling (*Aplonis corvina*) and the crake originally described as *Rallus monasa* but now generally referred to the genus *Porzana*. Little is known of either species and both are represented in museums only by specimens collected during the 1820s.

F.H. von Kittlitz discovered the crake during a visit to the Carolines in December 1827 and January 1828 yet a description of the species does not seem to have been published until 30 years later. He collected two birds in marshes at sea level and these, the only known examples, are now in the Leningrad Museum. No other European scientist seems ever to have seen these creatures alive. According to S.D. Ripley (1977), who recently examined the skins, they show the species to have been a fairly typical dark crake set apart from its close relatives by slightly larger overall size and stouter bill. Von Kittlitz recorded the species as uncommon even in 1828 and remarked that individuals lived alone in wetter, shadowy parts of the forest from whence, occasionally, they were heard calling.

The name *satamanot* was given by von Kittlitz as the locally current term for these birds. A century later, W.F. Coultas, a member of the Whitney South Seas Expedition, recorded the native name as *nay tai mai not*. Both reporters obviously heard the same sounds but wrote them down in different ways. Coultas's interpretation, being more carefully noted, is presumably more accurate. Against the implication of this native name, which means 'to land in the taro garden', it has been suggested that this crake was flightless, or almost so, and therefore particularly vulnerable to the attacks of humans and introduced mammals. Curiously, it seems that the native population considered the species sacred and on this account never interfered with it.

Exactly when the species became extinct is not known but it now seems certain that it is. Kusaie, of course, has been visited only rarely by naturalists. Otto Finsch could not find these birds during a visit in 1880; nor did he hear them call. Coultas failed to locate any in 1931 despite a prolonged stay. Both men recorded that Kusaie was overrun with rats.

ANOTHER PUBLISHED ILLUSTRATION OF THE *KUSAIE ISLAND CRAKE*

Ripley, S.D. 1977. *Rails of the World*, pl. 29. Artist, J. Fenwick Lansdowne.

HAWAIIAN RAIL

Porzana sandwichensis

PLATE XV

Porzana sandwichensis
(Gmelin, 1789)

Rallus sandwichensis Gmelin, 1789 (Hawaiian Islands)
Pennula millei (misprint for *millsi*) Dole, 1879
Pennula wilsoni Finsch, 1898

Length 14 cm ($5\frac{1}{2}$ in)

Description
Adult: (*millsi*) lores fulvescent; top of head greyish brown; ear coverts grey; back and wing and tail coverts chocolate brown but feathers on lower back showing darker centres; concealed tail feathers blackish with chocolate-brown edging; chin whitish; cheeks, throat and breast reddish brown shading to dusky chocolate brown on abdomen, flanks and under tail coverts; iris reddish brown; bill bluish horn, yellow at base; legs and feet orange brown.
Immature: (*sandwichensis*) similar to adult but generally paler; feathers on the mantle and back have dark-brown centres edged with a much lighter brown giving a mottled appearance; throat paler brown.

Measurements
sandwichensis: wing 73 mm; culmen 19 mm; tarsus 30 mm.
millsi: wing 70 mm; tail 20 mm; culmen 20 mm; tarsus 29 mm.

Whether there was just one small species of rail formerly inhabiting the island of Hawaii or whether there were in fact two, is something of a mystery. A paler form and a darker one were described under the names *Rallus sandwichensis* and *Pennula millei*, respectively, but most recent authorities (Stresemann, 1950; Greenway, 1958; Berger, 1973; Ripley, 1977) agree that both should be united in one species with the pale birds considered immature. S.D. Ripley (1977) points out, in particular, the similarities in plumage developments between these forms and the young and adult of the Spotless Crake (*Porzana tabuensis*).

First collected during the eighteenth century, the paler birds were illustrated by William Ellis, surgeon and artist on Captain Cook's third voyage, described by John Latham and redescribed and named by the German scientist Johann Friedrich Gmelin. The whereabouts of just two specimens are today known – one in Leiden carrying the rather generalized locality data of 'the Hawaiian Islands' and the other, taken by a native hunter on Hawaii itself, now in New York. Ninety years after Gmelin's diagnosis, the darker form was described under the name *Pennula millei* – *millei* being, in fact, a misprint for *millsi*. Although more specimens of this form exist it is represented by only a very few examples. So desperate was Walter Rothschild for specimens to add to his collection of rare and curious birds that he sent Henry C. Palmer, his agent, with a dog especially trained for the purpose of finding the birds out, but to no avail. Neither of the genera used by the original authors is now considered applicable, the single species recognized being referred to *Porzana*.

Just as there is doubt as to whether one or two species existed, so there is doubt as to whether Hawaii alone was inhabited or whether these birds also occurred on other islands in the group. R.C.L. Perkins (1903), on the authority of a local resident, R. Meyer, stated that they once lived on Molokai, but it should be said that Hawaii itself is the only island from which examples are positively known to have originated.

The Hawaiian Rail appears to have become extinct shortly before the close of the nineteenth century, the last specimens taken being collected during the 1860s with a final sighting occurring in 1884 or, possibly, 1893. Reasons for its disappearance have not been positively determined but the predatory habits of imported rats were probably an important factor and it seems certain that dogs and cats preyed upon these birds. Mongooses (*Herpestes auropunctatus*), too, have been blamed but by the date of their first introduction (early 1880s), the rails must have been desperately close to extinction and the presence of additional predatory mammals can have made little difference to the overall situation.

The last haunts upon the island of this bird appear to have been on the sides of the volcano Kilauea and in undisturbed parts of the Olaa district. It is said to have inhabited the open country below the level of the rainforest or, sometimes, in scrub patches within the forest itself. Formerly quite widespread, Rothschild (1893–1900) mentioned that birds of this species were at one time served as delicacies at the tables of Hawaiian kings. They were called by the natives *moho*, a word apparently meaning 'bird that crows in the grass'. Curiously, the Maoris of New Zealand are said to have called the extinct North Island Takahe by the same name; there is, however, no connection between these birds.

LAYSAN RAIL

Porzana palmeri

PLATE XVI; FIGURES 22–3

Porzanula palmeri
(Frohawk, 1892)

Porzanula palmeri Frohawk, 1892 (Laysan, Hawaiian Islands)

Length 15 cm (6 in)

Description
Adult: forehead, area above and around eyes, sides of face, foreneck, breast and abdomen ash grey; remainder of upperparts sandy brown with centres of each feather blackish brown, but scapulars more rufous with some streaks of white on outer webs, and sides of neck, breast and flanks plain sandy brown unmarked apart from some white flecking towards rear of flanks; under wing coverts sandy brown, some feathers showing white tips; iris red; bill, feet and legs green. Sexes alike.
Chick: covered in blackish down; legs and feet black; bill yellow.

Measurements
Adult: wing 57–63 mm; tail 25–29 mm; culmen 17–19 mm; tarsus 24–27 mm.

22 Laysan Rail photographed on Laysan Island in the Hawaiian group by Alfred Bailey in December 1912. Courtesy of Denver Museum of Natural History.

The coral island of Laysan, small in size – just 3.2 km (2 miles) long by 1.6 km (1 mile) wide – and dreadfully remote (almost 1,280 km (800 miles) to the north-west of Hawaii) was, understandably, rarely visited by early mariners, or even by more recent ones. The small rail that lived on this lonely refuge was almost – but not quite – overlooked, yet its ultimate fate was to prove as poignant as that of any of the more celebrated extinct birds.

There can be little doubt that the Rail of Laysan, with a swift, restless manner and trusting nature, endeared itself to at least some of those few who came into contact with it. During the comparatively short period in which it came under observation, the species seems never to have learned to fear man. While one observer noticed how the Rails would approach to bathe within 1 m (3 ft) of his position, another claimed they were bold enough to hunt for flies between his feet – provided he remained still. A third mentioned a bird standing with one foot raised examining a pair of shoes and other individuals were seen quite happily entering buildings in search of food. Alfred Bailey (1956) of the Denver Museum recalled the keenness with which some Laysan Rails clambered over both his own legs and those of a companion, set upon reaching the yolk of albatross eggs that the two men were blowing. Bailey himself trapped many individuals for relocation on seemingly suitable islands and discovered the ease with which these could be caught.

> We merely took a little box and a six inch stick to hold up one side of it. A chicken egg – which the rails could not break – was placed for bait, and when half a dozen birds were inside, jumping off the ground to give more force to the beaks strike on the egg, we . . . pulled the string.

Despite their obvious charm, these little rails could be pugnacious and their desire for eggs sometimes provoked them into chasing away their small companion on the island, the Laysan Finch (*Psittirostra cantans cantans*), from eggs of noddies and terns that the passerines themselves had broken and were about to eat. The Rail's other food sources were varied. All kinds of insects and their larvae were consumed, these sometimes dug from the sand, the birds using beaks as shovels. Scraps were stolen from the corpses of dead seabirds and seed and other plant materials were eaten.

The species' tiny size (for a rail) made movement often seem rather mouse-like. Speed of reaction, said to be very quick, enabled Laysan Rails to snatch flies from mid-air. When alarmed, the birds could move surprisingly rapidly over the bare sand or through the tussock in which they generally lived. This turn of pace was occasionally necessary to avoid the unwelcome attentions of frigate birds – for Laysan Rails were flightless. Their wings were used, however, for jumping, as an aid to running, and also for fighting smaller birds.

A variety of call notes were described, among them warbling and chattering sounds. Two birds meeting sometimes produced a rattling noise with their throats becoming swollen and bills slightly parted. Most interesting of all was a peculiar evening chorus, which began soon after darkness fell but lasted for only a few seconds. As if by prior arrangement

23 Nest and eggs of the Laysan Rail photographed on Laysan Island on 10 May 1902 by Walter K. Fisher. Courtesy of Denver Museum of Natural History.

the birds started in unison, delivered their performance together and then all remained silent. F. Frohawk (1892) compared the sound to 'a handful or two of marbles being thrown on a glass roof and then descending in a succession of bounds'.

Breeding took place chiefly between the months of April and July. The nest, built on the ground among tussock or reeds, was constructed of grass and leaves, often arched over to create a roof, with an entrance at the side. A clutch typically numbered three and the eggs were pale olive buff closely spotted with greyish lilac and raw sienna. According to S.D. Ripley (1977), egg measurements were 31 mm x 21 mm. By curious coincidence, marbles were again used as an image to describe these birds – this time in connection with the chicks. F.C. Hadden (1941) compared one at 3 days with 'a black velvet marble rolling along the ground. Its little feet and legs are so small and move so fast they can hardly be seen'. By 5 days they were reported to be able to run as fast as the adult.

It was in 1828 that the species first came to attention when Russian sailors noticed it on Laysan. Interestingly, they claimed similar birds lived on Lisianski Island 184 km (115 miles) further west, but none were seen here again until man introduced them many years later. Specimens of the Laysan Rail seem not to have been collected until the 1890s when some were taken by members of an expedition backed by Walter Rothschild. At this time, and perhaps for the next 20 years or so, the Laysan Rail was plentiful, numbers being estimated at around 2,000. However, rabbits (*Oryctolagus cuniculus*) and guinea-pigs (*Cavia porcellus*) were introduced during the early years of the twentieth century, apparently in the expectation that a meat cannery would represent a commercially viable proposition. It did not, and the habitat deteriorated rapidly. By 1912 it had become apparent that escalating numbers of rodents were utterly destroying the vegetation. Ten years later the island was ruined. A. Wetmore (1925) described the landscape: 'On every hand extended a barren waste of sand . . . The desolateness of the scene was so depressing that unconsciously we talked in undertones. From all appearances, Laysan might have been some desert, with the gleaming lake below merely a mirage.' Just two rails were seen. By 1936, there were none.

XV Extinct rails. Top right: Hawaiian Rail (*Porzana sandwichensis*); middle: a dark form of *P. sandwichensis* or perhaps a separate species (*P. millsi*); top left: Tahitian Red-billed Rail (*Rallus pacificus*). Chromolithographs from W. Rothchild's *Extinct Birds* (London, 1907), Pl.26. Courtesy of The Hon. Miriam Rothschild.

XVI Laysan Rails. Hand-coloured lithograph by F.W. Frohawk from W. Rothschild's *Avifauna of Laysan and the Neighbouring Islands*, Vol. 1 (London, 1893), Pl.12. Courtesy of The Hon. Miriam Rothschild.

XVII Samoan Wood Rail by George Edward Lodge, from *The Proceedings of the 4th International Ornithological Congress*, 1905 (London, 1907), Pl.8.

But the Laysan Rail was not yet extinct. During 1891, birds had been removed to Eastern Island in the Midway Atoll, some 480 km (300 miles) to the north-west. More were relocated on this island in 1913 by Alfred Bailey and G. Willett who also released some on Lisianski. The Lisianski birds, like their counterparts on Laysan, suffered as this island's plant life was also depleted by rabbits, but the population on Eastern Island continued to thrive long after Laysan had become a wasteland and some individuals were transported to nearby Sand Island. These, too, flourished and at the start of World War II rails were common on both islands.

In 1943 a US Navy landing-craft drifted ashore bringing with it an invasion of rats to both islets. Within 2 years all trace of the rails was gone.

After the extermination of the rabbits on Laysan, vegetation on the island recovered in a most remarkable fashion. By 1945 the habitat on Laysan was almost fully restored.

SAMOAN WOOD RAIL

Gallinula pacifica

PLATE XVII

Two small, closely allied and rather mysterious species of rail, relatives of the ubiquitous Moorhen (*Gallinula chloropus*), have been recorded from widely separated islands in the South Pacific and both are known from little more than museum specimens. One, the San Cristobal Gallinule (*G. silvestris*), is represented by just the type, which was brought to members of the Whitney South Seas Expedition by a native hunter on 4 December 1929. This hunter claimed the bird had lived high in the interior above 600 m (1,970 ft), among dense undergrowth in tracts of mountainous forest; but with no further individuals being ever located, the species remains an enigma. As its home grounds in the Solomons were said to be mountainous and forest covered, the species perhaps survives still, although clearly very rare.

Gallinula pacifica (Hartlaub and Finsch, 1871)

Pareudiastes pacifica Hartlaub and Finsch, 1871 (Savaii, Western Samoa)

Length 15 cm (6 in)

Description
Adult: plumage generally blackish with olivaceous tinge on back and wings and bluish tinge on throat and breast; bill, frontal shield and legs probably reddish orange.

Measurements
Wing 115 mm; tail 58 mm; culmen (including shield) 38 mm; tarsus 36 mm.

Its nearest relative appears to be a rail once found many hundreds of miles to the east upon Savaii, one of the Samoan Islands. This similarly mysterious species is the Samoan Wood Rail (*Gallinula pacifica*), known from almost nothing more than the skins, perhaps eleven in total, deposited in the museums at Bremen, Hamburg, London and New York. *Gallinula pacifica* was first collected in 1869 by J. Kubary and the last record of its occurrence appears to date from 4 or 5 years later. Almost certainly, one George Brown presented a specimen, perhaps two, to naturalists of *The Challenger* Expedition. The species is now assumed extinct, although the possibility exists that it survives for many rails are by nature secretive and skulking birds. However, the Whitney South Seas Expedition, while ascertaining the existence of the San Cristobel Gallinule in the Solomons, had no luck in locating the bird once known to Samoans as *puna'e*.

This was a very small, dark rail, probably flightless or almost so, and a peculiarity noted by the original describers, G. Hartlaub and O. Finsch, were the exceptionally large eyes, a feature from which it has been inferred that the species was crepuscular or even nocturnal in habit. Two eggs, one now in the British Museum (Natural History), were found by a native in a nest situated on the ground and made from a few twigs and some grass. Other Samoans claimed to the contrary that these birds nested in burrows!

Why the species became extinct, assuming it is, remains uncertain. Large populations of cats and rats may have reduced numbers but perhaps the species was already rare even before these mammals were introduced. As far as can be definitely determined, the *puna'e* was restricted to the island of Savaii but it may have occurred on Opolu also.

LORD HOWE SWAMPHEN

Porphyrio albus

PLATE XVIII

An aura of mystery that is not entirely dispelled surrounds New Zealand's Takahe (*Notornis mantelli*), even though this vanishing species has been intensively studied during recent years. Described originally from subfossil remains, then identified from a very few skins all obtained during the nineteenth century, the species disappeared, presumed extinct, for almost 50 years. When rediscovered in 1948, the event was significant enough for the large and plump blue rail to acquire a celebrity far beyond the confines of its homeland.

The Takahe is closely related to the Purple Swamphen (*Porphyrio porphyrio*) – the species of gallinule inhabiting much of the Old World. During that remote past in which stocks of *Porphyrio*-like birds first colonized New Zealand, they were enabled, by the peculiar conditions pertaining there, to begin developing in extraordinary ways. Without competition from terrestrial mammals, predatory or otherwise, no great advantage necessarily accrued to creatures that could become airborne. As colonizing stock lost its will and then its ability to fly, changes in shape and size complementary to the new role took place – the head and beak became more massive, the body much bulkier, the legs and feet stouter and more robust. Takahes look very much like swollen Purple Swamphens because this is just what they are. Perhaps the most significant difference between these species lies in their respective rates of fertility: whereas the Purple Swamphen is a vigorous force, still in places expanding its range (it

Porphyrio albus (White, 1790)

Fulica alba White, 1790
Gallinula alba Latham, 1790
Notornis alba Salvin, 1873
Porphyrio stanleyi Rowley, 1875
Notornis stanleyi Rothschild, 1907

Length 55 cm (22 in)

Description
Adult: white, white and blue, or, perhaps all blue; beak and legs red; eyes probably brown; spur on wing light horn colour.
Chick: believed to have been black.

Measurements
Specimen in Vienna: wing 228 mm; tail 75 mm; culmen (including shield) 79 mm; tarsus 92 mm.
Specimen in Liverpool: wing 220 mm; tarsus 90 mm; tail and beak too badly damaged for accurate measurement.

appears, in fact, to have colonized New Zealand only recently, very possibly to the Takahe's detriment), the Takahe now seems reproductively almost bankrupt.

As swamphens are so widespread and such successful colonists, it might be expected that populations would find other mammal free island refuges, where, subject to influences that led to the Takahe's development, they might themselves evolve along similar lines. In fact, this seems to have happened at least once.

On Lord Howe Island, lying in the Tasman Sea 480 km (300 miles) or so to the east of the Australian mainland, a population of swamphens, apparently mostly white in colour, lived until the early 1800s. These birds, while not quite so extravagant, had begun following an evolutionary path parallel to, but, of course, completely independent from, that of the Takahe. They might almost be called 'White Takahes,' and, indeed, have been.

The evidence that this population lived is of three kinds: early written accounts, illustrations made at dates contemporary with the birds' existence and two very old and rather enigmatic skins. Despite all of these, there remain fairly fundamental unresolved mysteries surrounding the population. Although it seems certain that Lord Howe was the Swamphen's home, some evidence points to a similar occurrence on Norfolk Island about 880 km (550 miles) to the north-east; this possiblity is discounted by most researchers. Whether only birds that were white made up the population, or whether all-white birds represented simply a colour morph of a species containing also blue or blue-and-white individuals is not clear. The fact that the typically blue Australian race of the Purple Swamphen (*Porphyrio porphyrio melanotus*) is found on Lord Howe only serves to confuse this particular issue further; possibly the extinct population associated freely with it, thus clouding early records.

With these elements of the story remaining uncertain, assessing the exact status of the vanished population has also presented problems because doubts have been expressed over whether it constituted a valid species. Several authors have speculated that the birds merely formed a race of the Purple Swamphen, a hypothesis supported by the undoubted closeness of affinity, by the tendency to albinism sometimes shown in populations restricted to small islands and also by occasional records of partly albino individuals belonging to the familiar species.

Recent research appears to eliminate any doubt, however, and reasons for separating *Porphyrio albus*, the extinct Swamphen of Lord Howe, from any other species are summarized by S.D. Ripley (1977). Firstly, the retrices are soft not stiff as in *P. porphyrio*. Secondly, X-ray examination of bones by Dr G.F. van Tets in Australia indicates that the Lord Howe birds were flightless, having very stout legs with short toes. To this evidence can be added that length of the secondaries and wing coverts in relation to the primaries is nearer *Notornis* than *Porphyrio*. In other words, the Swamphen of Lord Howe seems in certain respects physically intermediate between the rare, endemic New Zealand species and the common and widespread Purple Swamphen.

The two very old specimens generally referred to the species *P. albus* in themselves present problems. Although very similar, they are not identical and in neither case is the full provenance known. One was evidently in the Leverian Museum, forming lot 2782 ('a white fulica from New Holland') at the sale by which the collection was dispersed. This assemblage of items connected with natural history was put together by

XVIII Lord Howe Swamphen. Hand-coloured lithograph by J.G. Keulemans from G.D. Rowley's *Ornithological Miscellany*, Vol. 1 (London, 1875), Pl.9.

Sir Ashton Lever who housed it in London's Leicester Square for a time in the late eighteenth century. He disposed of it by lottery in 1788. The collection was so vast that it took more than 2 months to disperse it when it was later sold. Although the sale was staged in London, the Swamphen was knocked down to the Imperial Museum, Vienna on Saturday 31 May 1806, the sum changing hands being fourteen shillings. Still in Vienna, at the Naturhistorisches Museum, this specimen is assumed to be the type of J. White's species *alba*, partly because the illustration in his book *Journal of a Voyage to New South Wales* (1790) shows a bird in similar attitude to the stuffed one (it is no longer in this position, having been reduced to a cabinet skin), partly because White remarked in a preface that models for his plates were deposited in the Leverian collection. The specimen's actual place of origin is not known and is only assumed to be Lord Howe Island. According to Walter Rothschild (1907), a label attached to the bird and present during its early years in Vienna revealed Norfolk Island as the true locality – but then labels are not necessarily always clear guides to the truth.

The Vienna example is usually described as pure white, a respect in which it differs from a specimen now in the Merseyside County Museum, Liverpool showing a blue sheen on the wings, but Rothschild (1907), contradicting most writers, mentioned that even in the Vienna/Leverian bird some blue was evident on the wings. The provenance of the Liverpool specimen can be traced back as far as a catalogue entry advertising the sale of the Bullock collection in 1819, and reading, 'Lot 60. White Gallinule. New Zealand, rare; brought by Sir J. Banks'. It is known that the bird sold for three guineas but whether Banks really did bring it to England, whether it actually came from or via New Zealand, or whether it originated on Lord Howe Island cannot be said. K.A. Hindwood (1940) doubted the attribution to *albus* but other workers have accepted it quite happily enough. G.D. Rowley (1875) believed this was a fairly young individual at the time of its death; in addition to blue glossing of the wings, it shows some brownish feathers on an otherwise white plumage.

Apart from the plate in White's book, which may or may not have been drawn from the specimen now in Vienna, several significant but fairly crude contemporary pictures of Lord Howe Swamphens exist. These are kept by the British Museum (Natural History) and either are, or purport to be, the works of Thomas Watling and George Raper. Watling, born in 1762, appears to have been a coiner who turned his hand to art during his enforced stay in New South Wales, producing a series of pictures acquired by the museum in 1902. Raper was a midshipman aboard HMS *Sirius*, which sailed with the First Fleet to Botany Bay in 1787. He recorded the images of many of the birds he saw in a primitive but quite charming manner and three sets of pictures alleged to be his work still exist – one in the British Museum, another in the Mitchell Library, Sydney and a third in the Alexander Turnbull Library, Wellington. In addition to the illustrations in the British Museum (Natural History) collection, one of those in the set kept at Wellington seems to show the Swamphen of Lord Howe.

The other major pieces of evidence indicating the previous existence of these birds upon Lord Howe are in the form of written accounts. One description is contained in the manuscript of the journal of Surgeon Bowes, held, like some of Raper's pictures, in the Mitchell Library. Another reference is contained in the anonymous work called *The Voyage of Governor Phillip to Botany Bay* (1789). Here it is noted that the birds were considerably larger than ordinary swamphens, with very stout bills and plumage pure white – although males carried some blue on the wings. They were described as very tame and quite common, not only on Lord Howe but on Norfolk Island also. No other contemporary chronicler mentions this presence on Norfolk so the reference is usually held to be an error.

Whatever the truth, it seems certain that men were able to kill these birds with sticks – perhaps an additional indication of their flightlessness. It is hardly surprising to find they did not long withstand the coming of man. Whalers, casual visitors to the island and even, perhaps, the first colonists can be presumed to have extirpated them and they seem to have been entirely gone by the 1840s, if not earlier.

MAURITIUS RED HEN

Aphanapteryx bonasia

FIGURES 24–6; 28

Aphanapteryx bonasia
(Sélys Longchamps, 1848)

Apterornis bonasia Sélys Longchamps, 1848
Didus herbertii Schlegel, 1854
Didus broeckei Schelgel, 1854
Aphanapteryx imperialis Frauenfeld, 1868
Aphanapteryx imperatoris Frauenfeld, 1868
Aphanapteryx broeckei Milne-Edwards, 1868
Aphanapteryx broeckii Milne-Edwards, 1869
Pezophaps broeckei Schlegel, 1873
Pezophaps herbertii Schlegel, 1873
Kuina mundyi Hachisuka, 1937
Pezocrex herberti Hachisuka, 1953

Length unknown (about the size of a farmyard hen)

Description
Adult (from contemporary descriptions and illustrations): bill long, downcurved, probably dark in colour; plumage reddish brown, perhaps fluffy with feather structure slightly decomposed; wings short and useless; tail short; legs long and powerful, perhaps dark in colour; iris yellow?

Lying in the Indian Ocean several hundred miles east of Madagascar are the Mascarene Islands – Réunion, Mauritius and Rodrigues, all three of volcanic origin. They were first visited, so it is believed, by Arab trading vessels but the islands, being then uninhabited, were doubtless of little interest to merchants. Centuries later, during the early 1500s, Portuguese seafarers chanced upon them, followed by the Dutch, the French and the English. Although Réunion is the largest of the three, only Mauritius has any kind of renown in the world outside but the restricted fame it enjoys rests upon two quite unrelated factors. Among stamp collectors it is known for a printing error that resulted in some of the rarest and most valuable stamps, and to those with an interest in natural history, Mauritius is noted as the home of the Dodo (see page 117).

By far the most celebrated, dodos were not, however, the only strange birds living upon Mauritius during historical times but doomed to vanish soon after European arrival on the island. Scattered among the accounts and illustrations of dodos brought back by early travellers were descriptions and pictures of very different flightless birds. These enigmatic glimpses of lost creatures were written or made by men lacking in both artistic skills and expertise in the field of ornithology, but to later researchers they were clear enough to suggest that the birds mentioned belonged to the rail family. Although the pictures drawn by those who actually travelled were hopelessly inadequate (but nevertheless, enormously charming), there are several more formal paintings – by three different artists – of a rather higher stamp. These are believed to be drawn from living models – birds brought home to Europe as objects of great curiosity and rarity.

One picture of a rail-like bird, completed about 1600 by the Dutch artist George Hoefnagel, came to the attention of ornithologists during the nineteenth century when it was found among paintings in the collection of the Emperor Francis of Austria. It is supposed to have been drawn from life in the Menagerie at Ebersdorf; a matching work shows a dodo.

A painting in the Prado, Madrid, known as *Noah's Ark* and completed before 1584 by Bassano, includes a figure of a bird identified by S.D. Ripley (1977) as the rail of Mauritius.

Three pictures attributed to Roelandt Savery – celebrated chiefly for his dodo portraits – are also thought to show the same bird; of these, the best

24 Three figures of the Mauritius Red Hen by Pieter van den Broecke, 1617 (top); Sir Thomas Herbert, 1634 (middle); and Peter Mundy, *c.* 1638 (bottom).

known is an oil in the British Museum (Natural History) probably dating from the 1620s. This painting features a dodo but various birds are grouped around it and one, depicted in the act of swallowing a frog, is assumed to be a Mauritius rail.

Nor are these the only indications of the former existence of a large, flightless rail upon the island of Mauritius. Bones were found in the Mare aux Songes swamp during the nineteenth century and several points of significance can be determined from these remains. Firstly, they certainly belong to rails, but to a species unassignable to any genus with living representatives. Secondly, there was a considerable size differential between individuals – possibly the result of sexual variation. Thirdly, details of the bone structure confirm that the birds were flightless.

From the written descriptions, from the pictures and from the skeletal material, has been built an image of a reddish-brown bird with long, stout legs and short wings. If a kiwi could be constructed with a light build, less hair-like feathering and all the more extreme features modified to create an altogether more bird-like aspect, then this, perhaps, would resemble the assumed appearance of *Aphanapteryx bonasia*.

Attention has been directed to discrepancies among contemporary accounts and illustrations and there is no doubt that these exist. A very extensive and confusing literature has grown up around the extinct rails of Mauritius with later commentators comparing original writings and pictures and founding hypothetical species upon real or supposed differences among them. Whether or not more than one species existed cannot be said with certainty. Differences found in the original sources may easily be accounted for by inaccurate observation or poor drawing skills. In an extensive review of the literature and evidence, S.L. Olson (1977) concluded that only a single species could be upheld.

Of contemporary accounts perhaps the most picturesque is that of Peter Mundy, an Englishman who visited Mauritius in 1638. He described

> A Mauritius henne, a Fowle as bigge as our English hennes . . . of which wee gotte only one. It hath a long Crooked sharpe pointed bill. Feathered all over, butte on their wings they are soe Few and smalle thatt they cannot with them raise themselves From the ground . . . They bee very good Meat, and are also Cloven Footed, soe that they can Neyther Fly nor Swymme.

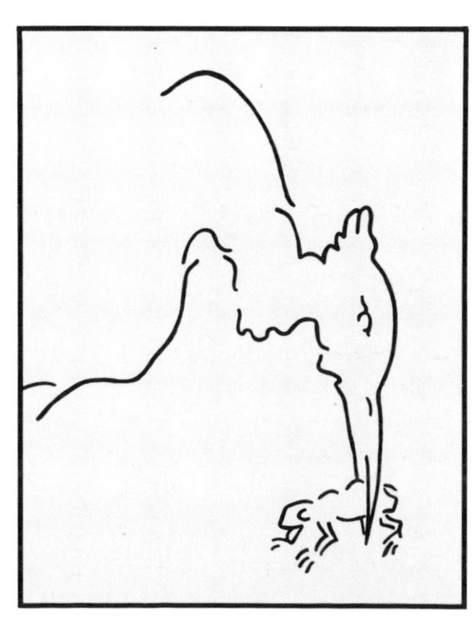

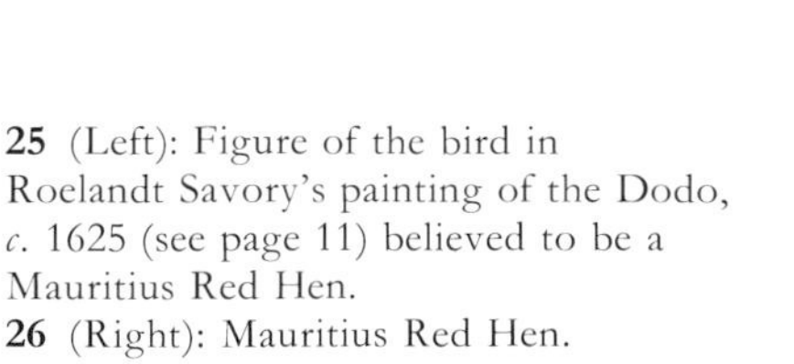

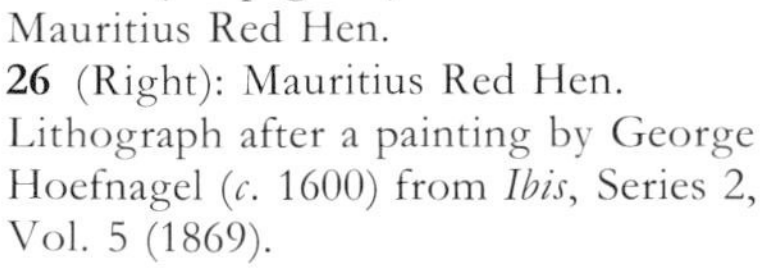

25 (Left): Figure of the bird in Roelandt Savory's painting of the Dodo, *c.* 1625 (see page 11) believed to be a Mauritius Red Hen.
26 (Right): Mauritius Red Hen. Lithograph after a painting by George Hoefnagel (*c.* 1600) from *Ibis*, Series 2, Vol. 5 (1869).

Other accounts mention the tendency of these birds to gather in flocks. Most describe the simplicity with which they could be caught, for a great part of the attraction that strange birds then held hinged around their potential suitability for the pot. The Mauritius Red Hen was said to be fatally attracted to the colour red. Mundy remarked of his bird: 'There is a pretty way of taking them with a redde Cappe, butte this was strucke with a sticke.' They apparently came running whenever a red rag was waved at them and once one was caught its squawks of alarm only served to draw others of its kind and, so, many could be taken together.

Clearly quite common in the early 1600s, this aberrant rail had become scarce by the second half of the century. François Leguat (1708) mentions that it was very rare in 1693 when he came to Mauritius and it seems likely that the species vanished entirely very soon after this date.

LEGUAT'S GELINOTE

Aphanapteryx leguati

FIGURES 27–8

Aphanapteryx leguati
(Milne-Edwards, 1874)

Erythromachus leguati Milne-Edwards, 1874
Miserythrus leguati Milne-Edwards in Haast, 1874

Length unknown (about the size of a small hen)

Description
Adult (from contemporary accounts): bill long ~ 50 mm (2 in), according to François Leguat), downcurved slightly (straight according to Leguat) and red in colour; orbital ring around eye red; plumage probably greyish, perhaps with white flecking; wings short and useless, feet red. Sexes probably alike.

Flying from persecution in their own land, a group of eleven French Huguenots left a Dutch port aboard a vessel bound for the Indian Ocean in the late summer of the year 1690. Their precise destination was the far-distant Mascarene island of Réunion, a territory they believed now abandoned by France. As the boat sailed into the waters of the North Sea, the Huguenots, under the leadership of one François Leguat, perhaps watched with mixed feelings the shores of northern Europe slipping further and further behind them. Weeks later, when the intended refuge finally came into view, any expectations were dashed, the French were still in possession, and the ship's captain changed course immediately for unoccupied Rodrigues, several hundred miles to the east. Here the refugees disembarked and stayed for 2 years until desperation (it has been said a lack of women) drove them to build a boat of their own and cross 480 km (300 miles) of open sea to Mauritius.

During the years spent on Rodrigues, Leguat made detailed records, published almost 20 years afterwards, of the birds he found there. In later years not only was the authenticity of his observations called into question, his own existence together with that of his little band of followers was doubted and his book pronounced a hoax. More recently, documentary evidence has come to light confirming that Leguat was, in reality, just what he had always seemed to be and it seems likely that his account can, generally, be relied upon. Certainly, descriptions of the creatures he encountered, sometimes considered fanciful, have in several instances been supported by the discovery of skeletal material.

For many years the case for the former existence upon Rodrigues of a bird seemingly closely related to the extinct rail of Mauritius (*Aphanapteryx bonasia*) rested solely on Leguat's account. Then, in 1874, A. Milne-Edwards, examining the bones of a flightless rail found in a cave on Rodrigues, felt able to connect them with the Huguenot's description. Subsequently, additional bones have turned up and from these it has been deduced that the Mauritius birds and those of Rodrigues were descended from a single ancestral stock that invaded both places during some period of remote antiquity. There is little doubt that the descendants had diverged to a considerable degree by the time of European arrival at the

Mascarenes. Although skeletal evidence led A. Günther and E. Newton (1879) to assign Rodrigues birds to the genus *Aphanapteryx*, J. Piveteau (1945) argued in favour of generic separation.

Bones were not the only evidence to support the word of Leguat, for just a year after the first of these were described, Alfred Newton drew attention to a mysterious and anonymous manuscript deposited in the archives of the Ministère de la Marine in Paris. This document was called the *Relation de l'Île Rodrigue* and it is thought to have been forwarded to the Compagnie des Indes about 1730. Within its pages is a description of a bird that clearly matches the one mentioned in Leguat's account. A creature about the size of a hen is spoken of, its plumage flecked with grey and white. The beak, coloured red like the feet, was in shape like that of a curlew but rather stouter and shorter. Wings were unfit for flight and the bird's cry was a continual whistling although in alarm they produced a kind of hiccup. It could run well but after having fed upon tortoise eggs was liable to become so fat that it was unable to perform effectively. The flesh was good to eat and the orange fat excellent for ailments.

Leguat himself referred to the birds as 'gelinotes', an old French word for hen. In the English translation of his book, published in 1708, this is translated (p.75) to Wood-hen:

> Our Wood-hens are fat all the year round and of a most delicate taste. Their colour is always of a bright gray, and there is very little difference in the plumage between the two sexes. They hide their nests so well that we could not find them out, and consequently did not taste their eggs. They have a red list about their eyes, their beaks are straight and pointed, near two inches long, and red also. They cannot fly, their fat makes them too heavy for it. If you offer them anything thats red, they are so angry they will fly at you and catch it out of your hand, and in the heat of the combat we had an opportunity to take them with ease.

27 Reconstruction of Leguat's Gelinote. Chromolithograph after a painting by F.W. Frohawk from W. Rothschild's *Extinct Birds* (London, 1907), Pl.30. Courtesy of The Hon. Miriam Rothschild.

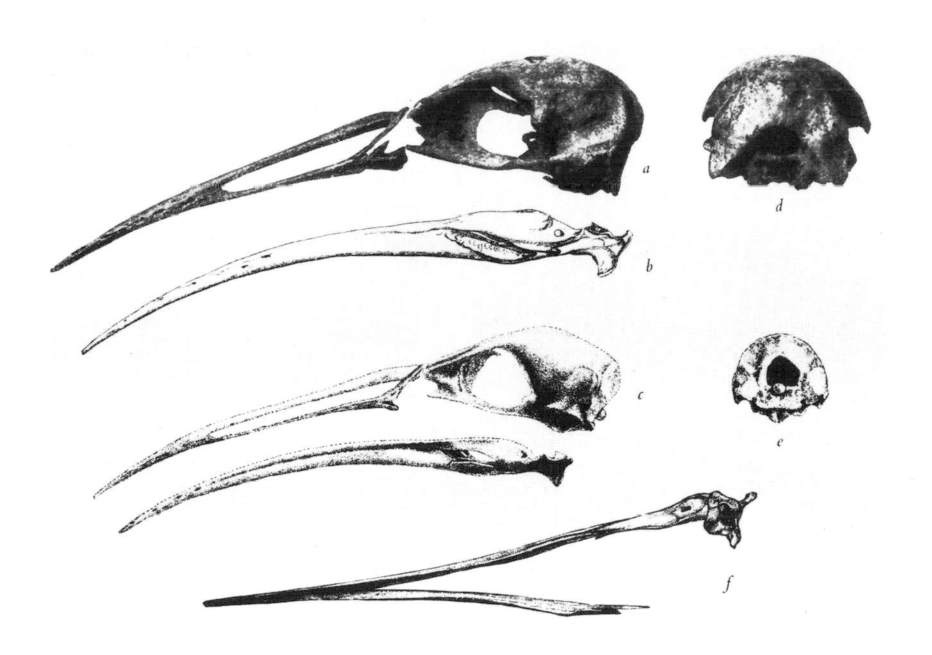

28 Cranial elements of *Aphanapteryx*: *A. bonasia*, the Mauritius Red Hen (a,b,d,f); *A. leguati*, Leguat's Gelinote (c,e). Arrangement derived from the work of S.L. Olson (after Günther and Newton, Piveteau and Milne-Edwards) in S.D. Ripley's *Rails of the World* (David Godine: Boston, 1977). Courtesy of S.L. Olson and S.D. Ripley.

J.C. Greenway (1958) believed this account related not to extinct rails but to wind-blown Madagascan Gallinules (*Porphyrio porphyrio madagascariensis*) or to a closely related species of which there is no evidence save this description. Quite why he believed this is not clear although he makes the point that grey and blue are sometimes synonymous in these writings. Blue would naturally be expected in a *Porphyrio* species. There seems little doubt, however, that the 'gelinotes' were the rails of which skeletal remains have been found and almost all commentators have accepted them as such.

It may be noted that Leguat described the beak as straight and pointed, whereas in the anonymous manuscript it was said to be downcurved. This discrepancy need not cause concern, for Günther and Newton (1879) demonstrated with skulls that a good deal of variability existed in the curvature of the bill, which sometimes did indeed appear almost straight.

It is interesting to read Leguat's remarks on the attraction that the colour red held for these birds. This corresponds with the preference alleged to have been shown by the related rails on Mauritius. There seems little reason to doubt Leguat but it is possible that he heard later of the Mauritian species' peculiarity and so applied it to his own account, perhaps intentionally to add colour or maybe through forgetfulness.

Leguat almost certainly arrived in Mauritius too late to see *A. bonasia* for himself. Neither did he nor his companions find the women for whom they are supposed to have been in desperate search. The flight from Rodrigues was only the beginning of a peculiar train of events that was to take them even further from their homeland. Soon after their landing the Dutch Governor of Mauritius was made aware of the Huguenots' arrival. Following several months of deliberation on the political implications of their presence, he banished Leguat's band to a small, rocky islet. After 3 years imprisoned thus, they were packed off, still prisoners, to Batavia in the Dutch East Indies but just a year later were released and shipped back to Holland. Here they arrived in June 1698, almost 8 years after leaving, having completed a round trip of some 38,400 km (24,000 miles) during one of history's least well-known, most pointless, but, for all that, quite remarkable voyages.

GULLS, WADING BIRDS AND MURRES

ORDER
Charadriiformes

EXTINCT SPECIES

White-winged Sandpiper (*Prosobonia leucoptera*)
Jerdon's Courser (*Cursorius bitorquatus*)
Great Auk (*Alca impennis*)

Species within this order fall into three major groups: (1) the waders; (2) the skuas, gulls and terns; and (3) the auks, guillemots, and puffins.

Three species have become extinct in historical times: the Great Auk, Jerdon's Courser and the White-winged Sandpiper.

Three races also seem to have vanished recently: the Canary Islands Black Oystercatcher (*Haematopus ostralegus meadewaldoi*), the Barrier Sub-antarctic Snipe (*Coenocorypha aucklandica barrierensis*) and the Stewart Island Sub-antarctic Snipe (*Coenocorypha aucklandica iredalei*).

Usually listed as a subspecies of *Haematopus ostralegus, meadewaldoi* is sometimes described as a race of *Haematopus moquini*. This black Oyster-catcher was last seen on Tenerife in 1968. Two populations of the rare Sub-antarctic Snipe (*Coenocorypha aucklandica*), a species from scattered localities in New Zealand waters, have disappeared. The Little Barrier race, *C. a. barrierensis*, is known only from the type captured alive on Little Barrier Island about 1870 and given to the Auckland Museum. One other bird was seen then, but none has been recorded since. Another subspecies, *C. a. iredalei*, from the Stewart Island group has also vanished.

Two problematical forms are the Auckland Island Shore Plover (*Thinornis rossi*) and Cooper's Sandpiper (*Pisobia (Tringa) cooperi*). Both are known only from the types. The specimen known as *T. rossi*, now in the British Museum (Natural History), was collected during the voyage of *Erebus* and *Terror* in November of 1840 and described and illustrated in G.R. Gray's account (1845). It is said to have been taken on the Auckland Islands south of New Zealand, but may have been taken on Campbell,

Table 11

RARE OR LITTLE-KNOWN GULLS, WADERS AND RELATED BIRDS

Black Stilt *Himantopus novaezelandiae*
Magellanic Plover *Pluvianellus socialis*
Javanese Wattled Lapwing *Vanellus macropterus**
Piping Plover *Charadrius melodus*
New Zealand Shore Plover *Thinornis novaeseelandiae*
Cox's Sandpiper *Calidris paramelanotus*
Spotted Greenshank *Tringa guttifer*
Tuamotu Sandpiper *Prosobonia (Aechmorhynchus) cancellatus*
Eskimo Curlew *Numenius borealis*
Bristle-thighed Curlew *Numenius tahitiensis*
Slender-billed Curlew *Numenius tenuirostris*
Snipe-billed Dowitcher *Pseudoscolopax semipalmatus*
Sub-antarctic Snipe *Coenocorypha aucklandica*
Chatham Islands Sub-antarctic Snipe *Coenocorypha pusilla*
Imperial Snipe *Gallinago imperialis*
Himalayan Snipe *Gallinago nemoricola*
Indonesian Woodcock *Scolopax rochusseni*
Dusky (or Lava) Gull *Larus fuliginosus*
Audouin's Gull *Larus audouinii*
Mongolian (or Relict) Gull *Larus relictus*
Damara Tern *Sterna balaenarum*
Chinese Crested Tern *Sterna bernsteini*†

*Previously known as *Xiphidiopterus* or *Rogibyx tricolor*
†Often listed as *Sterna zimmermanni*

which was visited at roughly the same time. This unique example is usually thought of as an aberrant, perhaps young, New Zealand Shore Plover (*Thinornis novae-seelandiae*), a bird now itself rare. J.C. Greenway (1958) suggested that *rossi* may have been a sympatric species now extinct.

The case of Cooper's Sandpiper is in some respects similar. It, too, was described from a single specimen, this one taken on Long Island, New York on 24 May 1833. Like *T. rossi*, the bird has been considered nothing more than a 'sport' or aberrant individual. Elliot (1895) believed it to be either a particularly large White-rumped Sandpiper (*Calidris fuscicollis*) or a hybrid between this species and the Pectoral Sandpiper (*C. melanotos*). The specimen is in the Smithsonian Institution.

Although unlikely that these forms constitute valid species, the possibility cannot be ruled out completely. Until 1965, the Mongolian or Relict Gull (*Larus relictus*) was known from a single specimen that several commentators believed to be a hybrid. It is only since similar birds were discovered breeding in several Russian and Mongolian localities that acceptance of the species' validity has become widespread.

Several species in the order are rare or on the point of extinction (see Table 11). The Javanese Wattled Lapwing (*Vanellus macropterus*), for instance, may very well now be extinct. Last seen in East Java in 1939, it is said to have once occurred in Sumatra, Timor and Java. Another very rare bird, the Eskimo Curlew (*Numenius borealis*) (Figure 29), has been on the verge of extinction for decades but individuals continue to be seen from time to time. During the nineteenth century this was an abundant species and huge numbers were shot as the birds migrated across the United States to their Canadian breeding grounds. By the turn of the century the population had been seriously depleted and it has never recovered.

A sandpiper from the Tuamotu Archipelago in the South Pacific, *Prosobonia cancellatus*, has been listed as extinct (Rothschild, 1907). It still exists, but only on predator-free islands in the group. On some of these the species seems reasonably plentiful but its continuing existence depends on the islands remaining free of cats and rats. This bird appears to be a close relative of the extinct White-winged Sandpiper. Another sandpiper, Cox's (*Calidris paramelanotus*), has recently been discovered and may or may not prove to be rare. It was described by Shane Parker (1982) from just two examples, both collected in South Australia during the 1970s.

29 The endangered Eskimo Curlew (*Numenius borealis*). Aquatint by J.J. Audubon and R. Havell the younger from Audubon's *Birds of America* (London, 1827–38).

WHITE-WINGED SANDPIPER

Prosobonia leucoptera

PLATE XIX

Prosobonia leucoptera
(Gmelin, 1789)

Tringa leucoptera Gmelin, 1789; based on Latham's White-winged Sandpiper, 1787 (Tahiti and Eimeo = Moorea, Society Islands)
Totanus leucopterus Vieillot, 1817
Calidris leucopterus Cuvier, 1829
Tringa pyrrhetraea Lichtenstein, 1844
Tringoides leucopterus Gray, 1871
Phegornis leucopterus Seebohm, 1888
Prosobonia ellisi Sharpe, 1906

Length 18 cm (6¾ in)

Description
Leiden specimen: crown blackish brown with nape and sides of face rather browner; lores and ear coverts slightly more reddish with a white spot behind eye; cheeks russet; chin and upper throat buffish white; back and wings blackish brown with crescent-shaped patch of white on lesser wing coverts continuing across leading edge of the wing; two central tail feathers blackish brown, the rest rufous banded black; underparts generally russet and unbarred; iris described as blackish; bill black; legs and feet greenish (Latham described legs and feet of one specimen as yellow).

Measurements
Wing 110 mm; tail 55 mm; culmen 20 mm; tarsus 32 mm.

Specimens of this species were collected by George Forster and William Anderson in 1773 and 1777 during James Cook's voyages to the South Pacific; the White-winged or Tahitian Sandpiper has never been seen since. Individuals were found on the island of Tahiti and also on Moorea (Eimeo). Three examples (there may never have been more) passed through the hands of John Latham, the celebrated writer and ornithologist, each one differing from the others. The whereabouts of just one, a bird in the museum at Leiden, is known today, so direct comparison of specimens is no longer possible. Pictures by Forster and William Ellis, however, exist in the British Museum, the Ellis illustration allegedly showing a bird from Moorea.

On the basis of plumage variation this bird was separated from its companions by R.B. Sharpe (1906) under the name of *ellisi*, but earlier workers, advantaged by actually handling specimens, evidently felt little doubt that all belonged to a single species. Presumably, plumage inconsistency can be accounted for by differences in age or perhaps a discrepancy in the season during which the birds were collected. One of Latham's birds differed from the Leiden example in that it had a rufous not whitish throat and a pale streak rather than a white spot behind the eye. The Moorea bird was redder with a ferruginous line above the eye.

Almost nothing is known of its habits. Forster found it close to small brooks and at the time of Cook's last voyage (1776–9) it was not uncommon. Rats may have played a major part in its extinction.

XIX White-winged Sandpiper. Hand-coloured lithograph by J.G. Keulemans from H. Seebohm's *Geographical Distribution of the Charadriidae* (London, 1888), Pl.18.

XX Jerdon's Courser. Hand-coloured lithograph by J.G. Keulemans from H. Seebohm's *Geographical Distribution of the Charadriidae* (London, 1888), Pl.13.

JERDON'S COURSER

Cursorius bitorquatus

PLATE XX

Cursorius bitorquatus
(Blyth, 1848)

Macrotarsius bitorquatus Blyth (*ex* Jerdon MS) 1848 (Eastern Ghats, India)
Rhinoptilus bitorquatus Strickland, 1850

Length 27 cm (11 in)

Description
Adult: narrow coronal streak white or pale buff dividing bands of brown stretching from (cont'd)

This little-known bird was first brought to the attention of naturalists by T.C. Jerdon in 1848 and last seen in 1900 when Howard Campbell spotted birds near Anantpur on the northeastern coast of India. Between these years, it was observed and collected, always within a very restricted area to the north of Madras, by several competent field workers. Jerdon and W.T. Blanford, both authors of classic works on Indian ornithology, managed to locate it at different times and it was recorded on various occasions during the last half of the nineteenth century.

Perhaps birds of this species still exist but, if so, it is difficult to understand why they have been overlooked in the years since 1900. Several times they have been searched for, without success, by experienced ornithologists, all of them well aware of the localities and the kind of habitat in which these birds had previously been sighted. Where they occurred, they are believed to have been permanent residents and records of them exist only from the Penner and Godavari valleys, more specifically

Cursorius bitorquatus (cont'd) forehead to hindneck; above each eye a broad white band extends from base of beak to hindneck, below each eye a broad band, grey streaked with black, follows a similar pattern; tail coverts white; tail feathers blackish; median wing coverts pale grey brown with broad white edging (forming a conspicuous wing bar); primaries black, outermost with broad white patch on outer webbing joining obliquely with similar patch on inner webbing, this feature decreasing until the fourth primary on which it forms a small spot; remainder of upperparts, scapulars and inner secondaries pinkish sandy brown; chin and upper throat white, looping round, with black edging, to separate rufous foreneck from brown breast; a second white band edged black stretches across lower breast; remainder of underparts creamy white; iris dark brown; bill pale yellow from gape to nostrils, blackish at tips of both mandibles; legs and feet pale yellowish white with flesh-coloured tinge; sole flesh-coloured; claws horny. Sexes alike.

Measurements
Wing 161–168 mm; tail 65 mm; culmen 19 mm; tarsus 68 mm.

in the neighbourhoods of Sironcha, Bhadrachalam, Nellore, Cuddapah and Anantpur.

The coursers are a closely knit group made up of eight species characterized by long legs, short wings and rather plain, although very beautiful, colours that afford the birds effective camouflage. They usually fly only when pressed, but if just disturbed will run off with shoulders hunched. In other circumstances they may be seen on their tiptoes, stretching upwards to inspect their surroundings. Like many other running birds, coursers have lost the hind toe and their feet have a curiously stunted appearance. Most often, they occur in dry even desert, regions where they feed chiefly on insects and their larvae.

Although Jerdon's Courser seems to have been closely related to the Indian Courser (*Cursorius coromandelicus*), a species distributed over much of peninsular India, it could be easily separated from this bird by a straighter bill and a double band across the breast. Its flight is said to have been faster than that of its more common relative and a different kind of terrain was occupied. Whereas the Indian Courser usually inhabits the dry, stony plains and areas of waste ground, this species was encountered only where scrubby jungle covered rocky, undulating country. It was most often found in pairs or small parties. Blanford, for instance, twice met two and once came across a group of three. Generally, they were quiet birds although, according to Jerdon, an occasional rather plaintive cry was uttered.

Breeding most probably occurred in June for which month there is a record of a male taken with enlarged gonads. A clutch of two eggs was described by an anonymous correspondent of the *Asian* newspaper in 1895. Whether this account can be considered reliable is open to question but the eggs are supposed to have been found on the ground in scrub jungle and were yellow stone in colour, heavily blotched and spotted with black.

Although the species is almost always referred to as Jerdon's Courser, it is occasionally called the Double-banded Courser. An African species (*Rhinoptilus africanus*) is, however, sometimes known by this name. Jerdon recorded that the local inhabitants identified the bird by the name of 'jungle empty purse', but, according to S. Ali and S.D. Ripley (1969), this rather strange name is not now locally current or even understood.

In the light of present knowledge it is not possible to account for the species' extinction. There appear to be no obvious factors that apply here, nor is any clue afforded by the extinct bird's relative, the successful Indian Courser.

GREAT AUK

Alca impennis

PLATE XXI; FIGURES 30–2

Lying to the west of the coast of Devon, England, shut off from outside sometimes for days on end by blankets of fog and swirls of grey rain cloud, is lonely Lundy – Isle of Puffins. This enormous, flat-topped, block of granite rises for some 122 m (400 ft) above the treacherous currents that surround it, the towering cliffs and rocky ledges providing shelter for thousands upon thousands of seabirds. Here, in or about the year 1835, a resident saw two birds the like of which he had never before set eyes upon.

Alca impennis (Linnaeus, 1758)

Alca impennis Linnaeus, 1758 ('Europa arctica')
Plautus impennis Brunnich, 1772
Alca borealis Forster, 1817

Length 75 cm (30 in)

Description
Adult (Summer): upperparts black but sides of head and wings dark brown and large oval white patch before eye; chin and throat blackish brown; remainder of underparts white; iris chestnut; culmen black with white grooves; feet black.
Adult (Winter): chin and throat white not blackish brown; oval white patch before eye lacking; a grey line through eye to ear; broad white band above eye. Sexes alike in both summer and winter but Rothschild (1907) suggested that females had a greyish tinge on flanks.

Measurements
Wing 160–178 mm; tail 83–95 mm; culmen 82–90 mm; tarsus 55–62 mm.

What they were will forever remain something of a mystery, but the man's claim was to have glimpsed the King and Queen of the Razorbill Murres standing 'up bold like'.

Razorbills (*Alca torda*), with their curious beaks and starkly defined plumage of black and white, are birds with which inhabitants of Lundy could hardly be unfamiliar. Individuals belonging to just one particular species might seem monarchs among them and of these creatures such a description is perfect. By the time of this encounter, the species in question had become so rare that to those who dwelt close to the rocky stacks and cliffs that line the shores of the North Atlantic, it was little more than myth passed on here and there in folktale and legend. These people spoke of the Garefowl, a murre so gigantic it was unable to fly, carrying a beak huge enough to make the bill of its smaller relative, the Razorbill, seem quite modest in comparison.

As the Garefowl passed into legend among the people of the North, the Great Auk was fast acquiring an almost legendary reputation among skin

30 Above: Eggs of the Great Auk (½ natural size). Lithographs in S. Grieve's *The Great Auk or Garefowl* (Edinburgh, 1885). Below: Stevens's Auction Rooms in London's Covent Garden – the venue of many Great Auk egg sales.

31 Alderman Robert Champley of Scarborough, an avid Victorian collector of Great Auk eggs, with four of his eggs.

and egg collectors. Its last refuges were quite deliberately sought out by seamen executing commissions for specimens from private individuals and from institutions. Amazingly high sums were offered and paid when examples changed hands, which seems all the more extraordinary because remains of Great Auks, while scarce, are by no means so rare as those of many other extinct birds.

Some eighty skins and seventy-five eggs have been recorded and throughout the nineteenth century examples were regularly offered for sale, many of them at Stevens's Auction Rooms in London's Covent Garden, a venue where various celebrated items connected with natural history were sold (Figure 30). Here, for instance, the Dresden Museum purchased the third specimen of the Takahe (*Notornis mantelli*) for around £100 and, in 1863, the historic collection of gorillas formed by the American explorer Paul Du Chaillu was disposed of. So close was the connection of this firm with the dispersal of remains of the Great Auk, that its telegraphic address was simply, 'Auks, London'. In early years, Great Auks and their eggs fetched just a few pounds each but by 1900 a particularly choice egg was able to reach the sum of £330.15*s*.0*d*., a fine skin having realized £350 5 years previously. Today, few remain that are ever likely to come onto the market, but a stuffed bird sold in London in 1971 was purchased by an Icelandic museum for £9,000.

The eggs of the Great Auk, for which there was such fierce competition, are a dirty yellowish white covered, particularly around the bulkier end, with an irregular tracery of pale grey or brown. They are large in size, an average one measuring 124 mm × 75 mm, and pyriform in shape, thereby lessening the likelihood of their rolling away and being smashed. Objects of this shape move in a tight circle when nudged, a feature of great importance to this species as it made no nest. Sometimes an accumulation of guano might have been used but, characteristically, the single egg was laid upon the bare stone of low and remote isles onto which the birds waddled to breed.

Almost nothing can be said of breeding behaviour. The inside of the Great Auk's mouth has been described as bright yellow in colour and this ornamentation may have been used to attract or, perhaps less likely, to deter. On the rocky platforms that provided the Auks with their home, territory would be defended by a few low croaks. The beak could be used to administer an unpleasant bite but, this apart, these Auks were more or less defenceless.

At breeding colonies, their great size set them apart from the other murres but they could also be immediately distinguished from the Guillemot (*Uria aalge*) and the Razorbill by the more upright stance adopted. Exact times of the breeding cycle are not known but eggs were found in June and young recorded in July. Fish, crustaceans and other marine invertebrates made up the diet and these the Auk could pursue through the water with tremendous speed and agility.

The species occupied much the same ecological niche in the waters of the Northern Hemisphere as some of the medium-sized penguins fill in the oceans of the south. The great proliferation of these birds in southern seas is remarkable when one considers that in the north only the Great Auk can be considered a direct counterpart. Presumably, this imbalance is due to the relative absence of major land masses below the Equator and the correspondingly larger area covered by water. It is an interesting fact that the name 'penguin' was originally applied to the Great Auk. Although other derivations are possible, the word is probably of Welsh origin, *pen*, meaning 'head' and *gwyn*, meaning 'white'. As European seafarers ventured with increasing frequency into southern seas, the word became transferred to the seemingly similar, and much more widespread, birds which they found there.

At one time the Great Auk was widely distributed across the North Atlantic, from the Gulf of St Lawrence in the west to Norway in the east. There are records of its occurrence from as far south as the Channel Islands and northern France and fossil remains have been unearthed in Gibraltar and Italy. Presumably, the advance of the great ice sheets pressed the Auks further to the south than has been the case more recently; contrary to the supposition of certain writers, these were not birds that could flourish under arctic conditions.

Despite their extensive distribution, Great Auks showed a preference for breeding in colonies at just a few select localities. Rather in the manner of the Northern Gannet (*Sula bassana*), this species left many seemingly suitable sites unoccupied. Breeding stations are known to have been situated on Bird Rocks, Gulf of St Lawrence; Funk Island, off the Newfoundland coast; Grimsey, the Geirfuglasker (sunk due to volcanic activity during 1830) and Eldey, all close to Iceland; and St Kilda, westernmost of the Hebrides.

When in water, the Great Auk could be caught only with much good fortune but on land it was clumsy and slow, waddling with little steps across the low, rocky shores onto which it managed to drag itself. Herded together at breeding stations, colonies of these flightless birds provided an easy target and man hounded them wherever they were exposed in

XXI Great Auk. Hand-coloured lithograph by Edward Lear from J. Gould's *Birds of Europe*, Vol. 5 (London, 1832–7), Pl.400.

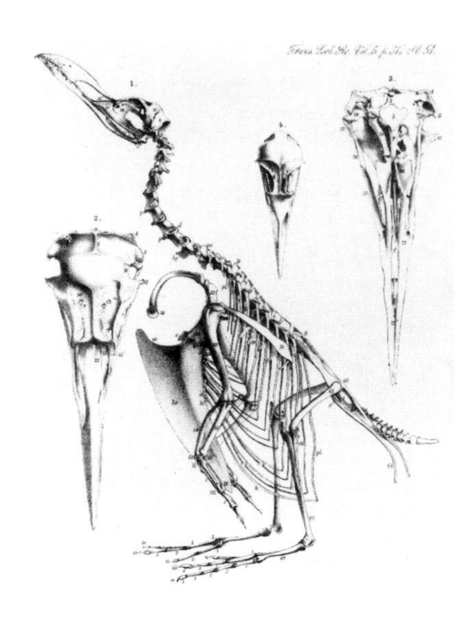

32 Skeleton and skull of a Great Auk together with the much smaller skull of a guillemot. Lithograph by E.A. Smith from the *Transactions of the Zoological Society of London*, Vol. 5 (1865).

localities that he could reach. Large size, suitability for the pot and its pathetic inability, at certain seasons, to evade capture, combined to make the Great Auk a quarry worth pursuing. To those who eked out a meagre living among the bleak isles and skerries of the far north, to sailors and fishermen anxious to replenish stocks before venturing into the icy waters of the Arctic, such a bird was irresistible. For a short time, vast numbers of birds were slaughtered for their feathers. On Funk Island, the unfortunate creatures were herded into pens, clubbed to death or immobility and tossed into vats of boiling water, which had the effect of loosening the plumage. The fires beneath these cauldrons were fuelled by the fat and oil from auks that had already met a similar fate.

By the end of the eighteenth century the Great Auk was all but gone. The few records of encounters with the species after this time make up a squalid list of human ignorance and cruelty. Many years after the event, an old resident of St Kilda claimed to have caught alive a Great Auk around the year 1840 with the help of four acquaintances. The bird was discovered asleep somewhere upon the rock known as Stack-an-Armine, and from here carried away to the islander's bothy. It was kept imprisoned but alive for 3 days and then the poor creature was beaten to death. It might have been a witch!

Some 20 years before this, during the early summer, another individual had been taken alive on St Kilda. Two men and two boys spotted it from a boat, sitting upon a low ledge. The men landed at either end of the shelf and began a steady approach while the boys rowed close in to the rock, just below the spot where the bird was resting. As the men drew nearer, the Auk, by now becoming increasingly alarmed, made desperately for the safety of the water but jumped straight into the arms of one of the waiting boys, who grasped his prize tightly and held on. This bird did not meet the fate usually kept in store for such hapless creatures, but was instead sent alive to Professor Fleming of New College, Edinburgh, who fattened it up with supplies of fresh fish and allowed it to exercise in the sea, preventing escape by means of a cord tied to one of the birds's feet. The precaution did not prove entirely adequate and the prisoner managed to escape into the Firth of Clyde during one of its daily dips.

One of the Great Auk's last strongholds was the Geirfuglasker, off the coast of Iceland, and when volcanic activity sunk it in 1830, a few birds continued to breed on the nearby island of Eldey. It was here, in 1844, that the species was seen for the last time. A party of sailors, commissioned by a collector to hunt for specimens, landed upon the island during the morning of the third of June. Among a mass of smaller seabirds, a pair of Great Auks was spotted and the birds were immediately attacked. As the frightened creatures made efforts to gain the water, waddling frantically forwards with their little wings outstretched, they were overtaken, one trapped between rocks and the other seized just a few steps from safety. Both were clubbed to death. There is a rumour that the female had been sitting on an egg and that this was crushed under foot in the race to reach the birds. The collector who commissioned the trip never took possession of his skins. They were sold, presumably for a better price, to a chemist in Reykjavik.

It is not realistic to suppose that this pair was actually the very last of their kind. Other individuals doubtless lingered here and there. But very soon after this date, somewhere in the deep waters of the North Atlantic or in a hollow upon a wind-lashed reef, perhaps sheltering from the fury of the waves, the last Great Auk died.

PIGEONS AND DOVES

ORDER
Columbiformes

EXTINCT SPECIES

Rodrigues Pigeon (?*Columba rodericana*)
Bonin Wood Pigeon (*Columba versicolor*)
Pigeon Hollandaise (*Alectroenas nitidissima*)
Forsters' Dove of Tanna (?*Gallicolumba ferruginea*)
Choiseul Crested Pigeon (*Microgoura meeki*)
Passenger Pigeon (*Ectopistes migratorius*)
Dodo (*Raphus cucullatus*)
Réunion Dodo (*Raphus solitarius*)
Rodrigues Solitary (*Pezophaps solitaria*)

Not being limited to any particular continent or climatic condition, pigeons and doves can be found in most parts of the world. They inhabit the tropics, the cold temperate regions and most of the environments in between. In general, they show themselves to be remarkably resilient birds despite relative defencelessness and the fact that they are favoured as food by man and a variety of predators. Many species, however, can be categorized as endangered and several are extinct. Among these latter are two of the most remarkable of all bird species: the extraordinary Dodo, by far the most aberrant member of the order, carries a name that has become virtually a synonym for extinction, and the Passenger Pigeon stands as an example of how even the most abundant and seemingly secure species may plunge catastrophically to extinction within an amazingly short period.

The species in the order which are thought to be extinct but of which there still remain either skins or skeletal material are: the Bonin Wood Pigeon (*Columba versicolor*), the Rodrigues Pigeon (?*C. rodericana*), the Passenger Pigeon (*Ectopistes migratorius*), the Choiseul Crested Pigeon (*Microgoura meeki*), the Pigeon Hollandais (*Alectroenas nitidissima*), the Dodo (*Raphus cucullatus*) and the Rodrigues Solitary (*Pezophaps solitaria*).

In addition to these extinct pigeons of which there is tangible evidence, several others have been named on the basis of written descriptions. These come properly within the category of hypothetical birds although two, the

Tanna Island Ground Dove of the Forsters (?*Gallicolumba ferruginea*) and the Réunion White Dodo (*Raphus solitarius*), can perhaps be accepted as valid extinct species. It has been alleged that other distinct dodo-like birds inhabited the Mascarenes; the history of these allegations is complex and is discussed later in the accounts of the various dodo species.

Another hypothetical extinct pigeon, supposedly a member of the Mascarene avifauna, is *Nesoenas duboisi* named by Walter Rothschild (1907) on the basis of a seventeenth-century description by 'Le Sieur' Dubois of an otherwise unknown bird living on the island of Réunion. This bird was said to be slightly larger than common European pigeons with a bigger beak, which was red towards the base. The plumage was russet red and the eyes were surrounded by a patch 'fiery red as in pheasants'. Nothing further can be said of this supposed species.

Another hypothetical form is a bird alleged once to have lived on Norfolk Island in the South Pacific and named by John Latham *Columba norfolciensis*. Latham described a bird 35 cm (14 in) long with head, neck and breast white, back and wings dusky purple showing darker markings, and underparts below the breast black. The tail was dullish purple, the bill black, and the legs red. Altogether there is considerable evidence to indicate that an extinct pigeon once lived on Norfolk Island but the issue has become very confused. A painting by John Hunter (see Hindwood, 1965) shows the 'Dove of Norfolk Island' but this does not match Latham's description. Still more confusing is the fact that Latham described two other birds under the name of *Columba norfolciensis* – one of these descriptions appears to apply to the Green-winged or Emerald Dove (*Chalcophaps indica*) and the other perhaps refers to the Australian White-headed Pigeon (*Columba leucomela*). Whether or not Latham really described a valid and now-extinct species cannot be satisfactorily determined unless additional evidence is forthcoming.

As well as the extinct species and the hypothetical extinct forms, several extant species have races that now seem extinct: the Madeiran Wood Pigeon (*Columba palumbus maderensis*), the Lord Howe White-throated Pigeon (*C. vitiensis godmanae*), the Seychelles Turtle Dove (*Streptopelia picturata rostrata*), the Cebu Amethyst Fruit Dove (*Phapitreron amethystina frontalis*), the Nukuhiva Red-moustached Fruit Dove (*Ptilinopus mercierii mercierii*) and the Norfolk Island Kereru (*Hemiphaga novaeseelandiae spadicea*).

Little can be said of the Madeiran Wood Pigeon (*Columba palumbus maderensis*) except that it vanished from Madeira during the early years of this century. Another bird belonging to the same genus but living at the other end of the world was the Lord Howe White-throated Pigeon (*Columba vitiensis godmanae*). There is no doubt that a now-extinct population of pigeons once lived on Lord Howe Island in the Tasman Sea. There are written descriptions of the bird dating from the last years of the eighteenth century and two paintings exist of the same vintage – one by George Raper in the British Museum (Natural History) and another by an unknown artist in the Alexander Turnbull Library, Wellington. It seems clear that the population was an isolated race of the widespread *C. vitiensis* but because no specimen remains it is not clear exactly how far it differed from the other subspecies.

On nearby Norfolk Island a race of the New Zealand Pigeon or Kereru (*Hemiphaga novaeseelandiae*) had established itself but this subspecies, named *spadicea*, seems to have become extinct shortly after 1800. No individuals were reliably recorded after 1801; presumably, the birds were

Table 12

RARE OR ENDANGERED PIGEONS AND DOVES

Common name	Scientific name
Long-toed Pigeon	*Columba trocaz*
Laurel Pigeon	*Columba junoniae*
São Tomé Olive Pigeon	*Columba thomensis*
Silver Pigeon	*Columba argentina*
Andaman Wood Pigeon	*Columba palumboides*
Silver-banded Black Pigeon	*Columba jouyi*
Black Wood Pigeon	*Columba janthina*
Yellow-legged Pigeon	*Columba pallidiceps*
Chilean Pigeon	*Columba araucana*
Jamaican Band-tailed Pigeon	*Columba caribeae*
Salvin's Pigeon	*Columba oenops*
São Tomé Bronze-naped Pigeon	*Columba malherbi*
Pink Pigeon	*Nesoenas mayeri*
Andaman Cuckoo Dove	*Macropygia rufipennis*
Brown's Long-tailed Pigeon	*Reinwardtoena browni*
Crested Long-tailed Pigeon	*Reinwardtoena crassirostris*
Timor Black Pigeon	*Turacoena modesta*
Flock Pigeon	*Phaps histrionica*
Squatter Pigeon	*Petrophassa scripta*
Red-winged Rock Pigeon	*Petrophassa rufipennis*
White-winged Rock Pigeon	*Petrophassa albipennis*
Galápagos Dove	*Zenaida galapagoensis*
Buckley's Ground Dove	*Columbina buckleyi*
Blue-eyed Ground Dove	*Columbina cyanopis*
Purple-winged Ground Dove	*Claravis godefrida*
Pallid Dove	*Leptotila pallida*
Grenada Dove	*Leptotila wellsi*
Tolima Dove	*Leptotila conoveri*
Mindoro Bleeding Heart	*Gallicolumba platenae*
Negros Bleeding Heart	*Gallicolumba keayi*
Tawi-tawi Bleeding Heart	*Gallicolumba menagei*
Truk Island Ground Dove	*Gallicolumba kuboryi*
Society Islands Ground Dove	*Gallicolumba erythroptera*
White-throated Dove	*Gallicolumba xanthonura*
Friendly Quail Dove	*Gallicolumba stairi*
Santa Cruz Ground Dove	*Gallicolumba sanctaecrucis*
Thick-billed Ground Dove	*Gallicolumba salamonis*
Marquesas Ground Dove	*Gallicolumba rubescens*
Palau Ground Dove	*Gallicolumba canifrons*
Wetar Island Ground Dove	*Gallicolumba hoedtii*
Tooth-billed Pigeon	*Didunculus strigirostris*
Sumba Island Green Pigeon	*Treron teysmanni*
Pemba Island Green Pigeon	*Treron pembaensis*
São Tomé Green Pigeon	*Treron saothomae*
Black-banded Pigeon	*Ptilinopus alligator*
Red-naped Fruit Dove	*Ptilinopus dohertyi*
Marche's Fruit Dove	*Ptilinopus marchei*
Merrill's Fruit Dove	*Ptilinopus merrilli*
Palau Fruit Dove	*Ptilinopus pelewensis*
Rarotongan Fruit Dove	*Ptilinopus rarotongensis*
Marianas Fruit Dove	*Ptilinopus roseicapillus*
Rapa Fruit Dove	*Ptilinopus huttoni*
White-capped Fruit Dove	*Ptilinopus dupetithouarsii*
Red-moustached Marquesas Fruit Dove	*Ptilinopus mercierii*
Henderson Island Fruit Dove	*Ptilinopus insularis*
Blue-capped Fruit Dove	*Ptilinopus monacha*
Carunculated Fruit Dove	*Ptilinopus granulifrons*
Ripley's Fruit Dove	*Ptilinopus arcanus*
Yellow-headed Fruit Dove	*Ptilinopus layardi*
Topknot Pigeon	*Lopholaimus antarcticus*
Cloven-feathered Dove	*Drepanoptila holosericea*
Comoro Blue Pigeon	*Alectroenas sganzini*
Seychelles Blue Pigeon	*Alectroenas pulcherrima*
Mindoro Fruit Pigeon	*Ducula mindorensis*
Society Islands Pigeon	*Ducula aurorae*
Marquesas Pigeon	*Ducula galeata*
Christmas Island Imperial Pigeon	*Ducula whartoni*
Peale's Pigeon	*Ducula latrans*
Giant Imperial Pigeon	*Ducula goliath*
Black Imperial Pigeon	*Ducula melanochroa*
Timor Imperial Pigeon	*Ducula cineracea*

unable to cope with the coming of man to the island and the effects of colonization.

The Philippine island of Cebu has lost many of its endemic forms and one of these is a race of the Amethyst Fruit Dove (*Phapitreron amethystina frontalis*). Other races exist still in the Philippines but the Cebu bird does not seem to have been collected at all during this century.

A distinct race of the Madagascan Turtle Dove once occurred on the

Seychelles Islands and was given the name of *Streptopelia picturata rostrata*. These birds were characterized by vinaceous heads and short wings. The nominate race, *S. p. picturata*, was introduced to the islands from Madagascar during the last century and, as individuals from each subspecies mixed, the stock of the native Seychelles birds seems gradually to have been absorbed by the newcomers. By 1975 no pure birds could be found and the race can be considered extinct as the result of what W.B. King (1981) calls 'genetic swamping'.

A large number of still-extant species and races are either endangered or potentially at risk (see Table 12). Two that may already be extinct and are sometimes included in lists of extinct birds are the Silver-banded Black Pigeon (*Columba jouyi*) from the Ryukyu Islands (Nansei-shotō) south of Japan and the Red-moustached Fruit Dove (*Ptilinopus mercierii*), from the Marquesas group. Two subspecies of *P. mercierii* have been described, the nominate race from the island of Nuku Hiva and the race *tristrami* from Hiva Oa. The nominate race appears not to have been seen since its discovery in 1849 and almost certainly is extinct, but on Hiva Oa birds were still alive during the 1920s and there is a possibility that they yet survive. The Silver-banded Black Pigeon was still extant during the 1930s but may have failed to survive to the present time; it could not be located in 1945. A large, black bird, it is closely related to the extinct Bonin Wood Pigeon (*Columba versicolor*) and also to the Black Wood Pigeon (*C. janthina*).

RODRIGUES PIGEON

?*Columba rodericana*

?*Columba rodericana*
(Milne-Edwards, 1874)

Columba rodericana Milne-Edwards, 1874 (Rodrigues)
?*Alectroenas rodericana* Rothschild, 1907

Length ~ 22 cm ($8\frac{1}{2}$ in)

Description
Appearance in life unknown; plumage possibly grey in colour.

This species is known with certainty only from bones found on the Mascarene island of Rodrigues, although there is a description given in the writings of the Huguenot refugee François Leguat that seems to apply to it. This written account relating to the first years of the 1690s was published in 1708 and reads as follows:

> The pigeons here are somewhat less than ours [European pigeons] and all of a slate colour, fat and good. They perch and build their nests upon trees; they are easily taken, being so tame, that we have had fifty about our table to pick up the melon seeds which we threw them, and they lik'd mightily. We took them when we pleas'd, and ty'd little rags to their thighs of several colours, that we might know them again if we let them loose. They never miss'd attending us at our meals, and we call'd them our chickens. They never built their nests in the Isle, but in the little Islets that are near it. We suppos'd 'twas to avoid the persecution of the rats, of which there are vast numbers in this Island.

If Leguat and his companions really did see living examples of the species now called *Columba rodericana*, then they were probably the last people to do so. Whether the rats mentioned by Leguat brought about the destruction of these birds or whether some other factor was involved, is not likely to be determined now but the pigeons probably became extinct during the first half of the eighteenth century.

The bones from which the species was described did not come to notice until the 1870s when they were examined by A. Milne-Edwards, the

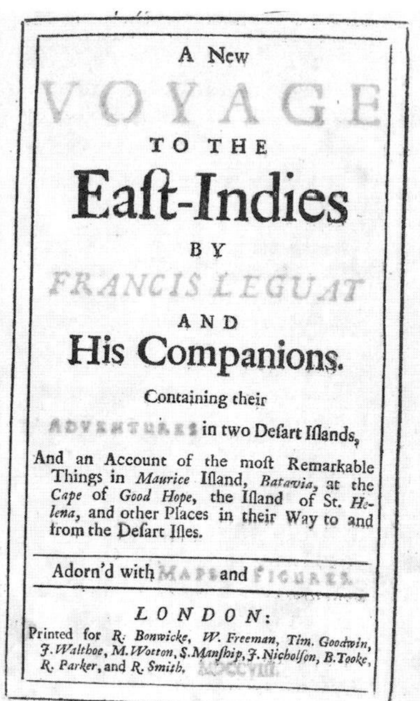

A New
VOYAGE
TO THE
Eaſt-Indies
BY
FRANCIS LEGUAT
AND
His Companions.
Containing their
ADVENTURES in two Deſart Iſlands,
And an Account of the moſt Remarkable Things in *Maurice* Iſland, *Batavia*, at the *Cape* of *Good Hope*, the Iſland of St. *Helena*, and other Places in their Way to and from the Deſart Iſles.

Adorn'd with MAPS and FIGURES.

LONDON:
Printed for *R. Bonwicke, W. Freeman, Tim. Goodwin, J. Walthoe, M. Wotton, S. Manſhip, J. Nicholſon, B. Tooke, R. Parker,* and *R. Smith.* MDCCVIII.

33 Françoise Leguat's *New Voyage to the East Indies* (London, 1708).

specialist on fossil birds at the Paris Museum. Primarily on the evidence of a sternum, Milne-Edwards deduced that a smallish pigeon had formerly inhabited Rodrigues and this bird he chose to assign to the genus *Columba*. Since then, G.R. Shelley (1883), Rothschild (1907) and M. Hachisuka (1953) have all aligned the form with *Alectroenas* – a genus found in Madagascar and islands of the Indian Ocean. Included in the genus are three living species – the Seychelles Blue Pigeon (*A. pulcherrima*), the Comoro Blue Pigeon (*A. sganzini*) and the Madagascar Blue Pigeon (*A. madagascariensis*) – and also the now-extinct Pigeon Hollandaise (*A. nitidissima*). From a geographical standpoint it is not unreasonable to associate the Rodrigues bird with these but in the light of present knowledge it may be safer to follow Milne-Edwards's original designation and retain the species within the genus *Columba*.

BONIN WOOD PIGEON

Columba versicolor

PLATE XXII

Columba versicolor (Kittlitz, 1832)

Columba versicolor Kittlitz, 1832 (Peel Island, now Chichi-jima, Ogasawara group)

Length 46 cm (18 in)

Description
Adult: upperparts greyish black generally but with (cont'd)

On the island groups to the south of Japan – the Borodinos (Daitō), the Ryukyus (Nansei) and the Bonins (Ogasawara) – bird populations have been heavily depleted; among those to suffer were three pigeon species. Together with a fourth that occupied an area much farther to the south, they make up a closely knit group within the large genus *Columba*.

Of the four, that with the most southerly range – the White-throated Pigeon (*C. vitiensis*) – is the only one with a seemingly secure future. This particular species is distributed over a vast area of the Pacific from the Philippines south and east through New Guinea to Fiji and Samoa but even though it is not at present endangered, one of its more isolated races – that formerly inhabiting Lord Howe Island and described under the name *godmanae* – is no longer extant.

Columba versicolor (cont'd) iridescent reflections of purple, amethyst and golden green – in particular, metallic green on top of head giving way to purple and green on hindneck, golden purple on back and green on rump, with wing coverts showing some golden tinges; face, neck and underparts greyish, glossed with golden green on breast; iris dark blue; bill greenish yellow; legs and feet dark red.

Measurements
Wing 265 mm; tail 205 mm; culmen 20 mm; tarsus 30 mm.

XXII Bonin Wood Pigeon. Oil painting by Errol Fuller.

Of the three species with natural ranges rather nearer to the shores of Japan, the Bonin Wood Pigeon (*C. versicolor*) is almost certainly extinct, the Silver-banded Black Pigeon (*C. jouyi*) may very well be so and the Black Wood Pigeon (*C. janthina*), even though still an inhabitant of many islands spread over a wide area, disappears wherever forest is interfered with. This last species, the only one of the three that has been studied to any extent at all, shows little fear of man – a trait not highly likely to secure survival. It seems probable that its two rather less well-known relatives were quite similar in this respect and certainly all three have shown the same inability to cope with the new conditions that develop when forested islands are tampered with.

The Wood Pigeon restricted to the Ryukyu Islands (Nansei-shotō) and Borodinos (Daitō-jima), *C. jouyi*, has not been reliably recorded since the 1930s; clearly it may not now survive. This large, black bird with a silver crescent on the hindneck was only ever seen in forest. It appears to have been present on several islands; while it may have vanished from the larger ones, it possibly clings to existence on some of the smaller.

It would be a great surprise to find that the pigeon endemic to the Bonin or Ogasawara group, *C. versicolor*, is other than extinct. The species was last located on 15 September 1889 when a male bird was taken on Nakondo-shima – one of the Parry (Mutojima) group – by a Mr Holst, collecting for the British ornithologist and writer H. Seebohm. Before this, the Bonin Wood Pigeon had been found only on Peel Island (Chichi-jima). Here, it was discovered by Captain Beechey in 1827 during the voyage of HMS *Blossom*. The results of this expedition, written up by N.A. Vigors, were not published until 1839 by which time F.H. von Kittlitz had visited Peel, encountered the species and published his own account of it; von Kittlitz's name, *Columba versicolor*, therefore takes precedence over any other.

Despite the absence of records, it is probably correct to assume that this Wood Pigeon occurred on other islands within the Ogasawara group. Practically nothing is known concerning habits, although it seems likely that, as with the species' close relatives, only well-wooded islands were occupied. It was a large and handsome bird, unmistakably bigger than *C. janthina* and rather paler in colour; from *C. jouyi* it could easily be distinguished by the lack of the silver crescent on the nape. Diet, almost certainly, was made up of fruit, seeds and buds.

The specimens of the Bonin Wood Pigeon that remain are divided between three museums: von Kittlitz's type is in Leningrad, a duplicate is in the museum at Frankfurt and although, according to H. Seebohm (1890), Vigors's type has been lost, another example is at the British Museum (Natural History).

PIGEON HOLLANDAIS

Alectroenas nitidissima

PLATE XXIII

This beautiful blue pigeon from Mauritius was first brought to attention by P. Sonnerat (1782); a specimen, now in poor condition, collected in 1774 during his voyage to various exotic parts of the world still exists in Paris. Sonnerat called his bird 'Pigeon Hollandais', its three colours – red, white and blue – reminding him of the flag of Holland.

Alectroenas nitidissima
(Scopoli, 1786)

Columba nitidissima Scopoli, 1786
(Île de France = Mauritius)
Columba franciae Gmelin, 1789
Columba batavica Bonaterre, 1790
Ramier hérissé Levaillant, 1808
Columba jubata Wagler, 1827
Alectroenas nitidissima Gray, 1840

Length 30 cm (12 in)

Description
Adult: feathers of head, mantle and throat white in colour, very long and pointed in shape; bare patch around eye extending to cheeks and lower mandible red; body, wings and thighs indigo blue; upper tail coverts dark red; tail dark red marked with black on outer webs; iris red; bill grey with yellow tip; feet probably slate-coloured.

Measurements
Wing 208 mm; tail 132 mm; culmen 25 mm.

While describing it he drew attention to its most obvious peculiarity – narrow, pointed, white feathers of the head, neck and breast having 'the polish, brilliancy, and feel of a cartilaginous blade'.

A second specimen exists in the Royal Scottish Museum, Edinburgh. Its early history is difficult to trace although the subject has been reviewed by A. Milne-Edwards and E. Oustalet (1893). The skin probably arrived in Edinburgh as early as 1816 having come from the collection of M. Dufresne; before this it is rumoured to have belonged to the Empress Josephine. Greenway (1958) claimed the bird was purchased from Julien Desjardins, a dealer in natural history specimens who spent part of his life in Mauritius, but this seems incorrect! Desjardin's name is, however, connected with the only other surviving specimen of this species – a skin in the museum at Port Louis, Mauritius. It came from an individual more recently taken than the others and the record of its shooting by one E. Geoffroy in the Savane Forest during 1826 constitutes the last definite note of the species. Geoffroy gave the skin to Desjardins some 3 years later and Desjardins presumably left it in Mauritius when he left for Paris.

Desjardins has, in fact, left the only eyewitness account of the bird in

XXIII Pigeon Hollandaise. Coloured engraving from F. Levaillant's *Histoire Naturelle des Oiseaux d'Afrique*, Vol. 6 (Paris, 1799–1808), Pl.267.

life. He claimed the Pigeon Hollandais lived alone near river banks and fed on fruit and freshwater molluscs. Some authors have doubted that such a species would eat aquatic animals but various pigeons do indeed find molluscs a delicacy and so the observation may well be accurate.

Why the species became extinct is unclear. It probably succumbed partly as a result of the early colonists' attitude to birds. Sonnerat (1782) records a bounty on 'vermin', a heading that seems to have covered almost every living creature, and a large striking pigeon would have made a very natural target anyway. Perhaps the growing presence of introduced Crab-eating Macaques (*Macaca fascicularis*) was just as destructive.

FORSTERS' DOVE OF TANNA

?Gallicolumba ferruginea

PLATE XXIV

?Gallicolumba ferruginea
(Wagler 1829)

Columba ferruginea Wagler, 1829 (Island of Tanna, Vanuatu, S. Pacific)

Length ~ 27 cm (11 in)

Description
Female: head and breast rusty brown; back dark reddish purple; wing dark green, primaries brownish grey with narrow pale edges; abdomen grey; bill black with slightly swollen cere; iris yellowish; feet red.

The Forsters, Johann Reinhold and his son George, sailed rather unexpectedly as naturalists aboard HMS *Resolution* on James Cook's second voyage around the world in 1772–5. Their opportunity to participate in this epic adventure followed a fit of pique from Joseph Banks (Figure 34). After his collecting successes on Cook's first circumnavigation (1768–71), Banks intended to make spoils from the second voyage even more spectacular. To this effect, he arranged for an elaborate extension to be built onto the *Resolution*, greatly increasing the potential scale of his intended operations. Shortly before the sailing date, the Admiralty announced that the 'improvement' rendered the vessel less than seaworthy! Banks refused to compromise his expectations and, with the Admiralty unable to change its own position, declined to join in the venture at all. Very much in the role of late reserves the Forsters were recruited but by the end of the mighty journey they had left an invaluable legacy to science. Poorly known species and also hitherto unknown ones were brought back with them or recorded in their notes and illustrations.

At this period, the serious exploration of Pacific islands had barely

34 (Left): Joseph Banks. Mezzotint by J.R. Smith after a painting by Benjamin West, *c.* 1775.
35 (Right): Man of Tanna. Engraving in the *Atlas* to James Cook's *Voyage towards the South Pole and Round the World* (London, 1777).

begun and so, when, in the late summer of 1774, the *Resolution* anchored off Tanna Island in the New Hebrides (now the Vanuatu) group, the shore-party perhaps expected to make interesting finds (Figure 35). Describing the events of 17 August, the older Forster, Johann Reinhold, notes simply, 'I went ashore, we shot a new Pigeon and got a few plants and the fruit of the wild nutmeg.'

Never again was the pigeon seen; even the specimen obtained has disappeared. Fortunately, a George Forster painting shows a dove not assignable to any known species. In its margin is written, 'Tanna, ♀, 17th August 1774'. There is little doubt, then, that this is a portrait of the mysterious Dove of Tanna. When and why it died out cannot be said.

CHOISEUL CRESTED PIGEON

Microgoura meeki

PLATE XXV

Microgoura meeki
(Rothschild, 1904)

Microgoura meeki Rothschild, 1904 (Choiseul, Solomon Islands)

Length 30 cm (12 in)

Description
Adult: forehead and front of face black, rest of face sparsely feathered showing a reddish hue; top of head, crest, mantle and breast dark bluish grey shading to brown on lower back; wings and rump olive brown; tail dark brown glossed purple; abdomen buffish chestnut; upper mandible black, lower mandible red; frontal shield whitish blue; iris brown; legs purplish red. Sexual differences, if any, unknown.

Measurements
Male: wing 195–197 mm; tail 100–105 mm; culmen 34 mm; tarsus 60 mm.
Female: wing 180–190 mm; tail 100 mm; culmen 33 mm; tarsus 60 mm.

One of the most remarkable collections of natural history items ever assembled was that gathered together by the English naturalist Walter Rothschild in the last years of the nineteenth century and the early part of the twentieth. During the period in which he was active, Rothschild amassed a vast stock of study skins, stuffed birds and animals, skeletons, fossils, preserved insects and other invertebrates. These he housed in a private museum at Tring in Hertfordshire along with his wonderful library on natural history subjects. He poured out enormous financial reserves on both the acquisition of items and on the staff needed to arrange and supervise them; it hardly seems surprising, therefore, that he ran into the monetary difficulties that forced him to part with some of his treasured possessions. At the end of his life, however, a magnificent array of material remained and this he most generously bequeathed, along with the property to house it, to the British people.

In addition to simply collecting, Rothschild founded his own scientific journal, *Novitates Zoologicae*, published papers in other journals of the day and produced several wildly eccentric but nevertheless important books. Among his special interests were kiwis, cassowaries, birds of paradise, and extinct or excessively rare birds. To secure examples of these, his agents, either dealers or explorers, scoured the world.

None, perhaps, was more successful than the intrepid English naturalist A.S. Meek who, during the years around the turn of the century, ventured into the formidable island of New Guinea on Rothschild's behalf. It was not in New Guinea itself, however, that he made his most sensational ornithological discovery but on the not far distant archipelago of islands known as the Solomons. Here, on the island of Choiseul during January of 1904, Meek collected for Rothschild a series of skins and a single egg belonging to a large and striking ground pigeon, which has, in fact, been described as the most magnificent and distinctive Solomons' bird.

The reputation of Choiseul's natives was very ugly and, doubtless, this had inhibited previous exploration. Meek, a veteran of such situations, was well prepared for attack. He wrote (1913):

> The danger there was thought to be so great that it was not considered wise for all the party to work ashore at the same time. I established my camp on a small island, so as to be in a better position for defence in case of attack, and

XXIV Forsters' Dove of Tanna. Watercolour by George Forster; Pl.142, Forster Portfolio, British Museum (Natural History). Courtesy of the Trustees.

> one of us always cruised about near the shore in a boat while a party was collecting on shore. The boat would have taken off the party to the island camp in case of danger threatening.
>
> The natives of Choiseul are not only savage to strangers, but are by no means friendly among themselves.

The skins obtained during what was evidently a hair-raising visit were despatched to Britain; in a covering letter dated 18 January 1904, Meek wrote to Ernst Hartert, Rothschild's curator, that six specimens had been sent. Interestingly, Rothschild and Hartert recorded that seven were actually received, six being retained for the Tring Museum. If a seventh skin existed, its whereabouts today remains unknown; Shane Parker (1972) suggests that it may have been presented to one of Rothschild's royal friends. Of the six skins that are fully documented, five were sold by Rothschild to the American Museum of Natural History during the period of financial difficulty and the sixth was passed to the British Museum

XXV Choiseul Crested Pigeon. Hand-coloured lithograph by J.G. Keulemans from *Novitates Zoologicae*, Vol. 11 (1904), Pl.21.

36 Blue-crowned Pigeon (*Goura cristata*), a close living relative of the extinct Choiseul Crested Pigeon. Engraving from *Cassell's Natural History* (London, 1889).

(Natural History) collection together with the cream-coloured egg. This egg, according to Meek, was found on the ground without any evidence of nesting material. Although Meek was told by natives that his *Microgoura* also occurred on the islands of Santa Isabel and Malaita, no one, other than he, ever managed to find it on Choiseul or anywhere else. This splendid species is therefore assumed extinct.

Individuals participating in the Whitney South Seas Expedition obtained information about Choiseul's Crested Pigeon when they called in 1927 and again in 1929, but no concrete evidence of its existence was forthcoming. In the north-west of the island, Rollo Beck – a member of the Whitney Expedition and an experienced bird collector – was given by natives to understand that the birds were alive but rare. In some other parts only the older inhabitants remembered the species, remarking that introduced cats had killed many individuals and recalling how easily the birds could be caught by hand at their roosts. Some of the natives imitated the birds' call by low, trilling sounds.

During the 1960s, Shane Parker, well known for his interest in lost birds and destined later to become the rediscoverer of the Australian Night Parrot (*Geopsittacus occidentalis*), embarked on an unsuccessful search for *Microgoura*. Studying reports of the Whitney Expedition, he noticed that localities at which native recognition of *Microgoura* occurred were always close to riverine flats and swamps – terrain rather unattractive to collectors. Indeed, field notes of the Whitney investigators show that little collecting was actually done in the swamps. Choiseul Bay, almost certainly the area from which Meek collected his specimens, no longer provided suitable habitat; about 1920, its original vegetation was almost completely destroyed to make way for coconut plantations. It was Parker's hope that he might find birds on the recently inhabited island of Wagina or the uninhabited island of Rob Roy.

The search proved fruitless, apart from a small body of information built up from reminiscences of older inhabitants. The natives of Choiseul knew the birds by the name of *kukuru-ni-lua* (pigeon-belong-ground) and those few who still remembered suggested they were resident in lowland – swampy forest but not mangrove. Meek's *Microgoura* was mostly terrestrial but roosted in small parties (twos, threes or fours) on low branches, the sites of which were betrayed by the droppings piled below. Two natives recalled that stones were often found in the birds' gizzards, a not unlikely occurrence as the presence of stones is reported in other pigeons. All seemed agreed that cats, introduced to eliminate rats, had exterminated the pigeons that clearly were very vulnerable to any form of predation. The most recent sighting that Parker could find word of appears to have occurred during World War II in the Kolambangara basin where a small roost is alleged to have gathered.

Although aberrant in several respects this pigeon is quite obviously related to the crested pigeons of the genus *Goura* (Figure 36). Parker (1967) pointed out that the position of the crest shown in the *Novitates Zoologicae* (1904) colour plate may be misleading and results from an inaccurate interpretation of Meek's museum skins. In the process of preparation the crest has been flattened back across the nape and the artist has followed this. Meek, to the contrary, remarked upon the similarity of the crests belonging to his new species and those crests carried by *Goura* pigeons. The likelihood is that in life *Microgoura* wore its beautiful crest in a much more erect fashion than indicated by the plate opposite, but we shall never know for certain.

PASSENGER PIGEON

Ectopistes migratorius

PLATE XXVI; FIGURES 37–41

Ectopistes migratorius
(Linnaeus, 1766)

Columba migratoria Linnaeus, 1766 (Carolina)
Columba canadensis Linnaeus, 1766
Ectopistes migratoria Swainson, 1827

Length 40 cm (16 in)

Description
Male: head slaty blue, some black blotches around eyes; back of neck iridescent bronze, green or purple according to light; back slate grey tinged with olive brown; lower back and rump greyish blue becoming greyish brown on upper tail coverts; two central tail feathers brownish grey, the rest white; wing coverts brownish grey with irregular blackish markings; primaries and secondaries much darker greyish brown, secondaries edged with white; throat and breast pale cinnamon rufous becoming paler on lower breast merging to white on abdomen; under tail coverts white; iris red; bill black; naked orbital ring purplish flesh colour; legs and feet red.
Female: similar to male but colouring much duller and tail considerably shorter; iris orange red; naked orbital ring greyish blue; legs and feet red but paler than those of male.
Immature: similar to adult female, but scapulars, wing coverts, feathers of foreneck and breast tipped with white giving scaled appearance; iris brownish with narrow ring of carmen; legs and feet pinkish brown.

Measurements
Male: wing 196–215 mm; tail

The Passenger Pigeon was once, perhaps, the most numerous bird on earth. Uncounted numbers blackened North American skies when, rank after rank, flock after flock, birds of the species passed overhead in search of feeding grounds or places to breed. So vast was the population, so seemingly complete its grip upon survival that the supply of birds appeared inexhaustible. In a hunting competition once organized, over 30,000 dead birds were needed to claim a prize. Yet, within a century, incredible multitudes were reduced to small, dwindling bands.

From the millions upon millions of Passenger Pigeons that thundered across North America at the start of the nineteenth century, by the year 1914 just one solitary descendant remained alive – an individual housed in a cage at the Cincinnati Zoological Gardens. When this last bird died, people still lived who could remember the tremendous massed flights of the Passenger Pigeon; the final decline from apparent security to extinction had occurred well within the lifespan of an average human being. A.W. Schorger, who, long after the extinction wrote a detailed monograph of the Passenger Pigeon, believed that in its heyday the species perhaps accounted for between 25 and 40 per cent of the total landbird population of the United States.

The earliest written mention of Passenger Pigeons appears to relate to 1 July 1534 when the French navigator Jacques Cartier saw them on Prince Edward Island, but so noticeable did the species make itself that there are many very early references. One of the more picturesque was written by T. Morton in 1637, describing birds he saw in New England: 'Millions of Turtledoves on the greene boughes, which sate pecking of the full ripe pleasant grapes that were supported by the lusty trees, whose fruitfull loade did cause the armes to bend'. In 1759, Pehr Kalm wrote:

> In the spring of 1749, on the 11th, 12th, 15th, 16th, 17th, 18th, and 22nd of March . . . but more especially on the 11th, there came from the north an incredible multitude of these pigeons to Pennsylvania and New Jersey. Their number, while in flight, extended 3 or 4 English miles in length, and more than 1 such mile in breadth, and they flew so closely together that the sky and the sun were obscured by them, the daylight becoming sensibly diminished by their shadow.
>
> The big as well as the little trees in the woods, sometimes covering a distance of 7 English miles, became so filled with them that hardly a twig or a branch could be seen which they did not cover; on the thicker branches they had piled themselves up on one another's backs, quite about a yard high.
>
> When they alighted on the trees their weight was so heavy that not only big limbs and branches of the size of a man's thigh were broken straight off, but less firmly rooted trees broke down completely under the load.
>
> The ground below the trees where they had spent the night was entirely covered with their dung, which lay in great heaps.
>
> As soon as they had devoured the acorns and other seeds which served them as food and which generally lasted only for a day, they moved away to another place . . .
>
> Extremely aged men stated that on three, four, five, or several more occasions in their lifetime they had seen such overwhelming multitudes in

Ectopistes migratorius (cont'd)
175–210 mm; culmen 15–18 mm; tarsus 26–28 mm.
Female: wing 175–210 mm; tail 140–195 mm; culmen 15–18 mm; tarsus 25–28 mm.

> these places . . . so that 11, 12, or sometimes more years elapse between each such unusual visit of pigeons.

Three-quarters of a century later (in the 1830s) John James Audubon was still able to describe a Pigeon migration in the most extravagant terms: 'The air was literally filled with pigeons; the light of noonday was obscured as by an eclipse; the dung fell in spots, not unlike melting flakes of snow . . . pigeons were still passing in undiminished numbers, and continued to do so for three days in succession.'

Since the colossal journeyings were occasioned by the need to find food, there was no particular time at which birds could be relied upon to arrive in any given district. Sometimes they might remain absent from a certain area for years on end before returning – perhaps for a prolonged stay, perhaps for a short one. Not only were migrations made spectacular by the vast numbers involved, they were notable for the sheer speed at which the birds travelled – possibly as fast as 96 km (60 miles) per hour. Passenger Pigeons were perfectly fitted for rapid flight. The breast was powerfully muscled for a bird of this size and the streamlined body was complemented by a small head with a long tail and pointed wings. In addition to stylish proportions, the plumage was also elegant, giving an overall effect of speed, grace and beauty.

As the columns of birds moved swiftly across the sky, each individual, or group of individuals, would follow exactly the movements of those in front. If one group dipped, so would those following behind; if individuals swerved or turned to avoid the pursuit of a hawk, others coming on would describe similar motions on reaching the spot. Passing over impoverished country Pigeons would fly high in the sky until their keen vision picked out more promising areas. Then would occur a descent to much lower levels, heights from which terrain could be examined in more detail. When suitably rich territory was spotted the flocks would wheel about in circles and descend in a blaze of flashing, iridescent colours as light caught massed ranks of birds. Once alighted and foraging for food, ranks at the back would still rise up, pass over the main part of the flock and land in front. These manoeuvres were accomplished in such rapid succession that the mass of birds appeared continually on the move. J.M. Wheaton (1882) described such a scene:

> This movement soon became continuous and uniform, birds from the rear flying to the front so rapidly that the whole presented the appearance of a rolling cylinder having a diameter of about fifty yards, its interior filled with flying leaves and grass. The noise was deafening and the sight confusing to the mind.

37 Passenger Pigeon (adult male).

Because of the enormous reserves of energy used in their spectacular flights, Passenger Pigeons were voracious feeders. Their meals consisted principally of acorns, chestnuts and beechnuts and all kinds of fruits when available. In addition to these foods, grain and a variety of other cultivated plants were subject to attack. Insects and, particularly, worms were also great favourites. One interesting peculiarity of the feeding routine was an ability to disgorge food from the crop to make way for more desirable foodstuffs, if and when these became available.

During mass-feeding sessions, ground was rapidly cleared and, on finishing, Pigeons settled in the trees where they usually spent much of the remainder of the day, towards evening departing for their roosting places.

XXVI Passenger Pigeons: female (above); male (below). Aquatint by J.J. Audubon and R. Havell the younger from Audubon's *Birds of America* (London, 1827–38), Pl.285.

Characteristically, these were located in parts of the forest where trees were large with little undergrowth. Beneath the roosts dung might lie several inches deep. Audubon (1831) described how he waited at such a site on the Green River, Kentucky. The time was dusk and the backwoodsman was in the company of pigeon hunters:

> Everything was ready, and all eyes were gazing on the clear sky, which appeared in glimpses amidst the tall trees. Suddenly there burst forth a general cry of 'Here they come!' The noise which they made, though yet distant, reminded me of a hard gale at sea passing through the rigging of a close-reefed vessel. As the birds arrived, and passed over me, I felt a current of air that surprised me. Thousands were soon knocked down by the pole men. The birds continued to pour in. The fires were lighted, and a magnificent, as well as wonderful and almost terrifying sight presented itself. The pigeons, arriving by thousands, alighted everywhere, one above another, until solid masses as large as hogsheads, were formed on the

38 Egg of the Passenger Pigeon.

> branches all round. Here and there the perches gave way under the weight with a crash, and falling to the ground, destroyed hundreds of the birds beneath, forcing down the dense groups with which every stick was loaded. It was a scene of uproar and confusion. I found it quite useless to speak, or even to shout to those persons who were nearest to me. Even the reports of the guns were seldom heard, and I was made aware of the firing only by seeing the shooters reloading.

Typically, Passenger Pigeons bred in large colonies, arriving in a district a little before they actually began to breed. Nestings might cover thousands of acres or just a few. Where colonies were of great size, and some were colossal, they tended to be long and comparatively narrow. Nestings more than 64 km (40 miles) in length were recorded and claims for 160 km (100 miles) were made, even as late as 1871. A.W. Schorger (1955) considered that 16 km (10 miles) long by 5 km (3 miles) broad might be fairly characteristic. Very long colonies were almost certainly discontinuous and may best be described as loosely associated gatherings of flocks. Within each concentrated nesting area every suitable tree would be occupied and only after all potential sites were filled would birds begin to extend the limits of their chosen area.

The nest itself was scrappily made from small twigs. Many observers recorded that construction was so flimsy the egg could be seen from beneath, yet nests did have a certain permanence for their remains could be found at sites of nestings several years old. By choice, they would be positioned on strong branches close to tree trunks.

Regarding the courtship of Passenger Pigeons, Audubon's account is again most picturesque:

> The male assumes a pompous demeanour, and follows the female whether on the ground or on the branches, with spread tail and drooping wings, which it rubs against the part over which it is moving. The body is elevated, the throat swells, the eyes sparkle. He continues his notes, and now and then rises on the wing, and flies a few yards to approach the fugitive and timorous female . . . they caress each other by billing . . . the bill of the one is introduced transversely into that of the other, and both parties alternately disgorge the contents of their crop by repeated efforts.

39 Chick of the Passenger Pigeon.

Egg laying occurred between March and September with a peak reached in April and May. Although several writers claimed that two were usually laid, it seems certain that the general rule was for each pair to produce a single egg, white in colour and similar in appearance to that of a domestic pigeon, having average measurements of 38 mm × 27 mm. Both parents took turns at the nest and incubation lasted for 12 or 13 days, a remarkably short time for a pigeon. After hatching, the chick was cared for by the adults for around 2 weeks, then it was left. *En masse*, the old birds would depart. For a while, the chick, fat and unable to fly properly, sat crying in the nest, then it dropped to the ground. Within a few days the baby could itself take to the air. From start to finish the nesting cycle would have taken around 30 days. Some writers claimed Passenger Pigeons nested several times during the season, others that they bred only once in summer.

Because they were so often encountered in vast multitudes, the sound each bird produced joined together to swell to an overwhelming racket, audible for miles. When building, their call was likened to the croaking of wood frogs, but at other times shrieks, clucks, tweets, twitterings and

40 Nest and egg of the Passenger Pigeon.

cooings were heard, and it is claimed that during mating series of bell-like notes were produced. While a flock was feeding, certain individuals, perhaps especially posted as sentinels, were likely to give a call of alarm if danger threatened and the note was rapidly taken up by others as they rose.

During the early years of European colonization the breeding range of the species stretched from the Great Plains in the west to the Atlantic Ocean in the east, from southern Canada in the north as far to the south as northern Mississippi. The great deciduous forests of this area formed their home but stragglers occasionally turned up very far afield. There are records of Passenger Pigeons in British Columbia and northern Canada and in the south the birds sometimes wandered to Mexico and islands of the Caribbean.

Although the species was so plentiful in Audubon's time, and even much later, it is possible that numbers were in decline from the beginning of European colonization. As the nineteenth century wore on the population dropped more rapidly; perhaps it is truer to say more noticeably. The decrease probably reached its most critical phase during the 1870s. At the start of this decade huge flocks still existed, by its end the remnants were scattered and the species' hold on survival had broken for ever.

A variety of reasons have been advanced to account for the disappearance. Some, like forest clearance or imported avian disease, may well have been significant contributory factors. Others, such as mass drownings, or the curse of a Christian minister, are altogether more fanciful. The principal factor seems to have been, quite simply, the vast quantities of birds killed by pigeon hunters. Hundreds of thousands were slaughtered season after season, decade after decade. Eventually, the multitudes were thinned. Once the huge flocks were broken and dispersed, hunting on a major scale lost its commercial attraction, so persecution can hardly be used to explain the final dip to extinction. In a land so vast as the United States there could be no mopping-up operation for creatures as small as pigeons. Given the species' predilection for associating in vast numbers, it may be assumed that once the population had plummeted to certain levels the species was doomed even though many individuals remained alive. The evolution of the Passenger Pigeon seems to have occurred in such a way that it could exist only in huge flocks.

By the last decade of the nineteenth century there were very few Passenger Pigeons left. Reports of birds sighted or shot during the last few years of the old century and the first few of the new, are listed by Schorger (1955). Many are probably genuine, some false. Schorger accepts as the very last legitimate record of a wild Passenger Pigeon, one of a bird taken at Sargents, Pike County, Ohio, on 24 March 1900. The stuffed skin of this individual is now in the Ohio State Museum.

A few still lingered in captivity, some in Milwaukee, Chicago, and some in the Cincinnati Zoological Gardens. By the spring of 1909, only the Cincinnati birds were left alive, two males and a female. By the end of the summer of 1910, only the female, the famous Martha, survived (Figure 41). Since she was almost certainly the last of the Passenger Pigeons, her final years and death attracted attention, but not all that has been written of her is reliable. Her age at death was claimed as 29 but, as befits a lady, in reality it was unknown. Various myths surround Martha's last hours. It has been suggested that she expired before a hushed team of eminent ornithologists, but this is incorrect. The moment of truth found Martha alone! She was found dead on the floor of her cage at 1.00 p.m. precisely on 1 September 1914, having been checked by her keeper just a short while

41 Martha – the last Passenger Pigeon, who died on 1 September 1914 aged about 29 in Cincinnati Zoological Gardens. Her stuffed skin can be seen at the Smithsonian Institution, Washington DC.

previously. This may be the only instance in which virtually the exact time of the extinction of a species is known, for it is most unlikely that any wild Passenger Pigeons survived to 1914. The small body of Martha was suspended in water and the water frozen. In a huge block of ice she was transported to the Smithsonian Institution in Washington DC. Her carcass was minutely examined and her skin stuffed and eventually placed on show.

Of the Passenger Pigeon Audubon wrote:

> When an individual is seen gliding through the woods and close to the observer, it passes like a thought, and on trying to see it again, the eye searches in vain; the bird is gone.

DODO

Raphus cucullatus

PLATE XXVII; FIGURES 42–5

Raphus cucullatus
(Linnaeus, 1758)

Struthio cucullatus Linnaeus, 1758 (India = Mauritius)
Didus ineptus Linnaeus, 1766

Length ~ 100 cm (3 ft 3 in)

Description
Adult (from contemporary accounts and pictures): plumage generally of a greyish colour, darker on upperparts, lighter on throat and abdomen; tail feathers whitish; primaries much reduced and of a light colour; thighs blackish; bare part of face probably ash-coloured; feet and legs yellow; iris probably whitish; beak green or black.

> About 1638, as I walked London streets, I saw the picture of a strange fowle hung out upon a clothe and myselfe with one or two more then in company went in to see it. It was kept in a chamber, and was a great fowle somewhat bigger than the largest Turky Cock, and so legged and footed, but stouter and thicker and of a more erect shape, coloured before like the breast of a young cock fesan, and on the back of dunn or dearc colour. The keeper called it a Dodo, and in the ende of a chymney in the chamber there lay a heape of large pebble stones, whereof hee gave it many in our sight, some as big as nutmegs and the keeper told us that she eats them (conducing to digestion), and though I remember not how far the keeper was questioned therein, yet I am confident that afterwards shee cast them all again.

With these words the English theologian and historian Sir Hamon L'Estrange introduced readers of his memoirs to the Dodo of Mauritius, the bird with a name whose original meaning is obscure but a name which has now become virtually a synonym for extinction. As far as extinct creatures are concerned, only dinosaurs can match dodos for celebrity; and not only does the peculiar name stick in the mind, the bird's extraordinary appearance – made vividly familiar even to children through Sir John Tenniel's remarkable illustrations for Lewis Carroll's *Alice in Wonderland* – leaves an indelible impression.

The particular individual that Sir Hamon saw – some 350 years ago – seems to have had an interesting subsequent history. Several quite fanciful ideas have been put forward concerning it (Hachisuka, 1953) but in all probability the bird was finally stuffed and took a place in the illustrious museum established by John Tradescant, naturalist and gardener to Charles II, in Lambeth, London. The stuffed creature and the rest of Tradescant's collection passed in 1659 from this museum to that of Elias Ashmole at Oxford. One of Ashmole's statutes – number eight – reads: 'That as any particular grows old and perishing the keeper may remove it into one of the closets or other repository; and some other to be substituted.'

In 1755 the Dodo was examined, found wanting in certain aspects and – presumably under statute eight – was ordered out for destruction. The instruction appears to have been obeyed but not quite to the letter;

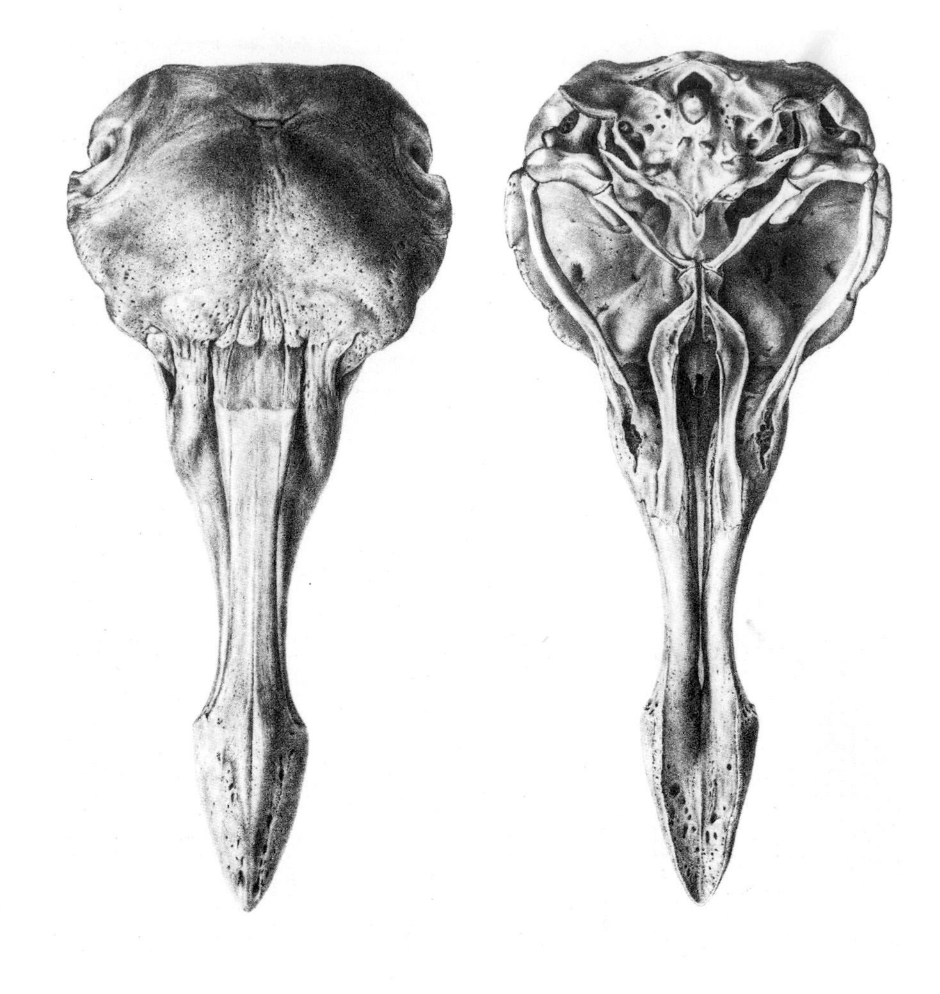

42 Dodo skull from above, below and side. Lithographs from H.E. Strickland and A.G. Melville's *The Dodo and its Kindred* (London, 1848).

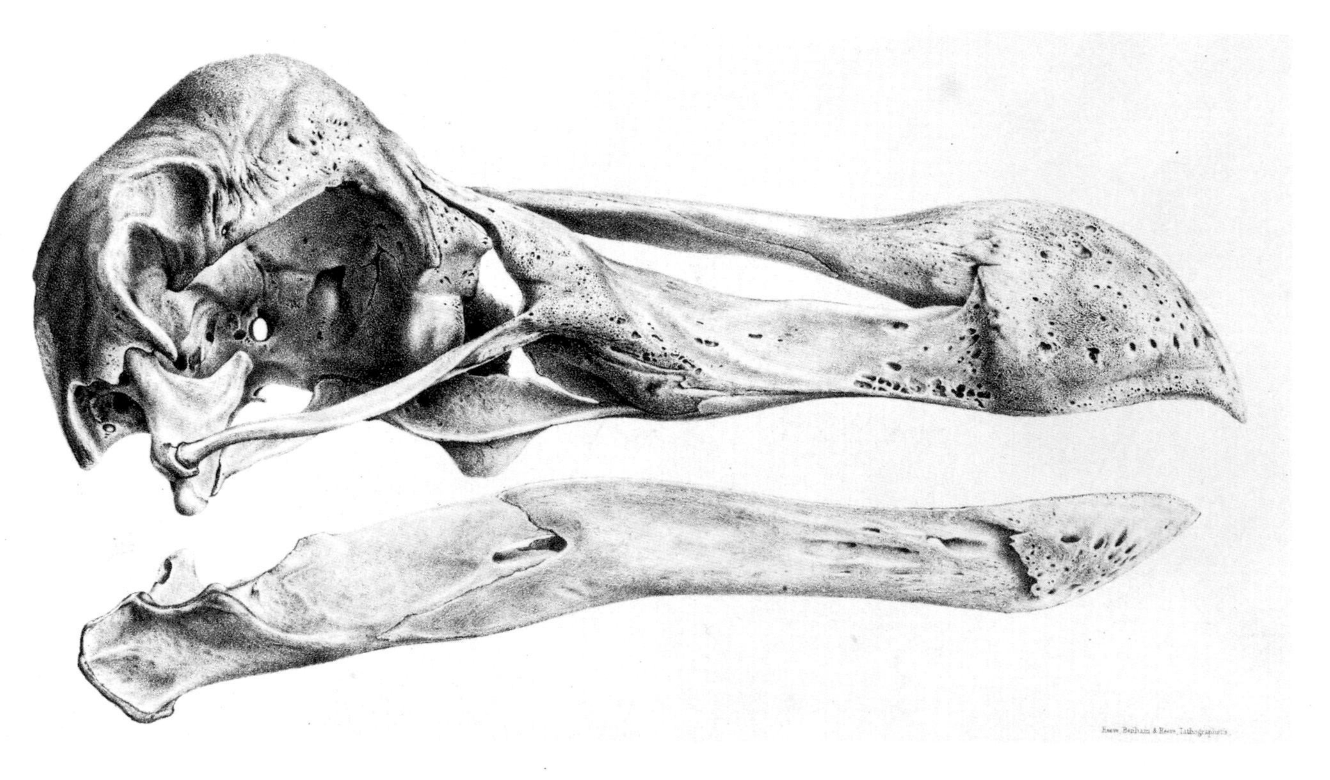

someone cut off the head and right foot, thus preserving something of Tradescant's Dodo for posterity. Regarding 'some other to be substituted' – none could be. By 1755 the living Dodo was long gone from Mauritius and no other stuffed example can be proved to have outlasted this Tradescant/Ashmole one. Aside from the remains in the Ashmolean Museum, Oxford (Figure 43), the only tangible relics that have come down to us of human encounters with living Mauritius Dodos are a few scraps of skin and bone in some European museums – a left foot in the British Museum (Natural History) (Figure 44), a head in Copenhagen and a fragment of a head in Prague. The dodos – apparently stuffed – that can often be seen in museums are all fakes, made up from the feathers of chickens or other birds; their heads and feet are made of plaster or other suitable material.

By the early part of the nineteenth century, with just these bits and pieces plus a few pictures remaining, the unravelling of the puzzle concerning the Dodo's true nature was rather a controversial issue. At first, naturalists seem to have thought of the bird as a kind of short-legged ostrich, but gradually other associations were considered – among them gallinaceous birds, albatrosses and plovers! By the late 1830s, the word 'vulture' was in vogue, supported by the eminent opinion of Richard Owen.

What now seems the truth was first hit upon by a Professor Reinhardt of Copenhagen. Following on from Reinhardt's work, Hugh Strickland – co-author of an early book on didine birds, *The Dodo and its Kindred* (1848) – examined the available evidence, looked at the paintings that survived and agreed with the Danish professor that the Dodo was perhaps a giant dove – grotesque in its development, but a member of the pigeon order nevertheless. So wild and unlikely seemed the opinion that several of Strickland's contemporaries felt inclined to ridicule it, but not long was to pass before the idea of a connection with pigeons and doves gained

43 The Dodo head preserved in the Ashmolean Museum, Oxford. Lithograph from H.E. Strickland and A.G. Melville's *The Dodo and its Kindred* (London, 1848).

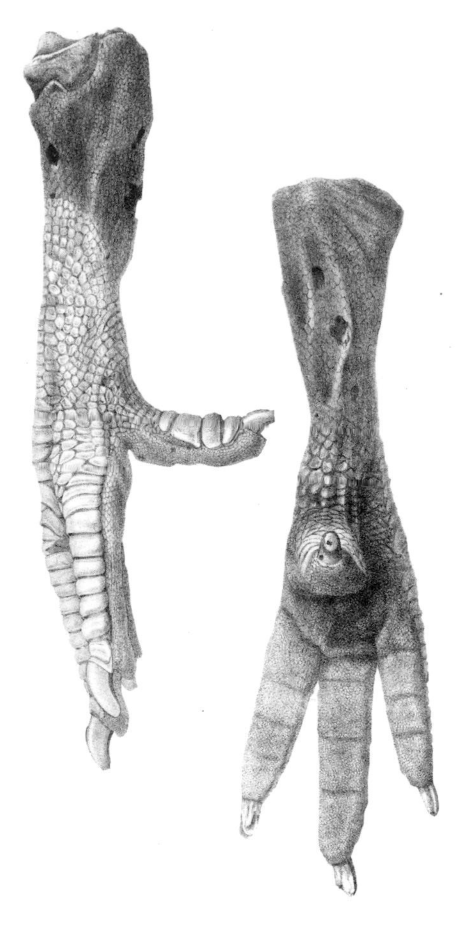

44 Side and back views of the Dodo foot preserved in the British Museum (Natural History). Lithograph from H.E. Strickland and A.G. Melville's *The Dodo and its Kindred* (London, 1848).

credibility, for the first specimens of the remarkable Tooth-billed Pigeon (*Didunculus strigirostris*) were arriving in Europe from Samoa (Figure 46). This pigeon species looks surprisingly dodo-like about the head and clearly illustrates the possibility of a dove developing into something of the dodo type. Poor Strickland, however, had not lived long enough to see his triumph; intent upon studying the geology of a newly dug railway cutting, he remained oblivious of an approaching train and paid for his carelessness with his life.

Not many years after this premature death, studying the Dodo in greater depth was to become very much of a possibility thanks to the endeavours of a Mr George Clarke, master of the Government School at Mahébourg, Mauritius. Clarke put together a large collection of Dodo bones that he uncovered at his own expense in the Mare aux Songes, Mauritius, the first of his discoveries being publicized after 1865. From the finds have been assembled some complete or almost complete skeletons and it is this material that provides the basis for all dodo research of an anatomical nature (see Figure 45).

The first written account of the living Mauritius Dodo is probably that of Jacob Cornelis-zoon van Neck, published in Holland in 1601; the last is almost certainly one given by Benjamin Harry some 80 years later in a manuscript kept today at the British Museum (Natural History). In the years between the writing of these two accounts, quite considerable numbers of travellers to Mauritius penned their descriptions of meetings with Dodos and from these, together with a sizeable collection of seventeenth-century paintings, can be formed a fairly good impression of the birds in life.

They were large and fat, even – perhaps – gross. Their swollen body shape together with much reduced wings rendered flight impossible and in all probability Dodos were rather sluggish in their movements. According to some writers, they were so clumsy that when they tried running to elude capture their fat bodies wobbled and their bellies scraped along the ground. Naturally, they were forced to nest at ground level and on the authority of Françoise Cauche (1651) it is said that a single white egg was laid, deep in the forest, on a bed of grass; interestingly, Dodos appear not to have been shore birds – as is often assumed – but inhabitants of the woods. It seems virtually certain that stones were used as an aid to digestion, Sir Hamon L'Estrange being by no means the only writer to mention this.

The call was never clearly described but one school of thought inclines to the opinion that the word 'dodo' – probably coined by Portuguese sailors – is simply a rendering of it. This may be so for 'dodo' seems to have no particular meaning in any language unless it is derived from the word *duodo*, meaning idiot! The derivation of other early names for the species (for example, *dronte* taken from a now-obsolete Dutch word meaning 'swollen' or 'bloated'; or *walghvogel* meaning 'disgusting bird') seems rather more obvious.

The most striking detail of appearance was perhaps the grotesque, hooked beak and there can be little doubt that with this weapon the otherwise defenceless bird could administer a fearsome bite. The rest of the head and body was almost as extraordinary. Probably the most picturesque – and also one of the most detailed – descriptions is that given by Sir Thomas Herbert (1634) who in later life was destined to accompany Charles I to the scaffold as the king's personal attendant. In his account of the island of Mauritius, Herbert says:

> First here and here only ... is generated the Dodo, which for shape and rareness may antagonize the Phoenix of Arabia: her body is round and fat, few weigh less than fifty pound ... her visage darts forth melancholy, as sensible of Nature's injurie in framing so great a body to be guided with complementall wings, so small and impotent, that they serve only to prove her bird.
>
> The halfe of her head is naked seeming couered with a fine vaile, her bill is crooked downwards, in midst is the trill, from which part to the end tis of a light green, mixt with pale yellow tincture; her eyes are small and like to Diamonds, round and rowling; her clothing downy feathers, her traine three small plumes, short and inproportionable, her legs suting to her body, her pounces sharpe, her appetite strong and greedy.

Although no skins of Dodos remain to prove the loose description of Herbert and others beyond doubt, many pictures by seventeenth-century artists exist – some by painters with quite considerable reputations. Among the most illustrious to whose hand is attributed a Dodo portrait is one of the d'Hondecoeter family in Holland, but the best-known series of paintings is that thought to have been produced by Roelandt Savery, an artist of not inconsiderable merit whose paintings have sold for in excess of £100,000.

While the expertise of some of these Dodo painters is not to be doubted, a problem exists with all of the pictures – it is not clear whether or not they were actually drawn from life. Several commentators (e.g. Oudemans, 1917; Hachisuka, 1953) assume they were but not on very well-defined grounds. Certainly, the better quality works were drawn directly from models of some kind but these models may not have been living birds – some of the Savery paintings (and also the d'Hondecoeter) look very much as though they were produced using crudely stuffed specimens as

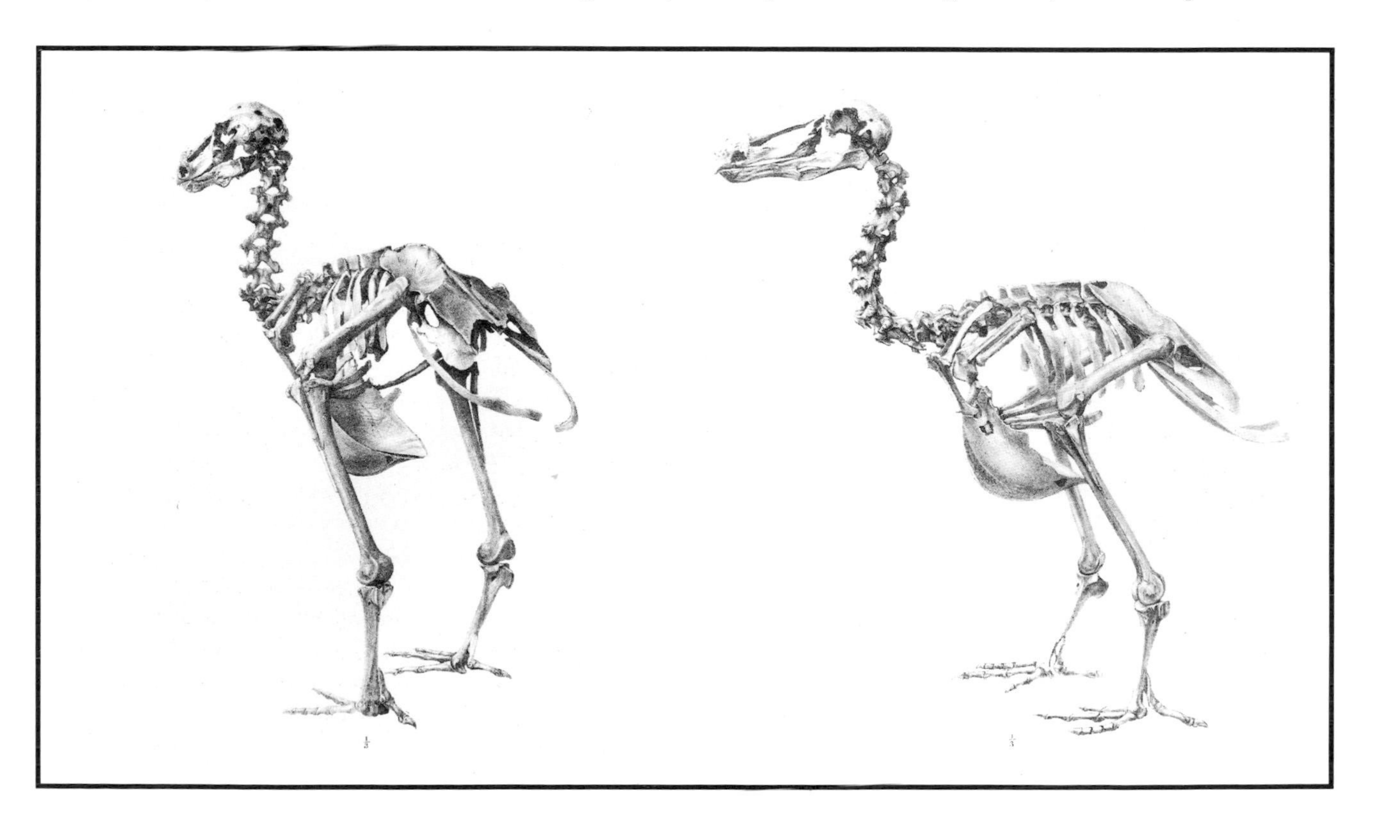

45 Rear (left) and side (right) views of a Dodo skeleton. Lithographs by James Erxleben in the *Transactions of the Zoological Society of London*, Vol. 7 (1871).

46 Tooth-billed Pigeon (*Didunculus strigirostris*), a dodo-like bird from Samoa. Engraving from R. Lydekker's *Royal Natural History* (London, 1893–6).

XXVII Three hundred years of Dodo illustration: from 1601 to 1894.
Centre: John Tenniel's illustration for Lewis Carroll's *Alice in Wonderland* (London, 1865).
Other illustrations from left to right and top to bottom:
Nineteenth-century engraving after a seventeenth-century painting attributed to Roelandt Savory in R. Lydekker's *Royal Natural History* (London, 1894).
Engraving after an aquarelle by J. Walther (*c*. 1657).
Engraving from Jacob Cornelius-zoon van Neck's *Het Tweede Boeck* (Amsterdam, 1601).
Watercolour by Cornelius Saftleven. Boymans Museum, Rotterdam.
Drawing from the journal of Pieter van der Broeke; early seventeenth century.
Engraving from Sir Thomas Herbert's *Relation of Some Yeares' Travels . . . into Afrique and the greater Asia* (London, 1634).
Seventeenth-century drawing by C. Clusius copied from an account (now lost) of a voyage by Jacob Cornelius-zoon van Neck.
Figure from a painting by Roelandt Savory. Zoological Society of London.
Painting by G. Hoefnagel (*c*. 1600). Österreichische Nationalbibliothek.
Painting by an anonymous artist. Museo Nazionale, Florence.
Engraving from *Variorum Navigationis* by T. and J.J. de Bry (Amsterdam, 1601).
Pen-and-ink sketch by Adrian van der Venne (*c*. 1626). University Library Utrecht.

reference and a number look as if they were simply copies of others. Despite the unquestionable talent of the artists concerned, it may therefore be unrealistic to suppose that complete accuracy in points of detail can be relied upon. Against this it can be argued that there are pictures of quality (most particularly Cornelius Saftlevens's at the Boymans Museum in Rotterdam; see Plate XXVII) which do seem drawn from life and perhaps very accurately reflect the spirit of the living bird.

The uncertainty regarding these pictures' origins would not be greatly important were it not that some writers have taken their contents far too seriously. Variation in detail led A.C. Oudemans (1917) and then M. Hachisuka (1953) to catalogue differences in plumage colour and other features according to sex, age or season. Upon these not very reliable foundations they built an elaborate superstructure of ideas which the base is quite unfitted to support. Hachisuka (1953) even attempted to assess the exact number of Dodos transported live from Mauritius to Europe on the basis of differences existing in surviving paintings!

All that can realistically be concluded from the evidence of the paintings is that on several occasions during the seventeenth century living birds were brought from the Indian Ocean to Europe and that some of these were exhibited to the public and also, presumably, made available to artists of the day – either in a living or a dead state. As well as these exports to Europe, it is established on the testimony of Peter Mundy (in Temple and Anstey, 1919–36, *The Travels of Peter Mundy in Europe and Asia, 1608–1667*) that two individuals were despatched to India and one other may even have gone as far afield as Japan.

Dodos were hunted mercilessly during the short time in which Europeans came into contact with them and for a period of upwards of half a century they proved a very useful source of fresh meat for travellers in the Indian Ocean. Whether or not they were good to eat is, however, something of a moot point. While many fell prey to hungry seafarers they may not always have been eaten with great relish. Herbert (1634) mentions that the Dodo, 'is reputed more for wonder than for food, greasie stomaches may seeke after them, but to the delicate they are offensive and of no nourishment'. Van Neck (1601) was even more insulting, calling them *walghvogels* apparently because the longer they were cooked the tougher and less palatable became their flesh; he noted, though, that the belly and breast had a pleasant enough flavour.

The extinction of the Dodo on Mauritius was probably accomplished by 1690. According to Benjamin Harry, chief mate of the *Berkley Castle* on its voyage to the 'Coste and Bay', 1679, individuals still survived on the island in 1681 but no more recent writer makes mention of these. In addition to direct persecution from man, recently introduced monkeys and hogs swarmed over the island and ground-nesting dodos – or at least their eggs – must surely have proved vulnerable to their depredations. The Huguenot François Leguat, who stayed on Mauritius in 1693, does not list dodos as inhabitants of the island and it seems likely that these wonderful birds had vanished by the time of his visit.

The Dodo was one of the most fantastic creatures ever to have lived. Even during the century in which it came to notice, only to become extinct, the species aroused great interest in Europe. Had Dodos survived for a few more decades, colonies might perhaps have established themselves in European parks and gardens; they were probably hardy enough creatures. Today, Dodos might be as common as peacocks in ornamental gardens the world over! Instead, all that remains are a few

bones and pieces of skin, a collection of pictures of varying quality, and a series of written descriptions enormously expressive of the age in which they were conceived yet curiously inadequate in the information they convey.

Recalling the Dodo, the Englishman Peter Mundy wrote:

> as I remember they are as bigge bodied as great Turkeyes, covered with Doune, having little wings hanging like short sleeves, altogether unuseful to fly withall, or any with them to helpe themselves. Neither Can they swymme, butt as other land Fowle Doe, on Necessity, Forced into the water, being Cloven Footed as they are.

RÉUNION DODO

Raphus solitarius

FIGURES 47–9

Raphus solitarius (Sélys Longchamps, 1848)

Apterornis solitarius Sélys Longchamps, 1848
Pezophaps borbonica Bonaparte, 1854
Victoriornis imperialis Hachisuka, 1937

Length ~ 100 cm (3 ft 3 in)

Description
Appearance in life not with certainty known.

That a dodo-like species once lived on the island of Réunion in the Indian Ocean can hardly be doubted. The supposition that such a bird existed is based on four seventeenth-century written accounts and several pictures that supposedly relate to a 'Dodo of Réunion'. There are also one or two tantalizingly brief mentions of dodos on the island given in eighteenth-century documents. The earliest account is that given by J. Tatton (1625), who wrote of:

> A great fowl the bigness of a Turkie, very fat, and so short-winged that they cannot flie, beeing white, and in a manner tame; and so are all other fowles as having not been troubled or feared with shot.

Six years before 1625, V.I. Bontekoe had visited the island but his journal was not published until 1646:

> There were also some Dod-eersen, which had small wings but could not fly; they were so fat they could scarcely walk, for when they walked their belly dragged along the ground.

A later observation of Réunion Dodos comes from the traveller M. Dubois (1674), who wrote about so many of the island's mysterious birds:

> These birds are so called [solitaires] because they always go alone. They are as big as a large goose and have white plumage with the tips of the wings and tail black. The tail feathers are like those of the ostrich, they have a long neck and . . . the legs and feet are like those of a turkey. This bird is caught by running after it, since it can scarcely be said to fly at all.

Finally, comes the written description of a M. Carré (1699):

> I saw a kind of bird in this place which I have not found elsewhere; It is that which the inhabitants call the Oiseaux Solitaire for to be sure, it loves solitude and only frequents the most secluded places; one never sees two or more together; it is always alone. It is not unlike a turkey, if it did not have longer legs. The beauty of its plumage is a delight to see. It is of changeable colour which verges upon yellow.

47 White Dodo and other birds. Watercolour by the seventeenth-century Dutch artist Pieter Witthoos formerly in the collection of Walter Rothschild.

> We wished to keep two of these birds to send to France and present them to His Majesty, but as soon as they were on board ship, they died of melancholy, having refused to eat or drink.
>
> The flesh is exquisite; it forms one of the best dishes in this country, and might form a dainty at our tables.

From these accounts a general image has been built up of a creature that can be aptly described as a 'lightly bleached dodo'.

The seventeenth-century pictures that are associated with the Réunion Dodo largely confirm this impression. This is hardly surprising as these pictures have been connected with the birds simply because they fit the image already established. It is an unfortunate fact that the illustrations, although contemporary with the Dodo's existence, have nothing positively to connect them with the birds of Réunion other than loose resemblance to the written descriptions.

Several pictures by Pieter Witthoos are supposed to have been drawn from life from an individual taken to Amsterdam (Figures 47 and 48); there is no irrefutable evidence that this is so, however, and certainly nothing to link the model, whatever it was, positively with Réunion. One aspect of interest is that the primary feathers point downwards and forwards. It is doubtful whether this would have been the position in life – perhaps the arrangement was due to some accident or deformity, or maybe it is an indication that Witthoos produced his painting not, as is supposed, from a living model but from a badly stuffed specimen.

48 Pieter Witthoos's White Dodo. Engraving from a watercolour (whereabouts unknown) in the *Illustrated London News*, 20 September 1856.

The three other pictures that have been associated with the Réunion Dodo (see Hachisuka, 1953) are all alleged to be the work of Pieter Holsteyn Senior, and to have been produced during the 1630s (Figure 49). As with the Witthoos pictures nothing is known of the circumstances surrounding the painting of them.

When all the paintings are matched up with all of the written descriptions inconsistencies can quite naturally be found among the details shown or described. The birds can be said to have been the possessors of horny sheaths to the upper mandibles that were either greyish, greenish or even patterned with yellow, black and white; these

49 Pieter Holsteyn's White Dodo. Aquarelle in Teylers Museum, Haarlem.

bills may have been hooked or they may have ended in a ball-like lump; the tail feathers perhaps consisted of ostrich-like plumes fairly tightly curled, or they may have been elongated, resembling in appearance more those of a Silver Pheasant. Attempts have been made to rationalize these discrepancies in terms of sexual differences (Oudemans, 1917; Rothschild, 1919) and with equal justification they could be regarded as indications of age or season; more likely, they are the result of individual variation, bad workmanship, careless observation or a combination of these.

Although it can be said with some degree of confidence that Réunion once had its own dodo, M. Hachisuka (1937b and 1953) used the discrepancies between the various written accounts and the associated paintings as evidence that not one but two distinct species inhabited the island. One he felt was similar to the Solitaire of Rodrigues (*Pezophaps solitaria*) and the other he believed to be closer to the Dodo of Mauritius (*Raphus cucullatus*), and this second species he named as *Victoriornis imperialis*. Although the hypothesis is not completely ridiculous, the grounds for supposing that two species existed are rather tenuous and some of Hachisuka's reasoning is remarkably eccentric. It might be remembered that the concept of even one dodo-like species inhabiting Réunion is not borne out by specimen material of any kind, so even the single species might justifiably be thought of as hypothetical.

What, then, can be said of the Dodo of Réunion? From the testimony of Bontikoe (1646), who called the birds *dod-eersen*, it may be deduced that there was a considerable likeness between this species and the Mauritian Dodo. This impression is borne out by the paintings of Witthoos and Holsteyn – assuming that these can be relied upon in every sense. However, it can hardly be imagined that flightless birds could have lived in isolation on two islands as widely separated as Réunion and Mauritius without there being differences between them, but these differences can only be guessed at. Presumably, these two species, developing from the same or very similar original stock, were following independent evolutionary paths, although the factors determining their progression were, of course, quite alike on both islands. It seems likely that, despite their differences, they retained much in common.

Réunion Dodos appear to have become extinct at a comparatively late date, particularly surprising when it is remembered how little is known of them. They probably survived until at least 1750. Hachisuka (1953) found a record to the effect that dodos were still living while M. de la Bourdonnaye was Governor of Mauritius and Réunion; this was between the years 1735 and 1746. M. de la Bourdonnaye is, in fact, supposed to have sent a dodo from Réunion back to France. If it ever arrived and what subsequently became of it is not known.

RODRIGUES SOLITARY

Pezophaps solitaria

FIGURES 50–1

In caves on the small, isolated and rather mysterious island of Rodrigues (or Rodriguez) in the Indian Ocean have been found large numbers of bones. From these, naturalists have pieced together several more or less complete skeletons to establish that a bird bearing great affinity to the Mauritius Dodo, although by no means so extreme in development, once

Pezophaps solitaria (Gmelin, 1789)

Didus solitarius Gmelin, 1789
Pezophaps solitaria Strickland and Melville, 1848

Length ~ 90 cm (3 ft)

Description
Male: general colour grey and brown; iris black.
Female: generally either the colour of blonde hair or sometimes brown; above beak a band of black feathers velvet-like in texture; on lower neck two breast-shaped elevations lighter in colour than rest of underparts. The feathers of both sexes possibly rather hair-like.

lived upon this lonely tropical paradise in the Mascarenes; and the records of some of the very few early travellers who reached Rodrigues confirm that such a bird did indeed live there – a bird they named the Solitary because it was most often found alone. In this tranquil kingdom generations of Solitarys must have lived enjoying extraordinary peace and seclusion until their world was shattered by the coming of man.

The two principal sources for information concerning living Solitarys are the accounts of the Huguenot refugee François Leguat (published in 1708) and the description contained in the anonymous document known as the *Relation de l'Île Rodrigue* – probably written around 1730. Leguat and his small band of followers were marooned upon Rodrigues in 1691 and stayed for 2 years, eventually becoming so desperate in their isolation that they constructed a boat and in this home-made craft crossed several hundred miles of open sea to Mauritius.

At least some of the desperation that drove them to such a foolhardy venture was a result of a lack of female companionship and, certainly, parts of the female anatomy appear to have been playing on Leguat's mind when he penned his description of the female Solitary. A peculiarity of the lower neck he could not avoid comparing to the beautiful curvature of a woman's breasts: 'They have two elevations upon the crop, of which the feathers are whiter than the rest, and which resemble, very marvellously, the beautiful bosom of a woman'. The feature is clearly shown in the picture Leguat carefully drew to illustrate his book but the comparison is neatly sidestepped in some translations – they refer instead to the 'fine neck of a beautiful woman'.

That Leguat's description of the Solitary can in general be relied upon is evident because some of the anatomical details he mentions can be traced in the skeletal material now available for scientific study. Thus the 'little round mass as big as a musket ball', which he claimed occurred on the wing, exists in skeletons as a bony knob on the metacarpal.

It is perfectly obvious from Leguat's account that he was quite enchanted with the Solitary. Following the glowing comparison already given he adds: 'They walk with such stately form and good grace that one cannot help admiring and loving them.' Of mating and nesting he says:

> All the while they are sitting upon it, or are bringing up their young one, which is not able to provide for its self in several Months, they will not suffer any other Bird of their Species to come within two hundred yards round of the Place: But what is very singular, is the Males will never drive away the Females, only when he perceives one he makes a noise with his wings to call the Female, and she drives the unwelcome Stranger away, not leaving it till 'tis without her Bounds. The Female do's the same as the Males, whom she leaves to the Male, and he drives them away. We have observed this several times, and I affirm it to be true. After these birds have rais'd their young One, and left it to its self, they are always together, which the other birds are not, and tho' they happen to mingle with other Birds of the same Species, these two Companions never disunite. We have often remark'd, that some days after the young one leaves the nest, a Company of thirty or forty brings another young one to it; and the new fledg'd Bird with its Father and Mother joyning with the Band, march to some bye Place. We frequently follow'd them, and found that afterwards the old ones went each their way alone, or in couples, and left the two young ones together, which we called a *Marriage*.
>
> This peculiarity has something in it which looks a little Fabulous,

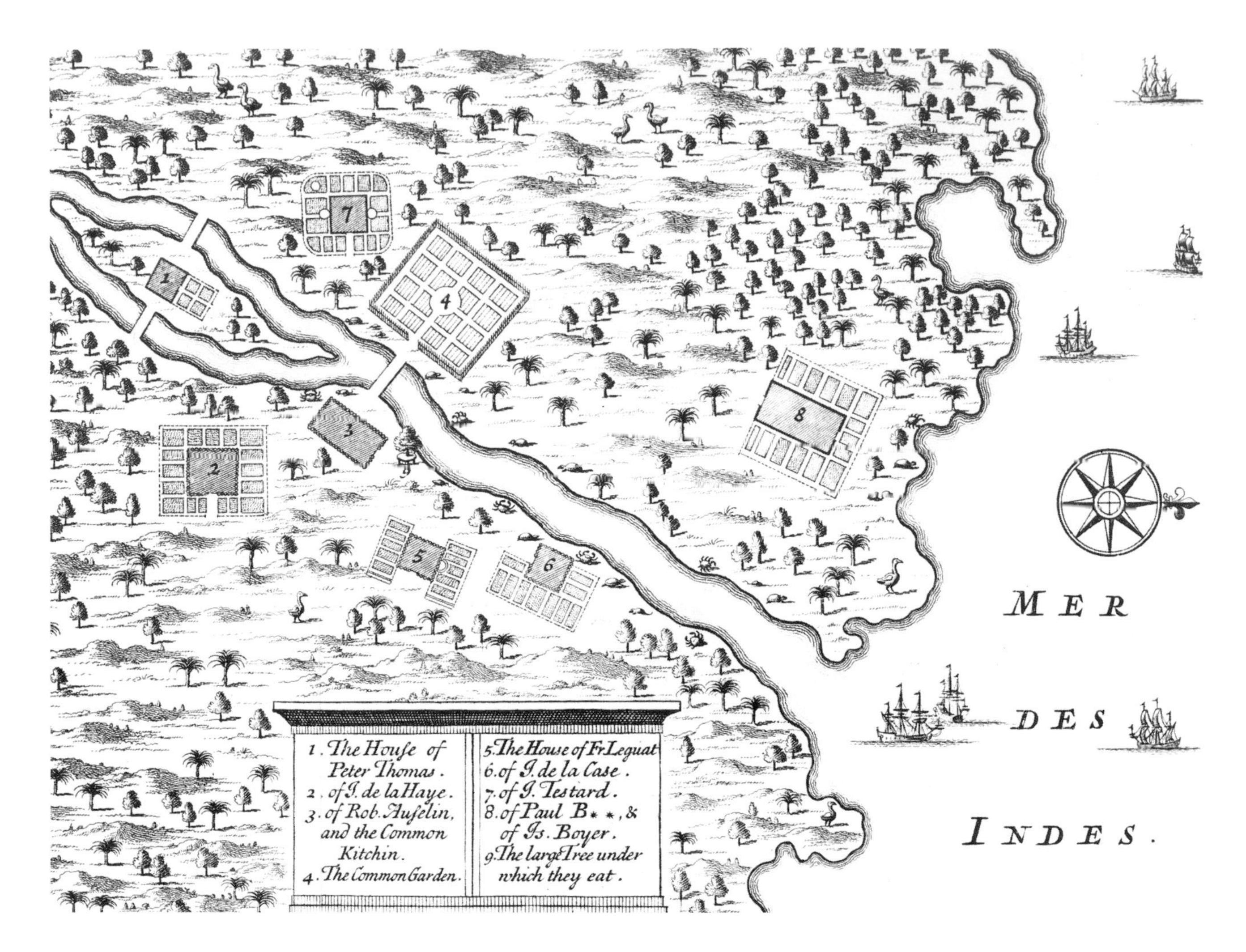

50 François Leguat's 'Plan of the Settlement' on Rodrigues, decorated with Solitarys. Foldout engraving in Leguat's *New Voyage to the East Indies* (London, 1708).

> nevertheless, what I say is sincere Truth, and what I have more than once observ'd with Care and Pleasure: neither cou'd I forbear to entertain my Mind with several Reflections on this Occasion.

Even so, the admiration and affection felt does not seem to have stopped the Huguenots from tucking in to roast Solitary from time to time, for in the *Voyages et Aventures* it is stated, 'From the month of March to September they are extremely fat and they are of excellent taste, especially when they are young'.

While Leguat's account is quite full, so too is that given by the author of the *Relation de l'Île Rodrigue*:

> The Solitary is a large bird, which weighs about forty or fifty pounds. They have a very big head, with a sort of frontlet, as if of black velvet. Their feathers are neither feathers nor fur; they are of a light grey colour, with a little black on their backs. Strutting proudly about, either alone or in pairs, they preen their plumage or fur with their beak and keep themselves very clean. They have their toes furnished with very hard scales, and run with quickness, mostly among the rocks, where a man, however agile, can hardly catch them. They have a very short beak, of about an inch in length, which is sharp. They nevertheless do not attempt to hurt anyone, except when they find someone before them, and when hardly pressed try to bite him. They

> have a small stump of a wing which has a sort of bullet at its extremity, and serves as a defence. They do not fly at all, having no feathers to their wings but they flap them and make a great noise with their wings when angry and the noise is something like thunder in the distance. They only lay, as I am led to suppose, but once in the year, and only one egg. Not that I have seen their eggs, for I have not been able to discover where they lay. But I have never seen but one little one alone with them, and if anyone tried to approach it, they would bite him severely. These birds live on seeds and leaves of trees, which they pick up on the ground. They have a gizzard larger than the fist, and what is surprising is that there is found in it a stone of the size of a hen's egg, of oval shape, a little flattened, although this animal cannot swallow anything larger than a cherry-stone. I have eaten them; they are tolerably well tasted.

At about the same time as the anonymous author visited Rodrigues (1730), the Solitary came to the notice of a one-time Governor of Réunion by the name of d'Heguerty. In 1754 he penned his recollections of Rodrigues and its inhabitants:

> One also finds birds of different species, which can be caught by running after them, and among them the Solitarys, which scarcely have tails or wings; this bird, as big as a swan, has a sad face; in captivity one sees him always in the same line, no matter how much room he has, and returning the same way, without variation.

51 Rodrigues Solitary. Engraving in F. Leguat's *New Voyage to the East Indies* (London, 1708).

Presumably, these poor captured birds did not leave the Mascarenes, or perhaps not even Rodrigues. As far as is known, none was ever brought to Europe and no contemporary paintings or other illustrations by European artists are believed to exist. The only representations of the Solitary appear to be in the drawings in editions of Leguat's journal. Although crude, Leguat had obviously gone to a great deal of trouble over the picture he published and it certainly has enormous charm (Figure 51). Probably it conveys a perfectly good, albeit very generalized, impression of the bird, but in fairness to Leguat his own little apology may be repeated, 'I desire he [the reader] wou'd pardon my deficiency in Designing'.

Although the species seems to have been still common enough in the 1730s, it must have declined rapidly after this. In 1761, the Abbé Pingré made the long trip to the far-off Mascarene island to observe the transit of Venus. He arrived in time to see the wonder in the Heavens but not in time to see the Solitary – though he was assured that individuals still lived. In 1831, a resident of the island for some 40 years claimed that he had never seen a bird large enough for the Solitary. Presumably, the last of the species died during the second half of the eighteenth century; Rodrigues has an area of only 104 km^2 (40 sq. miles) and it seems unlikely that large birds could survive there undetected.

The Rodrigues Solitarys, like dodos, were primarily birds of the woods – not the shores – and here they fed on dates (according to Leguat) or seeds and leaves (according to the anonymous author). Although their only defence was a nasty bite and their wings were useless for flight, they do not seem to have fallen prey to man very easily when disturbed in cover and they may have been quite fleet even in the open. But there is little doubt that they were caught, and caught often. What a sad sight it must have been to see them then; according to Leguat, they made no sound, they simply 'Shed tears'.

PARROTS

ORDER
Psittaciformes

EXTINCT SPECIES

Paradise Parrot (*Psephotus pulcherrimus*)
Society Parakeet (*Cyanoramphus ulietanus*)
Black-fronted Parakeet (*Cyanoramphus zealandicus*)
Newton's Parakeet (*Psittacula exsul*)
Mascarene Parrot (*Mascarinus mascarinus*)
Broad-billed Parrot (*Lophopsittacus mauritianus*)
Rodrigues Parrot (*Necropsittacus rodericanus*)
Cuban Red Macaw (*Ara tricolor*)
Carolina Parakeet (*Conuropsis carolinensis*)

Parrots are probably the most popular of all birds. Yet, human interest and favour do not seem to have done them very much good. There are a considerable number of recently extinct species and many of those that still survive are seriously threatened. The craze for taking young birds to rear as pets has certainly contributed to this position.

Of the parrots that have become extinct since 1600, seven species are represented by museum skins, so of these, at least, we can be fairly sure of the appearance in life. They are: the Paradise Parrot (*Psephotus pulcherrimus*), the Black-fronted Parakeet (*Cyanoramphus zealandicus*), the Society Parakeet (*C. ulietanus*), the Mascarene Parrot (*Mascarinus mascarinus*), Newton's Parakeet (*Psittacula exsul*), the Cuban Red Macaw (*Ara tricolor*) and the Carolina Parakeet (*Conuropsis carolinensis*). Two more species, *Lophopsittacus mauritianus* and *Necropsittacus rodericanus*, are known from bones and perhaps also from early travellers' accounts. It is not possible to be absolutely certain that these accounts relate to the birds in question but probably they do.

A number of what can only be termed 'hypothetical' extinct parrots have been scientifically described and named, mostly by nineteenth- and twentieth-century naturalists anxious to attach their own names to as many species as possible (see Table 13). These allegedly valid species are founded solely upon accounts contained in the writings of early travellers, and all are supposed to have been inhabitants of either the Mascarenes or the West Indies. Since none were named from actual specimens, none can be accepted as proper species; but they have caused some confusion.

The original sources that provide the evidence for the former existence of these parrots are very variable. Some 'hypothetical' parrots were written of in detail, others were named from entries in journals that are hardly more than mentions.

One of the major problems in interpreting these early writings, most of which were assembled by people untrained in natural history, is the complication that man's interference with parrot populations can create. Humans have moved parrots around fairly indiscriminately. If a seventeenth-century writer with just a passing interest in ornithology claimed to have seen, say, a blue and yellow macaw on a West Indian island, there is no way of determining whether he saw a bird belonging to an otherwise unrecorded species or whether a familiar South American form, transported from its natural home, was observed. So many reports of anomalous parrots might be explained simply as observations of escaped pets.

Boddaert's description of *Psittacula eques* is too vague to be of real value. It is based on a plate, no. 215, in E.L. Daubenton's *Planches Enluminées* (1783), which in turn appears to be drawn from an account given by the traveller 'Le Sieur' Dubois (1674). This brief mention is assumed to indicate the former existence of a distinct parakeet belonging to the ring-necked group on the Mascarene island of Réunion.

Various species of Amazon parrots occur on islands in the Caribbean and evidence suggesting that a distinct population once inhabited Guadeloupe is quite convincing. Birds belonging to it were mentioned by both J.B. du Tertre (1667) and J.B. Labat (1742) who give fairly detailed references which agree in most particulars. J.F. Gmelin (1789) named the form *Psittacus violaceus* but although no specimens exist it quite clearly would today be referred to the genus *Amazona*. Labat mentioned more Amazons on Martinique; these may also have been distinct but his description of them is rather less detailed. A.H. Clark (1905) named this supposed species *Amazona martinicana*. The exact status and validity of these two populations are unlikely ever to be satisfactorily determined and neither can properly be listed as other than 'hypothetical'.

Another mysterious, and perhaps distinct, bird that may have formerly inhabited Guadeloupe was named *Conurus labati*, by Walter Rothschild (1905a) but taxonomists would now almost certainly refer it to the genus *Aratinga* should the species prove valid. The original sources were again du Tertre and Labat who described smallish green parrots, Labat adding that these carried a few red feathers. Assuming that these birds did exist and were, as supposed, conures, there would still be no reason to believe that they were clearly differentiated from surviving conures on other West Indian islands. Coloured illustrations of the two 'hypothetical' Amazons and the Conure are given in Rothschild (1907).

Of the other 'hypothetical' forms, the validity of the two Mascarene parrots *Necropsittacus borbonicus* and *N. francicus* is discussed in the account of the Rodrigues Parrot (*N. rodericanus*) and the mysterious West Indian Macaws are referred to under the heading of the Cuban Red Macaw (*Ara tricolor*).

In addition to the extinct species of which we have certain knowledge and also the 'hypothetical' forms, various populations of parrots restricted to islands have disappeared. These are all races of otherwise extant species and those listed have each vanished at some time during the last 150 years. Of these, one of the most interesting, and also one with a rather confusing history, is the Norfolk Island Kaka (*Nestor meridionalis productus*). Often

Table 13

HYPOTHETICAL EXTINCT PARROTS
Psittacus eques Boddaert, 1783
Necropsittacus francicus Rothschild, 1905
Necropsittacus(?) *borbonicus* Rothschild, 1907
Ara martinica Rothschild, 1905
Ara erythrura Rothschild, 1907
Ara atwoodi Clark, 1905
Ara guadeloupensis Clark, 1905
Ara gossei Rothschild, 1905
Ara erythrocephala Rothschild, 1905
Anodorhynchus purpurascens Rothschild, 1905
Conurus labati Rothschild, 1905
Psittacus violaceus Gmelin, 1789
Amazona martinicana Clark, 1905

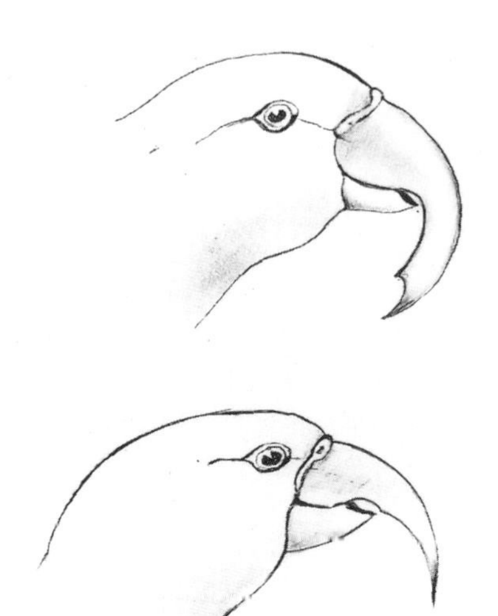

52 Comparison of the head of a typical Norfolk Island Kaka (*Nestor meridionalis productus*) and the freak that was named *Nestor norfolcensis* by A. von Pelzeln (1860).

53 The extinct Norfolk Island Kaka (*Nestor meridionalis productus*). The last member of the race died in captivity in London some time after 1851.

listed as a full species, it seems doubtful, however, that Norfolk Island birds had passed a point at which they could be recognized as forming a species distinct from the very variable New Zealand Kaka. It might be added, though, that the now-vanished population was very isolated from other subspecies of the Kaka on its Tasman Sea outpost.

These large and attractive parrots occupied both Norfolk and nearby Phillip Island, seemingly surviving on the latter for some time after their disappearance from the larger island. There seems to have been a tendency for some individuals to develop a monstrous abnormality of the bill and several freak specimens apparently came to the attention of naturalists (Figure 52). On the basis of this fairly obvious deformity, A. von Pelzeln (1860) named a new species, *Nestor norfolcensis*; in so doing he introduced an element of confusion because some ornithologists took him seriously. Rothschild (1907), for instance, believing the species valid, speculated that birds with highly specialized beaks might have been inhabitants of Lord Howe Island even though there are no good grounds for any such supposition. Greenway (1958) quite properly relegated *norfolcensis* to the level of a synonym for *productus*.

The Norfolk Island Kaka (Figure 53) was probably a bird of the woodlands that, like its relatives still surviving in New Zealand, made its nest in hollow trees. The very last member of the race is thought to have been an individual that died in a cage in London some time after 1851. They seem to have been quite tame and confiding birds and were often trapped as potential pets. The convicts and settlers who began to arrive on the islands in numbers during the early 1800s probably exploited them often as a convenient food source. In New Zealand, the Kaka is still widespread but numbers have declined along with suitable habitat and fears have been expressed for the species' future.

From the same general area of the South Pacific, two more races of a parrot species still common in New Zealand have vanished. These are *Cyanoramphus novaezelandiae erythrotis* and *C. n. subflavescens*, both island forms of the Red-fronted Parakeet. This species has adapted to a considerable range of habitats and one of the outposts of its colonizations was bleak and inhospitable Macquarie Island, far to the south of New Zealand proper. Here it nested among the tussock grasses and occurred in some numbers until sealers and their cats began to deplete the population. The race *erythrotis* finally vanished from here during the early years of this century. The other island race (*subflavescens*) proved something of a pest to settlers trying to cultivate parts of Lord Howe Island and disappeared, presumably under hunting pressure, during the years following 1869. Six more races of the Red-fronted Parakeet are recognized, all of them still extant.

Another small parrot with a Pacific distribution and a large number of recognized races is *Loriculus philippensis*, the Philippine Hanging Parrot. Like the Red-fronted Parakeet, this species, too, has lost two island representatives – *siquijorensis* from Siquijor Island and *chrysonotus* from Cebu. Both have disappeared during this century.

The Seychelles Parakeet (*Psittacula eupatria wardi*), a subspecies of the Alexandrine Parakeet, became extinct around 1900. Probably this was a result of extensive deforestation – land being required for plantations – and excessive shooting on account of the damage that these parakeets did to crops. J. Forshaw and W.T. Cooper (1973) listed the Seychelles Parakeet as a distinct species but this hardly seems justifiable.

Two island races have been lost from the West Indies, one belonging to

Table 14

RARE OR ENDANGERED PARROTS

Stephen's Lory *Vini stephani*	Orange-fronted Parakeet *Cyanoramphus malherbi*	Blue-throated Conure *Pyrrhura cruentata*
Tahitian Lory *Vini peruviana*	Horned Parakeet *Eunymphicus cornutus*	Rufous-fronted Parakeet *Bolborhynchus ferrugineifrons*
Ultramarine Lory *Vini ultramarina*	Orange-bellied Parrot *Neophema chrysogaster*	Brown-backed Parrotlet *Touit melanonota*
Blue-fronted Lorikeet *Charmosyna toxopei*	Turquoise Parrot *Neophema pulchella*	Golden-tailed Parrotlet *Touit surda*
Striated Lorikeet *Charmosyna multistriata*	Scarlet-chested Parrot *Neophema splendida*	Puerto Rican Amazon *Amazona vittata*
New Caledonian Lorikeet *Charmosyna diadema*	Ground Parrot *Pezoporus wallicus*	Red-spectacled Amazon *Amazona pretrei*
Red-throated Lorikeet *Charmosyna amabilis*	Night Parrot *Geopsittacus occidentalis*	Red-tailed Amazon *Amazona brasiliensis*
Long-billed Corella *Cacatua tenuirostris*	Kakapo *Strigops habroptilus*	Yellow-shouldered Amazon *Amazona barbadensis*
Salvadori's Fig Parrot *Psittaculirostris salvadorii*	Glaucous Macaw *Anodorhynchus glaucus*	St Lucia Amazon *Amazona versicolor*
Black-lored Parrot *Tanygnathus gramineus*	Lear's Macaw *Anodorhynchus leari*	Red-necked Amazon *Amazona arausiaca*
Masked Shining Parrot *Prosopeia personata*	Spix's Macaw *Cyanopsitta spixii*	St Vincent Amazon *Amazona guildingii*
Princess Parrot *Polytelis alexandrae*	Caninde Macaw *Ara caninde*	Imperial Amazon *Amazona imperialis*
Golden-shouldered Parrot *Psephotus chrysopterygius*	Red-fronted Macaw *Ara rubrogenys*	
	Golden Conure *Aratinga guarouba*	
	Yellow-eared Conure *Ognorhynchus icterotis*	
	Thick-billed Parrot *Rhynchopsitta pachyrhyncha*	

a species of conure and the other being an amazon. Like the macaws, amazons have suffered greatly during the last few centuries, the numbers of many species having seriously declined. With the possible exception of two 'hypothetical' amazons, only *Amazona vittata gracilipes*, however, has actually become extinct. This subspecies of the Puerto Rican Amazon, once an inhabitant of Culebra Island, is only poorly differentiated from the nominate form, which is itself now seriously threatened, only around thirty individuals remaining. Amazons on Culebra were last recorded about 1900.

The Conure of Mona Island disappeared at about the same time, and just like the Culebra Amazon this extinct form was only poorly differentiated from the nominate race of the species. Mona Island, the former home of *Aratinga chloroptera maugei*, lies between Puerto Rico and the large island of Hispaniola, which is made up of the Dominican Republic and Haiti. *Aratinga c. chloroptera* occurs on Hispaniola itself and is still quite common.

Many parrot species are seriously endangered (see Table 14) and among these is perhaps the most interesting, certainly the most aberrant, of all parrots, the Kakapo (*Strigops habroptilus*) (Figure 54). This New Zealand species was once widespread on both main islands and on Stewart Island, but numbers have declined to a mere handful and the Kakapo seems doomed to extinction. It is a large, fat parrot quite extraordinary in appearance with soft plumage of green, subtly streaked and barred with brown and cream and a facial disk set off by whiskers. Not only is the appearance strange, the habits are equally peculiar. Partly because of its

54 The Kakapo (*Strigops habroptilus*) of New Zealand, one of the world's rarest birds.

visual resemblance to an owl, but also on account of nocturnal inclinations, this bird is often called the Owl Parrot. During the day Kakapos hide in burrows or holes among tree roots and emerge from these at night. In former times their booming call must, at certain seasons, have been a characteristic sound of the New Zealand bush after nightfall. Although the wings seem quite well developed, Kakapos – for all practical purposes – are incapable of flight and the sternum is almost flat. They formed great attachments to everybody who was kind to them and pet individuals are said to have behaved with their owners more like dogs than birds.

Another remarkable and very rare species is the Night Parrot (*Geopsittacus occidentalis*) of central Australia. This strange bird looks very much like a miniature kakapo and, as its name suggests, is active after dusk. For many years it was one of Australia's 'lost' species but was recently rediscovered, ending speculation that it was extinct. In appearance it can be confused with another rare Australian species, the Ground Parrot (*Pezoporus wallicus*), but this bird inhabits only coastal areas. The Ground Parrot is a slimmer, longer tailed bird which frequents swamps, coastal flats, heath and grasslands.

Two members of the genus *Charmosyna* remain very mysterious birds. One, *C. diadema*, the New Caledonian Lorikeet, is often listed as extinct.

XXVIII Paradise Parrots: female (left); male (right). Painting by Peter Slater. From the collection of Vic and Lorna Martin

There are good grounds for supposing that it might be. Only two specimens, both females, were ever collected and these were taken before 1860. Over the years, rumours of the species' continuing existence have circulated and Forshaw and Cooper (second edn, 1978) seem to accept that it still survives. The Blue-fronted Lorikeet (*C. toxopei*) from Buru, Indonesia is known only from a handful of examples, all taken in 1931.

Of the large and attractive Amazon parrots, several are very close to extinction including two of the most striking. The Imperial Amazon (*Amazona imperialis*) inhabits the mountain rainforests of Dominica where perhaps a hundred individuals survive; the brightly coloured St Vincent Amazon (*A. guildingii*) has a larger population on its island home but is endangered by hunting and trapping for the pet trade.

A number of problematical species have been described, in each case from a single or a very small number of specimens. Stresemann's Lory (*Lorius amabilis*), the Blue-thighed Lory (*Lorius tibialis*) and the Rufous-tailed Parrot (*Tanygnathus heterurus*) are each known from just the type. Forshaw and Cooper (1978) expressed the opinion that all three are aberrant individuals of better known species, these being *Lorius hypoinochrous*, *L. domicellus* and *Tanygnathus sumatranus*, respectively. Another form known only from the type is Wallace's Hanging Parrot (*Loriculus flosculus*) from the Indonesian island of Flores, but this may be simply a subspecies of *L. exilis* from the Celebes. *Pyrrhura hypoxantha*, the Yellow-sided Conure, is represented by three examples, two taken from amongst a flock of Green-cheeked Conure (*P. molinae*) in South America. A suspicion exists, therefore, that these, too, are merely aberrant individuals. *Psittacula intermedia* is known from a small number of skins from northern India, which, as the name suggests, show some features intermediate between two better known Asian species (*P. himalayana* and *P. cyanocephala*) and so *intermedia* may be a hybrid.

PARADISE PARROT

Psephotus pulcherrimus

PLATE XXVIII; FIGURE 55

Until quite recently, Australia possessed two 'lost' bird species – both, curiously, members of the Psittacidae. Only a single specimen of the squat and rather strange Night Parrot (*Geopsittacus occidentalis*) had been taken in this century (1912) and the small but very elegant Paradise Parrot was last reliably recorded during November of 1927. Hopes for the Paradise Parrot's continuing survival in rarely trodden parts of its former range are frequently expressed and received a boost in 1979 when living Night Parrots were identified on Coopers Creek, South Australia.

That species sometimes linger long after supposed dates of extinction is not open to question and, quite apart from the Night Parrot, the list of 'rediscovered' Australian creatures seems remarkable. Two species of Grasswren, for instance, the Eyrean (*Amytornis goyderi*) and the Black (*A. housei*), were missed for very many years before being tracked down in the 1960s. The Noisy Scrub-bird (*Atrichornis clamosus*) remained unseen from 1889 until 1961 and several exceptionally rare marsupials, among them Leadbeater's Possum (*Gymnobelideus leadbeateri*), have been found again in recent decades following lengthy periods during which they left no trace. Rediscoveries of this kind, perhaps not completely surprising in a country

Psephotus pulcherrimus (Gould, 1845)

Platycercus pulcherrimus Gould, 1845 (Darling Downs, Queensland, Australia)
Psephotellus pulcherrimus Mathews, 1917

Length 27 cm (11 in)

Description
Male: forehead red; crown and nape blackish brown and around each eye a ring of greenish yellow; cheeks, sides of head, chin, throat and breast greenish blue; lower abdomen, thighs and under tail coverts red; upperparts mostly greyish brown but rump turquoise blue, outer tail feathers blue and median wing coverts red; iris brown; legs and feet greyish brown.
Female: forehead whitish yellow with feathers faintly tipped red; crown and nape brown; face, throat and breast yellowish olive green with orange-brown markings; abdomen pale blue sometimes flecked with red; upperparts similar to those of male but showing reduced area of red on wing coverts.
Immature: like adult female; males show varying amounts of bluish green on the face.

Measurements
Male: wing 121–135 mm; tail 143–182 mm; culmen 12–14 mm; tarsus 16–19 mm.
Female: wing 122–126 mm; tail 144–168 mm; culmen 12–14 mm; tarsus 16–18 mm.

---

so vast and yet so thinly peopled, certainly lend credibility to hopes that any 'lost' Australian species may not in reality be quite extinct.

It may be premature to consider extinct a species recorded as recently as 1927 but the disappearance of the Paradise Parrot was quite unlike that of other, formerly 'lost', Australian birds. These always were considered exceptionally rare, in each case being known from little more than a handful of specimens collected during the nineteenth century or the early years of the twentieth. Such rarity was far from the case with the Paradise Parrot, which, although perhaps never more than locally common, was a relatively well-known bird. Its decline was noticed and followed over several decades around the turn of the century. Various factors appear to have combined to reduce the population to critically low levels and the rather scattered nature of distribution indicates that the range of the species may have been retracting even before the arrival of Europeans to the Antipodes.

Paradise Parrots must have had few natural enemies and it is difficult to assess the effect that the spread of introduced predators (cats, rats, foxes, etc.) may have had. Cattle and sheep grazing the land with an intensity that increased year by year undoubtedly cut into the supply of grass seed available. These birds are said to have fed directly from standing stalks – they slid the strands sideways through their beaks stripping seeds in the process, rather than picking fallen ones from the ground. Possibly, they preferred grass varieties that had become in short supply through competition from livestock and introduced vegetation.

Several years of drought certainly aggravated any problem of food shortage and bush fires may also have taken their toll at a critical period. A number of Australian species of bird are nomadic and therefore able to take advantage of temporary and shifting food supplies or to minimize the effects of disadvantageous conditions, but there is no evidence that the Paradise Parrot was able to do this. Demand for cagebirds led to individuals being captured and frequently exported, particularly to England where they were highly prized for their breathtakingly beautiful appearance. They were comparatively tame so captors could approach fairly close and netting must have been simple near to the nest. Apparently, the species bred in confinement but no aviary stock seems ever to have built up.

From time to time it is suggested that Paradise Parrots still exist in the hands of aviculturalists but claims almost certainly refer to hybrids of the Golden-shouldered Parrot (*Psephotus chrysopterygius*) and the Mulga Parrot (*P. varius*). These are occasionally offered for sale as genuine Paradise Parrots. According to J. Forshaw and W.T. Cooper (1981), male hybrids do show a remarkable similarity to the male of *P. pulcherrimus*. Perhaps because of this similarity, H. Groen (1966) and others believed the Paradise Parrot to be no more than a naturally occurring hybrid, but this argument is strongly refuted by Forshaw and Cooper (1981) who note that hybrid females are quite different from those of the 'lost' species. Although in early years caged individuals were not uncommon, A.H. Lendon commented in 1973 that the *Aviculturalist's Magazine* contains no authentic record for this species in captivity later than its first date of publication (1894).

During the years preceding World War I, the Paradise Parrot became rapidly rarer until by 1915 it appeared to have vanished completely. Three years later, the Australian ornithologist A.H. Chisholm launched a newspaper campaign aimed at locating any existing population. To this

55 Paradise Parrot photographed at its nest by C.H.H. Jerrard. Courtesy of the National Photographic Index of Australian Wildlife.

appeal there was no positive result until December 1921 when Chisholm received a letter from a Mr C.H.H. Jerrard who felt almost certain he had seen a pair in the Burnett River area of Queensland. Keeping the birds under careful observation, Jerrard was able to follow their movements over several months and build a rough shelter close to the nest from which were taken a remarkable series of black-and-white photographs. Chisholm too was able to see the birds before they vanished once more. Jerrard saw them for the last time in 1927 since when only a handful of tantalizing but unsubstantiated sightings suggest the species may be still extant.

Paradise Parrots were first recorded by John Gould's collector John Gilbert on the Darling Downs in 1844. Delighted with his find, Gilbert apparently asked that the species be given his own name but Gould failed to comply with the request. Whether he regretted his decision or not is probably unrecorded, but it was given a poignant twist shortly afterwards when Gilbert, engaged in yet another collecting trip, died at the point of an aboriginal spear.

It seems the species was then distributed over a wide area of southern Queensland and northern New South Wales, from Rockhampton to Brisbane along the coast, and inland to the Darling Downs. Gilbert claimed he also saw large numbers of birds further north in the Mitchell River area of Queensland but he may not have been close enough to make a correct identification and perhaps confused this species with the rather similar Golden-shouldered Parrot. It may be noted, however, that from this area have often come rumours of a red-shouldered parrot using termite mounds for nesting.

Typically, Paradise Parrots frequented river valleys, savannah woodland and scrub grassland. They were usually seen in pairs or small groups, spending most of their time on the ground feeding. On the wing, flight was swift and undulating. Breeding has been recorded at various times between September and March. Although nests were sometimes hollowed out in a steep or vertical river bank, birds usually burrowed in termite mounds to produce a narrow tunnel ending in a chamber perhaps 30 cm (1 ft) in diameter. No nesting material was introduced, a clutch of three to five rounded white eggs being laid on a carpet of loose earth. Incubation lasted for 3 weeks and was exclusively the responsibility of the female. Although males were observed to visit their mates frequently, they seem not to have entered the burrow at all during incubation. According to Chisholm (1922), 'The male would alight on the mound, and, looking into the hole, emit soft, sweet chirps until the faithful little home-keeper answered by coming out and flying off with him.'

The song of the species appealed to those who heard it and Jerrard in Chisholm (1922) supplies a vivid description of the male call and the impression it made:

> He has a very mixed and very animated song . . . his whole body vibrated with the force and intensity of his musical effort, imparting an agitated motion to the long tail which bore adequate testimony to the vim of the performance. It all seemed to indicate a very intense little personality under the beautiful exterior.

Whether Jerrard remains the last to observe this intense little personality only time will tell. The Night Parrot still survives in waste places of Australia and so, too, may this species, but until authenticated sightings are made it is probably correct to consider the Paradise Parrot extinct.

SOCIETY PARAKEET

Cyanoramphus ulietanus

PLATE XXIX

Cyanoramphus ulietanus
(Gmelin, 1789)

Psittacus ulietanus Gmelin, 1789 (Ulietea, now Raiatea, Society Islands)
Platycercus tannaensis Finsch, 1868
Psittacus fuscatus von Pelzeln, 1873

Length 25 cm (10 in)

Description
Adult: head dark chocolate brown; back and wings brown but outer webs of primaries greyish purple; rump and upper tail coverts dark brownish red; central tail feathers olive brown, outer tail feathers greyish purple; breast and abdomen dusky yellow ochre; thighs brown; bill grey; legs and feet greyish brown.

Measurements
Wing 130–145 mm; tail 132–134 mm; culmen 19 mm; tarsus 22–24 mm.

The fact that two distinct and now extinct species belonging to the genus *Cyanoramphus* existed in close proximity to each other in the Society Islands is rather curious, for the *Cyanoramphus* parakeets have a distribution centred today very much on New Zealand, 3,220 km (2,000 miles) to the south-west.

As far as can be determined, each of these extinct species was restricted to a single island in the group, the Black-fronted Parakeet (*C. zealandicus*) occupying Tahiti, the Society Parakeet (*C. ulietanus*) being confined to Raiatea. These two islands are separated by only 320 km (200 miles) of open ocean, Raiatea lying to the west of Tahiti; yet vast distances need to be covered before the habitations of any other *Cyanoramphus* species are reached.

It is generally assumed that the affinity of the Society Islands parakeets is very close; probably both species were descended from the same colonizing stock. But the development of two distinct kinds – so geographically close yet so remote from any related species – surprising though it may be, is not entirely without parallel among *Cyanoramphus* parakeets: the coexistence of the widespread Red-fronted Parakeet (*C. novaezelandiae*) and an endemic species (*C. unicolor*) on tiny Antipodes Island several hundred miles south-east of New Zealand is equally peculiar.

Both Society Islands species appear to have been extinct for a considerable length of time. Although the Tahitian Black-fronted Parakeet is a little-known bird, its relative from Raiatea is perhaps even more mysterious. Whereas the Tahitian species was located several times, there is no evidence to show *ulietanus* was found more than once. It is generally stated (Stresemann, 1950; Greenway, 1958) that this encounter took place when naturalists of Captain Cook's second voyage around the world explored the island of Raiatea (then called Ulietea) in September 1773 or, later, during the summer of 1774.

D.G. Medway (1979) could find nothing to support this contention, however, and suspects that the two specimens taken were collected during the visit of Cook's third expedition, which called at Raiatea on 3 November 1777, staying until 7 December. A specimen said to be from the third voyage was in the Sir Joseph Banks collection and this, presumably, is the skin now kept at the British Museum (Natural History), having probably passed through the collections of William Bullock and Victor Masséna (see page 140). A second skin was deposited in Sir Ashton's Lever Museum in London and is probably the type from which John Latham in 1781 described his 'Society Parrot', and on which J.F. Gmelin founded *Psittacus ulietanus* 8 years later.

At the dispersal of the Leverian Museum, this bird passed into the Imperial Collection at Vienna, being sold as lot 5616 '*Psittacus fuscatus*, South Seas' in 1806 on the forty-seventh day of the Leverian sale. It cost eleven shillings and has stayed ever since in Vienna, now being kept at the Naturhistorisches Museum.

Nothing more is known of the species but, amazingly, a recent claim has been made that it still survives. In 1981, D. Day reported the claim but it has not proved possible to uncover further details about this supposed rediscovery and Mr Day cannot himself remember where his information

XXIX Black-fronted Parakeet (above) and Society Parakeet (below). Oil painting by Errol Fuller.

came from. If the species really does still exist, the survival would certainly constitute a remarkable chapter in the history of zoological rediscoveries, but at present the rumour can only be discounted.

As with the Black-fronted Parakeet, it is difficult to advance reasons for the extinction of this species, but mammals introduced by Polynesians and, later, by Europeans – especially whalers who brought rats with them on board their ships – probably had a detrimental effect.

BLACK-FRONTED PARAKEET

Cyanoramphus zealandicus

PLATE XXIX

The genus *Cyanoramphus* with its exclusively South Pacific distribution is composed of several interesting, some even rather mysterious, species.

Both the Red-fronted Parakeet (*C. novaezelandiae*) and the Yellow-fronted Parakeet (*C. auriceps*) are widespread in New Zealand and occur on various of its offshore islands, *novaezelandiae* ranging as far afield as New Caledonia, Norfolk Island and formerly to a distant southern outpost on the island of Macquarie.

Cyanoramphus zealandicus (Latham, 1790)

Psittacus zealandicus Latham, 1790
Psittacus erythronotus Kuhl, 1820
Psittacus pacificus Vieillot, 1823
Platycercus erythronotus Stephens, 1826
Conurus phaeton Des Murs, 1845
Platycercus phaeton Des Murs, 1849
Cyanorhamphus forsteri Finsch, 1868

Length 25 cm (10 in)

Description
Adult: forehead blackish brown extending onto crown; cheeks and area just above eye bright green; red stripe extending from nostril to eye; lower face and rest of head bright olive green; upperparts olive green except for outer webs of primaries, which are violet blue, and rump, which is red; underparts olive green with bluish tinge; bill bluish grey; legs and feet greyish brown.
Immature (from Rothschild, 1907): 'Young differ in having a dull bluish-black forehead, brownish head, back mixed brown and green, rump and eye stripe chestnut red, and the underside greyish green'.

Measurements
Adult: Wing 135–145 mm; tail 135–146 mm; culmen 15–18 mm; tarsus 25 mm.

Eight races of the Red-fronted Parakeet have been recognized and these occupy a surprisingly varied range of habitats from tropical forest to barren subantarctic isles, some showing traits very unusual in parrots; throughout most of their range these are birds of the treetops but on some of the more isolated islands they live in bushes and low scrub. On Antipodes, Macquarie and the Kermadec islands Red-fronts have assumed an almost exclusively terrestrial mode of existence, living among clumps of tussock and foraging for food in heaps of cast-up seaweed. The race *subflavescens*, which once inhabited Lord Howe Island, is now extinct and so, too, is the extraordinary population *erythrotis* that managed to survive at least briefly upon Macquarie.

That parakeets were able to colonize so southerly an island and one subject to such hostile climatic conditions is remarkable. Equally remarkable is the fact that a still-extant race, *hochstetteri*, shares another of New Zealand's remote southerly islands with a very closely related but quite distinct species, the Antipodes Island Green Parakeet (*C. unicolor*). The coexistence of these two kinds on an isolated and windswept land less than 40 km² (15 sq. miles) in area is difficult to account for. Presumably, their requirements are somewhat different, for red-fronts and the endemic Antipodes Green Parakeets are common over the whole island.

One of the most puzzling of the *Cyanoramphus* parakeets is the form described under the name *C. malherbi*. Usually called the Orange-fronted Parakeet, the status of this allegedly distinct species is quite uncertain. Orange-fronts were present in mixed flocks that invaded the Canterbury district of New Zealand's South Island in the antipodean summer of 1884–5, and Walter Buller (1888) claimed they were common in the woods of Nelson at around this time. Since then, they have hardly been seen at all and it is now proposed (Holyoak, 1974a; Nixon, 1981) that these birds constitute a rare colour morph of *C. auriceps* because they are quite similar to typical individuals of the Yellow-fronted Parakeet. Apparently, a small population is currently under study so perhaps the uncertainty will soon be resolved.

Despite the mystery surrounding New Zealand's Orange-fronted Parakeet, it is from the Society Islands some 3,200 km (2,000 miles) to the north-east that the most enigmatic species in the genus have come. The Black-fronted Parakeet caused confusion among nineteenth-century naturalists partly because it was so elusive, partly because of the considerable difference in plumage between young birds and adults mentioned by Walter Rothschild (1907).

It was and remains a most mysterious bird. The species was first described (Latham, 1790) from specimens brought to Europe from Tahiti, presumably by naturalists who sailed with Captain Cook; according to Rothschild (1907), both William Ellis and one of the Forsters had examples. A drawing made by Sydney Parkinson, who was with Cook during his first voyage of exploration, exists in the British Museum (Natural History) but only a very few specimens of the species remain intact – two in the Merseyside County Museum, Liverpool, one in the British Museum (Natural History) in London and one in the Muséum National d'Histoire Naturelle, Paris.

It is unlikely that many more were ever collected. Of the two kept in Liverpool, one, supposed to have belonged to Sir Joseph Banks, may be Latham's type but the origin of the other is less definite. The example in London apparently came via the Masséna collection. This was a vast private assemblage of stuffed birds put together by Victor Masséna, Duc

56 Title page of the catalogue prepared at the disposal in 1846 of Victor Masséna, Duc de Rivoli's collection of stuffed birds.

CATALOGUE

DE LA

MAGNIFIQUE COLLECTION D'OISEAUX

DE M. LE PRINCE D'ESSLING, DUC DE RIVOLI,

DONT LA VENTE AURA LIEU AUX ENCHÈRES PUBLIQUES

DANS SA GALERIE, RUE DE LILLE, 98,

le 8 *Juin* 1846

JUSQU'AU 25 DU MÊME MOIS ET JOURS SUIVANTS, S'IL Y A LIEU,

A MIDI PRÉCIS,

PAR LE MINISTÈRE DE M. MAGIET,

COMMISSAIRE-PRISEUR A PARIS, CITÉ TRÉVISE, 10.

Exposition publique, du dimanche 31 mai jusqu'au 7 juin, de midi à 4 heures.

LES ACQUÉREURS PAYERONT, EN SUS DU PRIX D'ADJUDICATION, 5 CENT. PAR FRANC, APPLICABLES AUX FRAIS DE VENTE.

Le catalogue se distribue chez MM. CANIVET, rue Saint-Thomas-du-Louvre, 24 ; et PARZUDAKI, rue du Bouloi, 2, qui se chargeront des commissions des personnes qui ne pourraient pas assister à la vente.

S'ADRESSER, POUR RENSEIGNEMENTS PRÉALABLES, A M. CANIVET.

PARIS,

IMPRIMERIE SCHNEIDER ET LANGRAND,

RUE D'ERFURTH, 1.

de Rivoli, and Prince d'Essling – the son of one of Napoleon's favourite comrades in arms. In 1846, the collection was offered for sale (Figure 56) and acquired by the Academy of Natural Sciences, Philadephia, where it remains. The circumstances of its acquisition through an agent – J.E. Gray of the British Museum – are quite entertaining. Gray wrote (see de Schauensee, 1957):

> On my arrival in Paris, I put up at Meurice's and at once sent a messenger with a note to the Prince Masséna saying that I was willing to purchase the collection of birds at the rate of four francs per specimen, and that I was prepared to pay for it in ready money. While sitting at dinner at the *table d'hôte*, an aide-de-camp came in, all green and gold with a cocked hat and a

XXX Newton's Parakeet. Chromolithograph after a painting by J.G. Keulemans from W. Rothschild's *Extinct Birds* (London, 1907), Pl.19. Courtesy of The Hon. Miriam Rothschild.

XXXI Mascarene Parrot. Coloured engraving by Jacques Barraband from F. Levaillant's *Histoire Naturelle des Perroquets*, Vol. 2 (Paris, 1801–5), Pl.139.

> large white feather, to inquire . . . what I intended with ready money, and, when I explained, to inquire if I was ready to pay the sum that evening. I said no, that I had only just arrived in Paris . . . but I would be ready to pay as soon as the bank opened the next morning. He said the bank opened early, and would I come to the Prince at seven o'clock? to which I assented . . . I kept my appointment; the Prince met me, declared the collection agreed with the catalogue, on which I gave his Highness a check on Messrs. Green; and he gave me a receipt and handed me the keys to the cases, and I sealed them up, the affair being settled in a few minutes.

Parrots were not included in this sale but were subsequently negotiated for and bought; in recognition of Gray's part in the transaction, the Philadelphia Academy very generously presented the British Museum with some of these. If the Black-fronted Parakeet was among them – and presumably it was – then this was most certainly a handsome gift!

The very last record of Black-fronted Parakeets appears to date from 1844 during which year a Lieutenant de Marolles collected three birds, taking at least one skin back to Paris where it is now at the Muséum National d'Histoire Naturelle. What became of the other skins is unknown.

Virtually nothing is known of the habits of these birds. By the natives they were called *a'a* but, according to Anthony Curtiss (in Greenway, 1958), traditions of them are no longer current. Probably, the species ceased to exist soon after Lieutenant de Marolles collected his examples.

The Black-fronted Parakeet does not seem to have been encountered on any island other than Tahiti. It is not known exactly why it became extinct but several species that once inhabited the island no longer occur there and introduced mammals have doubtless taken their toll of the original avifauna.

NEWTON'S PARAKEET

Psittacula exsul

PLATE XXX

Psittacula exsul (Newton, 1872)

Palaeornis exsul Newton, 1872 (Rodrigues)

Length 40 cm (16 in)

Description
Male (perhaps immature): generally coloured greyish blue tinged with green, darker above than below; head bluer and darker with fine black line running from cere to eyes; chin black with broad black collar extending backwards becoming narrower towards nape; undersurface of tail greyish; upper mandible dark reddish brown, lower mandible black; legs grey; iris yellow.
Female: similar to male but crown more greyish and black lines on forehead much less discernible; black collar not extending further back than the sides of the neck; bill black.

Measurements
Male: wing 198 mm; tail (abraded) 206 mm; culmen 25 mm; tarsus 22 mm.
Female: wing 191 mm; tail 210 mm; culmen 24 mm; tarsus 22 mm.

The parakeets of the genus *Psittacula* have long been sought after as cage birds. Among the trophies of conquest brought back from India to Europe by the Romans were Rose-ringed Parakeets (*Psittacula krameri*), and these must have been among the earliest parrots to have been kept by Europeans as pets. Their slim, elegant proportions, long, finely shaped tails and their bright colours combine to make them popular. In addition, they are comparatively hardy and learn to speak well. The genus is very widely distributed, with a range extending from Borneo and the islands of Indonesia through south-east Asia and the Indian subcontinent to Africa.

Parakeets of this genus live on several islands in the Indian Ocean and although most of these forms can be regarded as insular races of widespread kinds, more than one has been considered divergent enough to warrant separation as a distinct species. Of these, the most singular, Newton's Parakeet, is now extinct. It occurred only upon the Mascarene island of Rodrigues and survived to a considerably later date than several other species endemic to the island. In general form this bird was similar to other members of the genus, but in colour it was strikingly different. It retained the black collar so familiar in *Psittacula* species but was otherwise coloured a slaty blue rather than the more typical green.

Only two specimens appear ever to have been collected and both are now in the Cambridge University Museum. The type, a female originally preserved in spirit, was described in 1872 by A. Newton and a second example was shot on 14 August 1875 by a Mr Vandorous. Although the bird was sexed as a male by the collector, there are several reasons for supposing that it may not have been fully mature. The colouring of the beak was not suggestive of a fully adult bird and there was no red alar patch, a characteristic of several *Psittacula* species. Absence of red on the wings is not necessarily of significance for it is not common to all members of the genus, but the accounts of early travellers suggest that the wings of Rodrigues birds were decorated with this colour.

Both of the early chroniclers of Rodrigues birds, the Huguenot François Leguat and the anonymous author of the *Relation de l'Île Rodrigue,* mention parakeets presumably referable to this species. During a 2-year stay upon the island, Leguat with his small band of castaways taught some parakeets to speak. One that spoke in both Flemish and French, they took with them on their desperate voyage from Rodrigues to Mauritius.

Despite this evident fondness for the parakeets, Leguat's followers found them very good eating; because the birds were quite small, the tail accounting for much of the total length, several must have been needed to provide a meal. According to Leguat, the birds themselves fed chiefly upon the nuts of an olive-like tree.

Although clearly quite plentiful at the time of the Huguenots' visit, by the nineteenth century the species had become rare. During the 1870s a number of sightings were recorded. A Mr J. Caldwell saw several birds but was unable to approach them closely, while one Henry Slater observed a solitary individual in wooded country at the southwestern end of the island. It seems likely that the species became extinct soon after this time and had vanished by the early years of this century.

MASCARENE PARROT

Mascarinus mascarinus

PLATE XXXI

Mascarinus mascarinus
(Linnaeus, 1771)

Psittacus mascarin Linnaeus, 1771
Mascarinus madagascariensis Lesson, 1831
Coracopsis mascarina Wagler, 1832
Vaza mascarina Schlegel, 1864
Mascarinus duboisi Forbes, 1879

Length 35 cm (14 in)

Description
Adult: entire front of the face forwards from the eyes black; rest of head greyish lilac; primaries dark brown; back and rest of wings greyish brown; underparts similar but paler; tail dark brown but basal third of lateral feathers white; bill red; feet reddish brown.

Measurements
Wing 211 mm; tail 144–152 mm; culmen 32–36 mm; tarsus 22–24 mm.

In all probability the very last living Mascarene Parrot was kept, far from its natural home, in the Zoological Gardens of the King of Bavaria in Munich. Here, its portrait was drawn to provide plate 39 of C.W. Hahn's *Ornithologischer Atlas* (1834–41) and here it survived until at least 1834. Other individuals may still have lived at this time on the distant island of Réunion but the absence of records may be an indication that all had gone by this date. Several living birds seem to have been kept in European aviaries during the eighteenth century but clearly the species was always rare enough to be regarded as a great novelty. What became of the King of Bavaria's bird after death is not known; certainly it does not seem to have been preserved, for only two specimens survive today and neither of these can be correlated with it.

One of the still surviving examples is in the Muséum National d'Histoire Naturelle, Paris. A. Milne-Edwards and E. Oustalet, who wrote of it in 1893, remarked that certain deteriorations were evident in the general condition but added that this was only to be expected in a skin perhaps a century old. Its age has now doubled, of course, but M. Hachisuka described the condition as still reasonable in 1953 while drawing attention to damage on the wings and tail. There is some evidence to show that the other specimen, kept now in the Naturhistorisches Museum, Vienna, was once a captive bird but apart from this its origin remains unknown. It is slightly albinistic, showing some white feathers on the back, wings and tail. According to Hachisuka (1953), this specimen carries the date 1806.

Mascarinus mascarinus was a striking but rather aberrantly coloured parrot with a red beak comparatively massive in relation to the size of the body. Very little is known of it and even the species' place of origin has been the subject of some controversy, early writers stating that it came from Madagascar. There is no positive evidence to connect this parrot with Madagascar although captive individuals may well have come to Europe via that island. There is, however, an unmistakable written description of the species' presence on the Mascarene island of Réunion by 'Le Sieur' Dubois (1674): 'Parrots a little bigger than pigeons, with the plumage light grey, a black hood at the front of the face, a large beak the colour of fire'. Certainly, therefore, the species now known as *Mascarinus mascarinus* was once to be found on Réunion. It has not been established whether it also occupied the other two Mascarene islands, Mauritius and Rodrigues, but probably it was absent from these.

BROAD-BILLED PARROT

Lophopsittacus mauritianus

FIGURE 57

Lophopsittacus mauritianus (Owen, 1866)

Psittacus mauritianus Owen, 1866
Lophopsittacus mauritianus Newton, 1875

Length ~ 70 cm (28 in)

Description
Appearance in life not known with certainty but plumage was probably dark bluish grey.

Housed in the University Library at Utrecht is the manuscript of a journal kept during the years 1601 and 1602 by one Wolphart Harmanszoon. What gives this document especial interest to ornithologists is that it tells of an early voyage to Mauritius and, most particularly, contains a drawing of a large and strange parrot (Figure 57). This drawing shows a bird with a proportionately massive head and beak, and a striking crest of feathers rising immediately above the upper mandible. Text accompanying the picture indicates that the bird's colour was a bluish grey all over.

Although penned at such an early date, this account was apparently forgotten – or perhaps overlooked – until after nineteenth-century naturalists had found concrete evidence of an extraordinary, and by that time extinct, species. Richard Owen in 1866 described *Psittacus mauritianus* from imperfect elements of a lower mandible taken, together with dodo remains, from the Mare aux Songes, Mauritius. Other parts of the skeleton have since been discovered, confirming that large and big-headed parrots once existed.

Only Harmanszoon's journal and illustration, however, give real indication of the bird's probable appearance in life. No other early travellers mention birds that can be positively referred to the species but it perhaps did not go completely unnoticed. A handful of seventeenth-century writers mention parrots that may have been the same, but each of these references seems too vague to be definitely associated with the known skeletal remains.

Since these exceptionally large and aberrant parrots clearly required generic separation, the monotypic genus *Lophopsittacus* was created to accommodate them. The species was, quite probably, flightless or almost so. In the Harmanszoon drawing the wings appear rather short but this cannot, perhaps, be relied upon. More conclusive of limited power of flight is the reduced keel and the overall bulk. Although the lower mandible indicates that the beak was massive and especially broad, the configuration suggests that it was quite weak, so the species probably fed on soft fruits. X-ray examination of the mandible (Holyoak, 1971) revealed an open arrangement of the tissue, lending support to the concept of a fairly weak structure. Holyoak also pointed out that differences among bones referred to the species might indicate a quite extraordinary sexual dimorphism in terms of size, but this conclusion is, of course, to a large degree conjectural.

As early records are so rare, it may be assumed that extinction took place during the first decades of the seventeenth century and that therefore this was one of the very first casualities of the human colonization of the Mascarenes. If, as speculated, this parrot was not an accomplished flyer, it probably would have provided easy, as well as tasty, prey for sailors who were anxious to replenish or supplement their monotonous and meagre rations.

Although the external features of *Lophopsittacus mauritianus* are not with certainty known, both Walter Rothschild (1907) and M. Hachisuka (1953) had coloured reconstructions prepared based on Harmanszoon's sketch. It seems likely that these give a reasonably accurate indication of the bird's appearance in life.

57 Adaptation of the drawing of a Broad-billed Parrot in the manuscript of Wolphart Harmanszoon's journal (*c.* 1601) in the Bibliotheek der Rijksuniversiteit, Utrecht.

RODRIGUES PARROT

Necropsittacus rodericanus

Necropsittacus rodericanus (Milne-Edwards, 1867)

Psittacus rodericanus Milne-Edwards, 1867 (Rodrigues)
Necropsittacus rodericanus Milne-Edwards, 1873

Length ~ 50 cm (20 in)

Description
Appearance in life not known with certainty but plumage was probably uniformly green.

Bones found on the Mascarene island of Rodrigues (Figure 58), among them an almost complete skull, show that a species of parrot comparable in size to a large cockatoo and possessing an outsized beak once lived there.

It was probably this species mentioned by the anonymous author of a document known as the *Relation de l'Île Rodrigue*. This manuscript, presumed to have been written around 1730, reports one of the earliest recorded visits to the island. Apart from the information given in this single account contemporary with the bird's existence, nothing is known of *Necropsittacus rodericanus*. The anonymous writer described a long-tailed, large-headed parrot, bigger in size than a pigeon, and implied that it was uniformly green in colour. It is also mentioned that the population was concentrated on islets lying to the south of the main island. Here, the birds ate small black seeds from a tree with leaves that smelt of lemons. To obtain water, it was necessary for these parrots to fly across to the large island where they were sometimes seen perching in trees. At what date they became extinct is not known but it was probably some time during the eighteenth century.

Related forms may have occurred on the other Mascarene islands, Réunion and Mauritius. Two additional species have been proposed for the genus *Necropsittacus*, both by Walter Rothschild, but these can be regarded as no more than hypothetical. During a 3-year stay on Réunion (1669–72), the traveller Dubois made notes on the birds that he saw, one of them an account of a parrot mostly green in colour but with head and upperparts of the wings red. While assembling material for *Extinct Birds* (1907), Rothschild considered it necessary to attach scientific identities to

58 A nineteenth-century view of the island of Rodrigues looking west. Papyrograph in H.E. Strickland and A.G. Melville's *The Dodo and its Kindred* (London, 1848).

XXXII Cuban Red Macaw. Coloured engraving by Jacques Barraband from F. Levaillant's *Histoire Naturelle des Perroquets*, Vol. 1 (Paris, 1801–5), Pl.5.

several enigmatical entries in Dubois's journal and he proposed the name *Necropsittacus borbonicus* for the green and red parrot. The original description of Dubois may be accurate and refer to an otherwise unrecorded species, or it may, quite unintentionally, be misleading. J.C. Greenway (1958) pointed out that the pattern of colouring is suggestive of an escaped lory.

The island of Mauritius also might conceivably have had a population of parrots referable to the genus although there seems no direct evidence in support of this. *Necropsittacus francicus* was named by Rothschild in 1905 but this is one of the most dubious of all hypothetical species. It is not even possible to be certain on which original sources the alleged species is founded.

Coloured illustrations of the hypothetical extinct parrot *N. borbonicus* are given by Rothschild (1907) and Hachisuka (1953).

CUBAN RED MACAW

Ara tricolor

PLATE XXXII

Ara tricolor (Bechstein, 1811)

Psittacus tricolor Bechstein, 1811 (South America)

Length ~ 50 cm (20 in)

Description

Adult: forehead red, becoming more orange on crown and merging to yellow on hindneck; bare facial area white; sides of face, chin, throat, breast and abdomen, together with thighs, orange; upper back brownish red, feathers margined with green; rump and lower back pale blue; lesser wing coverts brown, feathers edged with red; webs of primaries and secondaries purplish blue; upper surface of tail dark red becoming blue towards tip but underside brownish red; under tail coverts blue; bill dark brown, paler towards tip; iris yellow; legs brownish. Sexes alike.

Measurements

Wing, 275–290 mm; tail 215–290 mm; culmen 42–46 mm; tarsus 27–30 mm.

There is a great deal of confusion over the recently extinct macaws of the West Indies. Various allegedly distinct species from several different islands were named by naturalists during early years of this century on the basis of old accounts given by travellers and settlers. As these accounts, perhaps truthful enough in spirit, are not supported by specimens, it is impossible satisfactorily to evaluate them.

One problem with early reports of parrots on islands is that there can be no certainty whether the birds referred to were actually endemic or whether they were individuals belonging to a widespread species that had been introduced by human agency. For instance, macaws described as blue and orange appear to have been observed on the island of Martinique by one Père Bouton, being mentioned by him in an account written during the first half of the seventeenth century.

In rather characteristic fashion, Walter Rothschild (1905a) considered that this account provided sufficient grounds for naming a new species, even though no other author had mentioned similar birds. He therefore founded *Anodorhynchus martinicus*, subsequently referring this purely hypothetical species to the genus *Ara*. What Bouton really saw will almost certainly never be known but he probably observed birds of a familiar species, perhaps the Blue and Yellow Macaw (*Ara ararauna*) taken to the island as pets. A second Rothschild species, *Ara erythrura*, seems to be a synonym for the first.

A third hypothetical species, *Anodorhynchus purpurascens*, was named by Rothschild simply because a macaw deep violet in colour is said to have once inhabited Guadeloupe. If any purple macaws did live on this island, it is likely that they were imported from the South American mainland and were members of the well-known species the Hyacinth Macaw (*Anodorhynchus hyacinthinus*). A large red and yellow macaw has also been reported from Guadeloupe. In 1667 J.B. du Tertre wrote of these and A.H. Clark (1905) united this record with others from Martinique and Dominica, attaching to all the name *Ara guadeloupensis*. The correct status of these populations cannot now be determined. Each island may have had its own endemic species, all may have been conspecific, or they may have been birds of familiar species imported from elsewhere.

Yet another macaw named by Rothschild is *Ara erythrocephala*. This dubious species is based upon the rumour of a specimen supposed to have been seen by the nineteenth-century naturalist P.H. Gosse. Whether Gosse or anyone else did actually see an example of an otherwise unknown bird cannot be said but none now seems in existence. This allegedly green and yellow parrot perhaps once inhabited mountainous districts of Jamaica. Another green and yellow macaw was mentioned by a certain Thomas Atwood in an account of the island of Dominica published in 1791 and Clark (1908) gave to this report the scientific identity of *Ara atwoodi*.

Of all the recently extinct macaws of the West Indies, only one form is known with certainty from skins. This is *Ara tricolor*, the Red Macaw of Cuba, a bird represented by examples in some dozen museums in North America, the Caribbean and Europe. These indicate that this was an attractive and spectacular bird although it was a third less in size than its largest relatives. As with many rare things, a certain amount of intrigue surrounds the history of one of the skins. T. Barbour (1943) dropped a very heavy hint that a specimen that disappeared under mysterious circumstances from the Academy of Sciences, Havana, was surreptitiously extricated on behalf of Walter Rothschild.

The last living bird of which there is a record is of one shot at La Vega, close to the Zapata Swamp, during 1864, but probably some individuals still survived in southern Cuba for another 20 years or so. Parrots of this species may have been quite easy to obtain during earlier years but rapidly became rare around the middle of the nineteenth century. Local residents hunted adults for meat and often took young birds as pets, two activities that must have been important factors in the macaw's decline.

Little is on record of the species' habits. Much time seems to have been spent in pairs or small parties. Eggs were neither described nor, seemingly, preserved. It is likely that feeding requirements were similar to those of other macaws; the food was probably nuts, seeds, fruits and shoots.

Tradition suggests that the species once inhabited the Isle of Pines (now called Isla de la Juventud) and occurred over the whole of Cuba with the exception of the province of Oriente. With certainty, however, it was recorded only from a comparatively limited area and none of the specimens came from the Isle of Pines.

A predominantly red macaw may also have existed on Jamaica, from where a report of an individual shot about 1765 has come. The skin was apparently stuffed and has been described in some detail by a Dr Robinson, but it can no longer be traced. Rothschild believed that the description was evidence enough of a species endemic to Jamaica. He christened it *Ara gossei* but other writers, notably Clark (1905), have united it with *A. tricolor*. The true status of the form remains obscure.

Although the Cuban Red Macaw is the only extinct species of which skins exist, there is material evidence in the form of bones to indicate that a now-vanished bird once inhabited St Croix. This was named *Ara autochthones* (Wetmore, 1937) but it is not known when the population died out.

Reconstructions of a number of hypothetical macaws are featured in Rothschild's *Extinct Birds* (1907) wherein can be found coloured illustrations, mostly by J.G. Keulemans, of *Ara gossei*, *A. erythrocephala*, *A. martinicus*, *A. erythrura* and *Anodorhynchus purpurascens*. Also pictured, of course, is *Ara tricolor*.

CAROLINA PARAKEET

Conuropsis carolinensis

PLATE XXXIII

Conuropsis carolinensis
(Linnaeus, 1758)

Psittacus carolinensis Linnaeus, 1758 (Carolina)
Psittacus ludovicianus Gmelin, 1789
Conuropsis carolinensis Salvadori, 1891

Length 30 cm (12 in)

Description
Adult: forehead, lores, area around eyes and upper cheeks orange; remainder of head, throat and upper part of neck yellow; outer webs of primaries marked yellow towards their base; bend of wing, carpal edge and thighs yellow; rest of plumage green, paler on underparts; bill horn-coloured; legs and feet pinkish brown.
Immature: forehead, lores and area around eyes brownish orange; rest of head green; no yellow at bend or edge of wing, or on thighs.

Measurements
Male: wing 180–200 mm; tail 120–160 mm; culmen 23–25 mm; tarsus 16–19 mm.
Female: wing 180–195 mm; tail 125–155 mm; culmen 22–24 mm; tarsus 16–19 mm.

XXXIII Carolina Parakeets. Aquatint by J.J. Audubon and R. Havell the younger from Audubon's *Birds of America* (London, 1827–38), Pl.26.

A consignment of sixteen small, green parakeets with bright yellow heads was delivered to the Cincinnati Zoo at an unspecified date during the 1880s. The cost per head was just $2.50, yet among the batch was a pair destined to become the last known representatives of their kind.

At the time of purchase, Carolina Parakeets could still be seen in various zoos and aviaries, even though the species was approaching extinction in the wild. Among the last reliable records of wild birds is one of six individuals taken at Padget Creek, Brevard County, Florida on 18 April 1901 and another of thirteen seen at Lake Okeechobee, Florida during April 1904; a handful of other reports – probably accurate – were made during the new century's first decade and the last bird collected was very likely a female taken at Orlando on 4 December 1913.

As the birds vanished in the wild, so did the captive population decrease. Breeding among caged Carolina Parakeets had never been particularly successful and, with so few now living, little hope remained for the species' survival.

Eventually only a pair was left; cage-mates at the Cincinnati Zoo for something like 32 years – and now seemingly long past breeding age – their names were Lady Jane and Incas. The London Zoo, alert to their great rarity, is supposed to have offered $400 for the pair but the offer was declined. During the late summer of 1917 Lady Jane died; Incas, perhaps the very last Carolina Parakeet, pined through the autumn and early part of the following winter. On the evening of Thursday 21 February 1918, or (the record is not clear) exactly 1 week earlier, he died in his cage surrounded by his keepers. They claimed that old age was not to blame; their bird had died of grief.

Three and a half years previously, by great coincidence, the last Passenger Pigeon – Martha – had expired in the very same zoo (see page 112). Her body had been sent to the Smithsonian Institution in Washington DC for preservation, and similar arrangements were apparently made for Incas. Curiously, the little body never arrived in Washington and no one now knows what became of it. From a purely scientific point of view it hardly matters. There are rather more than 700 skins in collections around the world; P. Hahn (1963) reported seventy-six in Washington alone.

A parrot flying above fields lying heavy in snow presents something of an anomaly. Yet a record of 1790 describes a flock of parakeets appearing in January, 40 km (25 miles) north-west of Albany, and 100 years before the death of Incas, Carolina Parakeets – the only parrots indigenous to the United States – ranged over much of the eastern half of the country from the Gulf of Mexico in the south, as far to the north as New York and southern fringes of the Great Lakes; few members of the parrot order have adapted to life in such potentially harsh environments. This unusual distribution did not long survive the systematic colonization of North America. As European influence upon the land became increasingly felt through the nineteenth century, so the area occupied by parakeets was steadily pegged back; in the west towards the Mississippi River, in the south towards Florida.

Commentators noted how dates of last sightings in particular areas coincided with settlement and forest destruction but although the nature

of much of the land certainly changed, tracts of seemingly suitable habitat remain and it is curious that the species declined so drastically; the fall of the Carolina Parakeet from its widespread and numerous state to a position of virtual extinction within a space of 100 years is not easy to account for. There is no doubt it was ruthlessly hunted during these years but whether this alone can explain the total disappearance is difficult to say. Perhaps this was one of those species only able to maintain itself in large populations. Clearly, these were highly social birds; it seems that once numbers dropped below certain levels, the decline to extinction was inevitable.

From being primarily feeders on seeds of various wild plants – in particular, the seeds of a cocklebur (*Xanthium* sp.) – Carolina Parakeets quickly modified their habits to take advantage of new sources of food that became available with the spread of organized cultivation. As a result, they soon established themselves as pests in the eyes of colonists. The liking for seeds of various fruits, together with communal feeding habits, could result in the ruination of orchards. Long before ripening, fruits would be torn from the stalks and ripped open; a flock of birds might destroy the produce of an apple or pear tree within a very short time. Growing corn might also be destroyed; John James Audubon (1831) described how grain stocks were attacked in the fields: 'Flocks of these birds ... cover them so entirely, that they present to the eye the same effect as if a brilliantly coloured carpet had been thrown over them.' All kinds of seeds were attacked and eaten along with buds, blossom and nuts.

As might be expected, the depredations of Carolina Parakeets did not go unpunished, and perhaps it was a certain lack of caution as much as anything else that led to their downfall. During the early 1880s, when still occurring in flocks of several hundred individuals, they could be approached with ease at feeding grounds and so densely did they pack together that several might be destroyed by a single shot. Early writers reported it characteristic of this species that individuals would fly squawking around and above dead or wounded companions rather than desert them. After firing off a shot or two, a hunter just had to wait for a few minutes and any birds remaining alive would settle again in the midst of their dead or injured fellows. As numbers became diminished and birds were seen in parties rather than flocks, so the possibilities for destruction became more limited but Audubon in 1831 was still able to record:

> I have seen several hundreds destroyed in this manner in the course of a few hours, and have procured a basketful of these birds at a few shots, in order to make choice of good specimens for drawing the figures by which this species is represented in the plate now under your consideration.

Yet they were quite wary birds in some respects – a flock would inspect a feeding site carefully before landing and if alarmed in a general way might fly off and not return.

In its heyday this was a bird of forested lowland showing preference for terrain close to water, either river or swamp. For its home it sought stands of buttonwood, cypress or sycamore and its roosting places – often hollow stumps – were in trees sheltered in the heart of swamp or woodland. Within the stump, parakeets would crowd together attached to the sides by beak and claw. Sometimes, where all could not cram inside, those unable to squeeze in would cling to the outside of the tree. Dawn would find a flock in the topmost branches, chattering and agitated but for much of the day it would remain quiet and comparatively inactive; late afternoon or early evening saw another period of great industry. While engaged in bursts of feeding, a steady chattering was maintained but in flight a loud and raucous *kee*, *kee*, *kee* was delivered. Flocks flew swiftly, darting in and out of the woodlands in which they lived and spiralling down to chosen feeding areas.

Comparatively little is recorded of the species' nesting habits. There are descriptions of groups of parakeets laying their eggs together in holes in trees but other reports tell of fragile nests made from twigs and sited in tree-forks. There was occasional breeding in captivity but, here again, little information is preserved. Eggs in museum collections are white with an average measurement of 27 mm × 35 mm.

Two fairly well-marked races were recognized. The nominate subspecies, *carolinensis*, occurred in the southeastern United States while a western form, *ludovicianus*, covered the rest of the range. This western subspecies was generally paler with the green of the back and rump bluish rather than tending towards yellow. Yellow areas of the wing, however, were more evident.

Incas may or may not have been the last of the Carolina Parakeets. As is often the case with supposedly extinct birds, rumours of this species' continuing existence circulated. In 1926 Charles E. Doe, Curator of Birds at the University of Florida, saw three pairs of parakeets in Okeechobee County. They may or may not have been Carolina Parakeets but if they were, Doe robbed them of their eggs, none of which can be satisfactorily identified; escaped foreign parakeets were recorded in the area around this time. Ten years later, Alexander Sprunt together with Robert Porter Allen investigated rumours coming from the Santee River country of South Carolina. They both initially claimed to have indubitably seen the alleged extinct birds flying fast overhead.

From the same area – in 1938 – a local woodsman by the name of Shokes reported to Sprunt that he had seen two yellow-headed parakeets circling above him, screeching wildly. A young bird flew uncertainly out to join its parents. Leaving cover, it tried to gain height as it crossed the Wadmacaun Creek towards the safety of Wadmacaun Island. As the parakeet struggled on, it dipped closer and closer to the surface of the water but finally, in company now with the adults, it successfully reached its chosen destination.

No further reports came from the Santee Swamp – now in any case destroyed by development. Sprunt continued to believe he had seen living Carolina Parakeets in South Carolina; Robert Porter Allan changed his mind – perhaps they were both mistaken, he decided.

CUCKOOS

ORDER
Cuculiformes

EXTINCT SPECIES

Delalande's Coucal (*Coua delalandei*)

This order, which is divided into two major groupings, the touracos (Musophagidae) and cuckoos (Cuculidae), contains only one extinct species, Delalande's Coucal (*Coua delalandei*).

No species within the order seems seriously threatened at present although fears have been expressed for several of the Madagascan Coucals in the same genus (*Coua*) as the extinct one. Other cuckoos and touracos that may be endangered are Prince Ruspoli's Touraco (*Touraco ruspolii*) in southern Ethiopia and the Red-faced Malkoha (*Phaenicophaeus pyrrhocephalus*) in southern India and Sri Lanka.

DELALANDE'S COUCAL

Coua delalandei

PLATE XXXIV

Delalande's Coucal, a very large and striking member of the cuckoo order, is almost certainly extinct. It was a ground-dwelling species that once inhabited the wet forests stretching along the northeastern coasts of Madagascar; very little else can be said of it.

Exactly when it became extinct is not known, reports of birds being trapped during the first decade of the twentieth century and even on into the 1920s are probably reliable but no specimens seem to have been taken by a European after 1834; in this year a Monsieur Bernier sent an example to Paris.

Madagascar is the home of the lemurs – those extraordinary primates many of which become rarer almost by the day. Environmentally, this vast island is one of the world's most seriously threatened areas as development proceeds and spoils many of its natural habitats. Forest destruction helped to account for the Coucal's extinction as it has undoubtedly done for several recently extinct mammalian species. Native hunting seems to have been another important factor, at least in the Delalande's Coucal's case, for

Coua delalandei (Temminck, 1827)

Coccycus delalandei Temminck, 1827 (Madagascar)

Length 56 cm (22 in)

Description
Adult: head blue with blue area of bare skin around each eye surrounded by black feathers; rest of upperparts blue but tail shows a greenish tinge and outer pairs of feathers have white tips; throat and upper breast white; abdomen chestnut; bill, legs and feet black; iris brown.

Measurements
Wing 220 mm; tail 300 mm; culmen 45 mm; tarsus 71 mm.

XXXIV Delalande's Coucal. Hand-coloured engraving by M. Huet from C.J. Temminck and Baron M. Laugier de Chartrouse's *Nouveau Recueil de Planches Coloriées d'Oiseux*, Livre 74 (Paris, 1838), Pl.440.

certain Madagascan tribes valued the feathers highly.

Specimens of Delalande's Coucal can be seen in museums in Leiden, London, New York, Paris, Philadelphia, Tananarive (Madagascar) and Cambridge (Massachusetts).

ANOTHER PUBLISHED ILLUSTRATION OF *DELALANDE'S COUCAL*

Grandidier, A. 1876–85. *Histoire physique, naturelle et politique de Madagascar*, vol. 13, pl. 50. Artist, J.G. Keulemans.

OWLS

ORDER
Strigiformes

EXTINCT SPECIES

Rodrigues Little Owl (*Athene murivora*)
Laughing Owl (*Sceloglaux albifacies*)

Owls, like parrots or penguins, form a very natural and easily recognizable group. Although there are more than 130 species ranging in size between the European Eagle Owl (*Bubo bubo*), more than 2 ft (60 cm) in length, and the tiny Pygmy Owl (*Glaucidium minutissimum*), just 5 in (12.7 cm) long, all are readily identifiable as owls. They bear no obvious relationship to any other group and even at a superficial level they display a number of features that separate them clearly from other birds. Big heads with large forward-facing eyes give a surprisingly human aspect, enhanced in many species by a pronounced facial disk and ear tufts. Soft plumage and a blob-like body shape give an illusion of a cuddly, almost teddy-bearish character but in reality all of these features are simply adaptations to hunting in poor light. The visual effect of the wickedly shaped beak is softened to a large degree by its partial covering of feathers but the cruel talons reveal that these birds take over the role of the diurnal raptors when the sun goes down.

As parrots have always been especial favourites with humans, owls have in some respects attracted the opposite emotions, arousing unease and suspicion around which a mass of superstition and folklore has grown. As creatures of the night, it is only natural that owls should become associated with the powers of darkness. Their reputation for great wisdom is less easily explained, but it may be connected with their capacity for sitting quite still while fixing any observer with a steady gaze, seemingly full of expression and confidence.

Owls have long been associated with mysterious powers, and it is not surprising that an aura of mystery surrounds many species too. Because of their largely nocturnal habits and also the nature of the terrain they sometimes inhabit, many owls have not been studied to the same degree as other birds. Remote forests after nightfall are not appealing to most researchers and, in the tropics particularly, some recorded species are hardly known at all.

At least two species are listed as extinct but other poorly known owls may be also. *Athene murivora* vanished from the Mascarene island of Rodrigues during the eighteenth century; only bones and, perhaps, a traveller's description remain. From New Zealand, the Laughing Owl (*Sceloglaux albifacies*) seems to have been lost in the early years of this century although rumours to the effect that it survives occasionally circulate. Two further species have lost races from certain islands even though both species are themselves still widespread. These races are the Antigua Burrowing Owl (*Speotyto cunicularia amaura*), the Guadeloupe Burrowing Owl (*S. cunicularia guadeloupensis*) and the Lord Howe Island Morepork (*Ninox novaeseelandiae albaria*).

The Burrowing Owl (*Speotyto cunicularia*) is found throughout much of South America and the western half of North America and is, in many places, a quite common bird. Isolated populations occur in Florida and islands of the Caribbean and two of these no longer exist. From Antigua and nearby islands the race *amaura* was lost around the beginning of the century – a disappearance that probably resulted from the introduction of the Small Indian Mongoose (*Herpestes auropunctatus*), to which their burrowing proclivity would have made them especially vulnerable. From the small island of Marie Galante the subspecies *guadeloupensis* went missing at around the same time, probably following a similar introduction.

The Morepork or Boobook Owl (*Ninox novaeseelandiae*) occurs throughout Australia and New Zealand and is also recorded from southern New Guinea. Distinct races developed on the South Pacific islands of Norfolk and Lord Howe. The Norfolk Island race *undulata* is considered to be seriously threatened, having not been seen for some time, but it is occasionally still heard. The population *albaria* that once inhabited Lord Howe Island, 885 km (550 miles) to the south-west, seems no longer to exist. W.B. King (1981) lists 1940 as the approximate date of its extinction.

Table 15

RARE OR LITTLE-KNOWN OWLS

Madagascan Grass Owl	*Tyto soumagnei*
Tanzanian Bay Owl	*Phodilus prigoginei*
Seychelles Owl	*Otus insularis*
Puerto Rican Screech Owl	*Otus nudipes*
Rajah's Scops Owl	*Otus brookii*
Sokoke Scops Owl	*Otus ireneae*
Biak Island Scops Owl	*Otus beccarri*
São Tomé Scops Owl	*Otus hartlaubi*
Rufous Fishing Owl	*Scotopelia ussheri*
Papuan Hawk Owl	*Uroglaux dimorpha*
Forest Little Owl	*Athene blewitti*
Hume's Tawny Owl	*Strix butleri*
Buff-fronted Owl	*Aegolius harrisii*

Of rare owls (see Table 15), one with a particularly interesting history is the Seychelles Scops Owl (*Otus insularis*), a chestnut-coloured bird first described by H.B. Tristram in 1880. For many years from 1906 onwards this species was considered extinct but it was rediscovered in the mountains of Mahé during 1959. At least eighty pairs are currently thought to exist and at present this population seems able to maintain itself.

Another very rare owl sometimes listed as extinct is the Forest Little Owl (*Athene blewitti*) from central India. Only a very few specimens have ever been collected, the last in 1914, since which time its forest home has been largely destroyed. Whether it actually survives is open to some doubt even though an individual was photographed in 1968.

Several hypothetical owls have been described as former inhabitants of the Mascarenes. *Bubo leguati* and *Strix newtoni* were both founded by Walter Rothschild (1907) upon inadequate skeletal material. The evidence for the former existence of two other Mascarene owls, belonging to distinct species, is better but they remain rather mysterious. *Strix sauzieri* is known from bones but, in the absence of evidence to the contrary, this form may be assumed to have become extinct before 1600. A large owl perhaps inhabited Mauritius until the middle of the nineteenth century and a fairly detailed description of it, written by Julien Desjardins, exists. Desjardins apparently made his description from a specimen killed in 1836 but unfortunately the account is in some respects confusing. The owl is alleged to have been a member of a species belonging to the 'Scops' group, but in view of its considerable size this is open to doubt; however, the name *Scops commersoni* was attached to Desjardins's description by E. Oustalet in 1896.

J.C. Greenway (1958) pointed out that the account could be applied better to a species of *Asio* but there are problems even with this designation for no birds in the genus have the unfeathered tarsi that Desjardins makes clear was the case with the specimen he saw. Whatever '*Scops commersoni*' was, it remains at present a mystery. All that can be said is tentative – that a now-extinct species probably did inhabit Mauritius in recent years, and that this bird was possibly related to the scops owls.

RODRIGUES LITTLE OWL

Athene murivora

Athene murivora (Milne-Edwards, 1874)

Athene murivora Milne-Edwards, 1874 (Rodrigues)
Carine murivora Günther and Newton, 1879

Length ~ 23 cm (9 in)

Description
Appearance in life unknown.

An anonymous manuscript discovered among the archives of the Ministère de la Marine, Paris, during 1875 and known as the *Relation de l'Île Rodrigue*, details a visit to this Mascarene island that took place about 1730. It contains important references to many of the birds that then lived there, including a brief mention of a small owl that preyed upon little birds and lizards and was, in fine weather, regularly to be heard calling. There is no evidence that other early chroniclers of the bird life of Rodrigues ever saw these birds. If they did, they seem not to have recorded the fact.

It is generally assumed that this solitary account of owls on Rodrigues can be correlated with the discovery during the 1870s of skeletal remains that were not completely fossilized. Although this matching is, of course, conjectural it is certain that owls did inhabit the island until comparatively recent times, for the remains, quite unequivocally, belong to birds of this group.

A. Milne-Edwards, the great authority on fossil birds at the Muséum National d'Histoire Naturelle, Paris, during the last half of the nineteenth century, first described a Rodrigues Owl in the year preceding the finding of the anonymous manscript. His description was based upon a tibiotarsus and a tarso-metatarsus uncovered on Rodrigues yet belonging to no species at that time known to occur there. The proportions of these remains led Milne-Edwards to suppose they came from a 'little owl' assignable to the genus *Athene*.

Further material, including part of the cranium and the pelvis, has since been found, and this provides additional evidence that in size the species conformed to what might be expected in a 'little owl'. Hardly anything else can be said of *Athene murivora*, although it can be pointed out that the bones indicate legs of particular strength and length for an owl of this overall size. Presumably, the species was similar in habits and appearance to living birds in the genus but in the isolation and security of Rodrigues it had adapted to a rather more terrestrial mode of existence. Perhaps it bore some resemblance to the Burrowing Owl (*Speotyta cunicularia*) of the New World, a species that very much resembles the Little Owls and spends much time on the ground.

If *Athene murivora* is in reality the same as the owl reported in the anonymous writer's manuscript, the species may be assumed to have died out at some time during the eighteenth century. Reasons for this extinction remain unknown.

LAUGHING OWL

Sceloglaux albifacies

PLATE XXXV

Sceloglaux albifacies (Gray, 1844)

Athene albifacies Gray, 1844 (Waikouaiti, South Island, New Zealand)
Sceloglaux albifacies Kaup, 1848
Ieraglaux albifaces Kaup, 1852
Athene ejulans Potts, 1870
Sceloglaux rufifacies Buller, 1904

Length 48 cm (19 in)

Description
Adult: facial disk whitish showing more grey towards centre but sometimes much darker or more rufous; forehead, throat, ear coverts and sides of head greyish white; sides of neck white, feathers having narrow central streak of black; crown and nape dark brown, feathers broadly margined with yellowish brown or, sometimes, white; primaries dark brown, outer webs with angular white spotting, inner webs with obsolete bands; secondaries dark brown with wide transverse white barring; scapulas dark brown with white spots; tail dark brown with five bars and a terminal margin of tawny white; rest of upperparts dark brown but lower back streaked, spotted and barred with tawny and white; lower foreneck and breast dark brown, feathers margined with bright tawny colouring or yellowish brown; abdomen, sides of body and under tail coverts yellowish brown, each feather being centred with dark brown; thighs and tarsi light tawny; iris yellowish brown; toes fleshy brown covered with coarse yellowish hairs; bill black, horn-coloured towards tip; claws black. (cont'd)

When evening settles on the New Zealand bush and the last melancholy notes of the Grey Warbler (*Gerygone igata*) have trailed away into the twilight, then, in the stillness, above any scufflings of kiwis, the strange high-pitched cry of the Morepork (*Ninox novaeseelandiae*) might be heard. From its cry comes its peculiar name and the call of this owl is the characteristic sound of the bush after nightfall.

A hundred years ago – on the South Island at least – the call of the Morepork and the occasional scream of a kiwi were not the only indications of night birds. On particularly dark or drizzly nights, the Morepork's cry would be joined by that of another owl. A loud mournful succession of notes sometimes broke the silence – a sound described by one who heard it as, 'A series of dismal shrieks frequently repeated', and by another as, 'A peculiar laughing cry, uttered with a descending scale of notes'.

Adult birds hailing each other throughout the evening were sometimes likened to two men cooeeing one another but it was the sound uttered when laughing owls were on the wing that inspired the name; and if drizzle was not already falling when the weird call was heard, the sound could safely be taken as a presage to imminent rain.

Hardly had the Laughing Owl come to European attention than it began to vanish rapidly from those haunts it had frequented – perhaps it was a species in decline even before its discovery. The little information that now exists about its habits comes largely from the records of two of New Zealand's most respected early field naturalists – T.H. Potts and W.W. Smith. Curiously, both men were also involved – though only slightly – in the stories of other extinct birds: Potts maintained a considerable interest in eggs of the Great Auk (*Alca impennis*), actually owning several at a time when these regularly changed hands for very large sums of money (see page 98); Smith recorded the very last unquestioned sighting of the Huia (*Heteralocha acutirostris*) (see page 229) in 1907.

According to the records of these naturalists, and in particular Smith's, the Laughing Owl made its home in open country. Here it would find fissures in rock outcrops, especially of limestone, or even in caves. The entrance to the chosen crevice was often surprisingly small, giving the impression that entry, in the first instance, needed to be forced. Sometimes the nesting area – very much in the dry and warm – was 4.5–5.5 m (5–6 yds) back from the entrance. The birds introduced to their holes small quantities of dried grass, adding this to a bed of fine earth or crumbling, powdered rock.

In September, breeding began. Two white eggs were laid and both parents took some share in the duties of incubation, although the female did the bulk of the sitting. After 25 days the yellowish or whitish chicks hatched. These young birds were fed on worms but according to Smith, who formed his opinion on the basis of castings, the adult diet consisted of beetles, lizards and native rats.

The Laughing Owl's decline is often attributed to the decline of the Maori Rat or Kiore (*Rattus exulans*). Certainly, owls regularly ate rats but whether their dependence upon them was total is open to question. With the arrival of Europeans the Brown Rat (*Rattus norvegicus*) moved in to

Sceloglaux albifacies (cont'd)
Nestling: sparsely covered with coarse, yellowish-white down; abdomen bare.

Measurements
Adult: wing 260–280 mm; tail 160–175 mm; culmen 25 mm; tarsus 65–68 mm

---

replace the Kiore but it is hardly probable that the latter's demise signalled the end for the owl; Laughing Owls would surely have turned their attentions to the newcomers. Furthermore, the Kiore was itself a fairly recent arrival having been taken to New Zealand shores just a few centuries earlier by Polynesian navigators; presumably, Laughing Owls fed heartily enough on other creatures before the mammalian invasion of their homeland.

The depredations of other introduced mammals is more likely to have brought about the owl's downfall. Short and feeble wings indicate that powers of flight were rather limited, and the comparatively long and powerful legs suggest that this was a more terrestrial bird than most owls – it walked quickly with long strides and stood very upright. Also, a seemingly quite natural docility seemed to afflict these birds in certain circumstances. Walter Buller (1905) recorded an instance of this apparent tameness:

> A man . . . travelling from Nelson to the West Coast . . . observed a large Owl squatting on the ground near the roadside. He dismounted from his horse and caught the bird. Then, selecting a retired nook in the adjoining woods, he drove a thick pole into the ground and secured his captive by the leg, allowing a sufficient length of flax to permit of the Owl moving freely over the ground. On his return by the same road two days later he found that the bird had snapped, or in some way had got disengaged from, the flax string, and was perched on the top of the pole, permitting itself to be recaptured without the slightest resistance. He took it on with him to Nelson, and, not knowing its value, sold it to the narrator for a few shillings. It now graces the collection in the Nelson Museum.

Laughing Owls, though indubitably fierce birds in certain situations, might easily therefore have succumbed to attack from Ferrets (*Mustela putorius*) and Weasels (*M. rivalis*). W.W. Smith, who knew the birds better than anybody, attributed disappearance to 'the introduction by the Government of the weasel . . . for suppression of the rabbit nuisance'.

Evidence of a total dislike of daylight and the languid state to which the birds sank when exposed to it was often supplied or commented upon. Concerning a living bird in his possession, Buller (1905) described a charming instance:

> During the day it had a listless, dazed look, and generally kept its eyes partly closed. The only occasion on which I saw it awake from this lethargy was when I brought a live hawk (*Circus gouldi*) near to the wire-netting of its enclosure. It did not then manifest any excitement or alarm, but slowly raised itself up to its full height two or three times in succession, with the feathers of the head puffed out and the eyes opened to their full extent, as if in silent wonderment at so strange an apparition.

The Laughing Owl was first described by G.R. Gray from specimens sent to the British Museum by Percy Earl. Gray was able to include a description in his report on *The Zoology of the Voyage of H.M.S. Erebus and Terror* (1844–5) although parts of the bird section had to be rewritten to accommodate this newly arrived material from the distant colony.

Even though the increasing rarity of the species was quickly noticed, specimens – both living and dead – seem to have been quite readily available for the European market. Various collectors in England owned living examples at different times; clearly, several consignments of live owls were despatched from New Zealand and more than one individual is said to have fetched in excess of £50.

George Dawson Rowley, a Brighton collector and compiler of the *Ornithological Miscellany* (1875–8), one of the most extraordinary and entertaining of all Victorian bird books, exhibited several living owls before a meeting of the Zoological Society of London on 3 November 1874. He later commissioned J.G. Keulemans to paint portraits of two of these birds and used the picture produced for plate 8 of his book, noting:

> In making the drawing Mr Keulemans has had the advantage of giving portraits of living individuals [artists of the time often were given only skins as reference material] by which means all the soft parts can be correctly ascertained . . . the pupils are prussian blue with a strong light upon them, at night looking quite black; . . . the . . . iris . . . is so faint and dirty as to be hardly worthy of the name, and it is only on a close inspection that you see it, it might be described as brown-yellow. . . . The bird is strictly nocturnal; and the two here depicted sit all day in a corner on the ground, close up together. More gentle animals could not be; they allow themselves to be handled without any resentment. On March 1st last I tied their legs together, and put each bird into the scales; the dark one weighed 1 lb $4\frac{1}{2}$ oz [581.2 g], the light one 1 lb $5\frac{3}{4}$ oz [616.6 g].

That certain variations occurred in plumage colouring is evident from Rowley's final sentence and elsewhere he is more specific: 'They vary somewhat in appearance. One is dark, having the facial disk more rufous, the other light.'

This colour discrepancy among Laughing Owls was something well known to all who handled series of skins or groups of living birds. Although the type specimen was characterized by a whitish face, other individuals were darker and some showed an altogether more reddish hue. Variation of this kind is by no means an uncommon phenomenon among owls, many species having red and grey phases. In the case of *Sceloglaux*, however, it caused a great deal of confusion – something that grew largely from the long-running feud that raged between Sir Walter Buller, the New Zealand writer, collector and ornithologist, and the Hon. Walter Rothschild, the celebrated ornithologist and patron of natural history (see Figure 59 and page 192).

Although Laughing Owls once occurred on both of New Zealand's main islands, almost all of the specimens were taken from the South Island. Only two have a North Island provenance. The first of these was shot about 1855, close to the volcano Mount Egmont, by a taxidermist named Martin, who was working on behalf of a Captain King. Along with the rest of King's collection it has either been destroyed or exists unrecognized in private hands.

The second North Island specimen, and the one that was subsequently to feature in the Buller/Rothschild feud, was taken by a settler in the Wairarapa district during the summer of 1866–9 (high summer in the Southern Hemisphere comes, of course, at the turn of the year) and sent freshly killed to what was then known as the Colonial Museum, Wellington; it was skinned and the sex – female – determined. Here, Buller first saw it and in due course came to know the specimen very well; certainly, he very quickly realized that the plumage was redder than usual. During the early 1870s he was authorized to take it to England (Buller's

J.G.Keulemans del. Mintern Bros. imp

SCELOGLAUX ALBIFACIES.

main interest in this trip was to supervise publication of the first edition of *A History of the Birds of New Zealand* (1872)) so that the skin could be turned into a stuffed bird. This work Buller entrusted to a Mr Burton, a taxidermist operating out of Wardour Street, London. On Buller's eventual return home the stuffed *Sceloglaux* was taken back to the Colonial Museum where it stayed for nearly three decades.

During these years Buller built up his association with Walter Rothschild, providing the young banker with many rare and curious New Zealand specimens for the museum being assembled at Tring in Hertfordshire. Buller doubtless profited enormously from the association; when Rothschild was sent by his father to Cambridge, he arrived with a flock of live kiwis that Buller had supplied. In the beginning, a genuine friendship probably existed between the two men – at least on Rothschild's part; but somewhere along the line their relationship soured, probably as a result of Buller's greed and Rothschild's rather questionable behaviour over the matter of the Stephen Island Wren (*Xenicus lyalli*) (see page 192).

Through all these years Buller appeared to take little notice of the red Laughing Owl with a North Island provenance, although he drew attention to colour variation among Laughing Owls on several occasions. Then, the specimen became available. Presumably the Colonial Museum, being in possession of other Laughing Owl skins, felt able to part with one in exchange for items lacking in the collection and Buller would have been easily able to provide these; the North Island Laughing Owl thus fell into Buller's hands.

Bearing in mind the circumstances, only one interpretation can be put on what happened next. The great New Zealand ornithologist, fired with enthusiasm or greed (or both), hatched a cynical plot to extract maximum benefit from the situation. Having been familiar with the specimen for more than 30 years, Buller suddenly reappraised it. In 1904, using its reddish coloration as justification, he named it as the type of a new, and to all appearances extinct, species, *Sceloglaux rufifacies*, closely related to the South Island form but quite distinct from it. Then he offered the bird to Walter Rothschild! Well acquainted as he was with the colour variation to which owls in general and Laughing Owls in particular are prone, his action can only have been a ploy deliberately designed to increase the bird's value.

Rothschild was placed in a difficult position: he badly wanted the bird, but recognized the proposed price was hugely inflated. Also, his father presented a serious obstacle to the acquisition of specimens and Buller was not always discreet. Although for all practical purposes the Rothschild wealth may have appeared limitless, Walter's father disapproved – perhaps understandably – of his son's massive outlay on natural history material; Walter himself went to great lengths to conceal his spending excesses from his father's notice.

While Buller looked for every excuse to raise the price, Rothschild (1907) sought to convince himself that the stuffed bird was not worth buying:

> This supposed 'species' is a very doubtful one. A close examination in the Tring Museum of the type (which was offered me for such a high price that I did not feel justified in buying it, fond as I am of possessing extinct forms, types and varieties) by Messrs Hartert, Hallmayr and myself proved beyond doubt to all three of us that the specimen was not fully adult.

XXXV Laughing Owls. Hand-coloured lithograph by J.G. Keulemans from G.D. Rowley's *Ornithological Miscellany*, Vol. 1 (London, 1875), Pl.8.

This last remark may not have been strictly true because Buller gathered equally expert testimony to the effect that his owl was indeed mature, provoking Rothschild, who in exasperation had obviously overstated his case, to add:

> If I said to Sir Walter Buller that it was an 'extremely young, hardly fledged *Sceloglaux*', this was certainly incorrect, and was perhaps just an exclamation after a hasty preliminary examination, for the bird is of course fully fledged and has passed, at least partially, through one moult of feathers . . . It certainly shows unmistakable signs of maturity, as noticed at once by Dr Gadow.

Quite possibly, Buller had the ammunition to win this rather irrelevant battle but he had already lost the war; Walter Rothschild had seen through the game and would not make a purchase. The argument had raged quite publicly, and almost as a serial, across the pages of their rival publications. But there was one final humiliation in store for Buller.

Rothschild, justifiably angry and stung by Buller's manipulations, had spotted a defect in the specimen and this he now chose to bring to notice. The tail was not genuine – it belonged to an owl of the genus *Ninox*. This

59 A characteristic photograph (*c.* 1895) of The Hon. Walter Rothschild, founder of the Tring Zoological Museum in Hertfordshire, 1895. Courtesy of The Hon. Miriam Rothschild.

really was embarrassing to Buller. He could only claim that perhaps another tail was substituted – for unknown reasons – when the bird had been in the hands of Mr Burton, the taxidermist. At this period the bird was still the property of the Colonial Museum, the implication being that either Buller had been a party to the subterfuge by noticing and saying nothing, or he had not noticed at all: both alternatives were obviously equally embarrassing!

Having offered the specimen to Rothschild for such a high price, Buller must have been appalled to find the deceit uncovered by his potential customer. In the *Supplement to the 'Birds of New Zealand'* (1905), he wrote defensively:

> Mr Rothschild discovered that the taxidermist had skillfully replaced the tail of this specimen with that of an Australian Owl . . . but nothing turns upon that, as I have not made the tail a specific character.

To which Rothschild (1907) replied:

> Buller described it as a 'new species' and mentions among the distinctive characters . . . the colour of the tail. The tail, however, is 'skillfully' (as Buller calls it, though I should use a less complimentary adverb) stuck in, and does not belong to a *Sceloglaux*, but to an Australian *Ninox*, and also some feathers on the neck are foreign.

Unacquired by Rothschild, the specimen went elsewhere and its present-day location, sadly, appears unknown. With the vital piece of evidence missing, the matter of the North Island Laughing Owl can be judged only by the circumstances that surrounded its history. In the light of this, Buller's species *Sceloglaux rufifacies* is certainly invalid and it is doubtful whether this form should even be accorded subspecies ranking.

As with many supposedly extinct species rumours persist that this one still survives. Unfortunately, none of the recent alleged sightings appear to stand up to close scrutiny and the last unquestionable record relates to an individual found dead at Bluecliffs, South Canterbury in July 1914 by Mrs Airini Woodhouse. This 'last' individual could be seen stuffed and placed beneath a large glass dome at the home of its finder.

Anyone still believing in the Laughing Owl's survival and hoping to find its last resting place, might do worse than learn to play the accordion. Buller (1905) recounted the story of a settler who claimed: 'It could always be brought from its lurking place in the rocks, after dusk, by the strains of an accordion. Soon after the music had commenced the bird would silently flit over and face the performer, and finally take up its station in the vicinity, and remain within easy hearing till it had ceased.'

OTHER PUBLISHED ILLUSTRATIONS OF THE *LAUGHING OWL*

Buller, W.L. 1872. *A History of the Birds of New Zealand* (first edn), pl. 3. Artist, J.G. Keulemans.

Buller, W.L. 1888. *A History of the Birds of New Zealand* (second edn), pl. 20. Artist, J.G. Keulemans.

Buller, W.L. 1905. *Supplement to the 'Birds of New Zealand'*, pl. 7. Artist, J.G. Keulemans.

Gould, J. 1851–69. *Supplement to the 'Birds of Australia'*, pl. 2. Artists, J. Gould and H.C. Richter.

GOATSUCKERS

ORDER
Caprimulgiformes

EXTINCT SPECIES

Jamaica Least Pauraqué (*Siphonorhis americanus*)

Five extraordinary but clearly related families make up the order Caprimulgiformes – the oilbirds (Steatornithidae), frogmouths (Podargidae), potoos (Nyctibiidae), owlet frogmouths (Aegothelidae) and nightjars (Caprimulgidae).

Birds in all of these five families are nocturnal; as a result, some are very little known and it is difficult to determine which, if any, have actually passed the point of extinction. The Jamaica Least Pauraqué (*Siphonorhis americanus*) certainly seems to have done so, being last recorded in 1859; its Hispaniolan relative (*S. brewsteri*) is now rare. From the owlet frogmouths can be selected another species that may very well be extinct – the New Caledonain Owlet Frogmouth (*Aegotheles savesi*). It was last collected in 1880 but, as with any nocturnal species coming from a remote area, true status is very difficult to determine; the species may still exist or it may not.

A rare subspecies, the Puerto Rican Whippoorwill (*Caprimulgus vociferus noctitherus*), shows just how easily these birds can be overlooked. It was thought extinct from the early years of the nineteenth century until the 1960s; probably around 1,000 individuals survive!

XXXVI (Opposite): Jamaica Least Pauraqué. Chromolithograph after a painting by George Edward Lodge from W. Rothschild's *Extinct Birds* (London, 1907), Pl.5A. Courtesy of The Hon. Miriam Rothschild.

JAMAICA LEAST PAURAQUÉ

Siphonorhis americanus

PLATE XXXVI

Siphonorhis americanus
(Linnaeus, 1758)

Caprimulgus americanus Linnaeus, 1758 ('America calidiore' = Jamaica)
Chordeiles americanus Bonaparte, 1850

The name of the mysterious Jamaica Least Pauraqué is derived from its call. For the identification tag of a nighthawk this is quite regular (Whip-poor-will, Nightjar, Poor-me-one, Potoo, etc.) simply because these birds are far more likely to be heard than seen; the case of Jamaica's lost pauraqué was rather more extreme than most – it was hardly ever seen at all.

Only a very few specimens were ever collected, the last of them in 1859,

Siphonorhis americanus (cont'd)
Siphonorhis americanus Sclater, 1861

Length 24 cm ($9\frac{1}{2}$ in)

Description
Adult: upperparts mainly brown mottled and streaked with grey and darker brown but a rufous collar on hindneck is spotted with black and white; primaries irregularly barred with light and dark brown; tail mottled and flecked with grey and brown, feathers tipped with white; underparts generally lighter particularly on lower breast and abdomen, which is flecked with arrow-shaped marks of a darker brown – white band separates lower throat from upper breast.

and these remains are now divided between the American Museum of Natural History in New York and the British Museum (Natural History) in London. Reasons to account for the bird's extinction are not particularly evident; it is just possible, although most unlikely, that the Least Pauraqué still survives in Jamaica. Nightjars are birds that are often overlooked – even individuals belonging to the more common species; their nocturnal habits make observation difficult and their rather beautiful camouflaging plumage keeps them well hidden even during the day-time.

It was a quite distinct species of nightjar, well differentiated from its allies on the American continent and the Caribbean Islands – most particularly by its possession of pronounced tubular nostrils about 2 mm ($\frac{1}{8}$ in) long. The Jamaica Least Pauraqué has, in fact, only one obviously close relative – it shares the genus *Siphonorhis* with the Hispaniolan Pauraqué (*S. brewsteri*). Smaller than other West Indian Goatsuckers, the Hispaniolan bird is a mere 19 cm ($7\frac{1}{2}$ in) long, with the extinct Jamaican species only a little larger at 24.1 cm ($9\frac{1}{2}$ in).

Virtually nothing is known of the Jamaica Least Pauraqué; presumably it lived much as other nightjars do. Its extreme rarity even at an early date is indicated by the fact that it was unknown to the celebrated naturalist P.H. Gosse, the chief nineteenth-century chronicler of Jamaican birds and author of *The Birds of Jamaica*, or, indeed, to any of his learned acquaintances.

SWIFTS, MOUSEBIRDS AND TROGONS, KINGFISHERS, WOODPECKERS AND RELATED BIRDS

ORDERS
Apodiformes, Trogoniformes, Coliiformes, Coraciiformes and Piciformes

EXTINCT SPECIES

Ryukyu Kingfisher (*Halcyon miyakoensis*)

These five orders – the Apodiformes (swifts and hummingbirds), Trogoniformes (trogons), Coliiformes (mousebirds), Coraciiformes (kingfishers, rollers, bee-eaters, hoopoes, motmots, todies and hornbills) and Piciformes (barbets, woodpeckers, jacamars, puffbirds, honeyguides and toucans) – are lumped together as they contain only one species, and that a somewhat doubtful one, that seems recently extinct. This is the Ryukyu Kingfisher (*Halcyon miyakoensis*) known from just the type specimen. As might be expected, however, there are several rare or little-known species in these groups (see Tables 16 and 17).

The hummingbirds contain among them several mysterious forms. Like birds of paradise, hummingbirds were at one time collected in vast numbers, their feathers being required for the nineteenth-century fashion trade; many species were initially described from 'trade' skins and some are known from little more than this material. Other species are still being

Table 16

RARE OR ENDANGERED ROLLERS
Crossley's Ground Roller *Atelornis crossleyi*
Pitta-like Ground Roller *Atelornis pittoides*
Short-legged Ground Roller *Brachypteracias leptosomus*
Scaled Ground Roller *Brachypteracias squamigera*
Long-tailed Ground Roller *Uratelornis chimaera*

Table 17

RARE OR ENDANGERED WOODPECKERS
Imperial Woodpecker *Campephilus imperialis*
Ivory-billed Woodpecker *Campephilus principalis*
Helmeted Woodpecker *Dryocopus galeatus*
Red-cockaded Woodpecker *Picoides borealis*
Okinawa Woodpecker *Sapheopipo noguchii*

discovered in the remote South American jungles. There is little doubt that severe deforestation will result in extinctions.

Similarly, several trogons will soon be threatened by forest clearance. The beautiful Quetzal (*Pharomacrus mocinno*), for instance, has declined drastically as the result of habitat destruction.

Among the rather diverse order Coraciiformes, the rollers – as a group – seem to be faring quite badly and several species are under threat. Also in this order the Helmeted Hornbill (*Rhinoplax vigil*) is under threat because its habitat in Malaysia, Sumatra and Borneo is being destroyed.

A kingfisher species, *Halcyon gambieri* from the Tuamoto Archipelago in the Pacific Ocean, has lost a race. As a species this bird still occurs on Niau Atoll but its nominate race, a former inhabitant of Mangaréva in the Gambier Islands, has, according to King (1981), been extinct since 1844.

In the last order of the non-perching birds, the Piciformes, is another extinct subspecies – a race of the Common Flicker (*Colaptes auratus rufipilous*). This race once inhabited Guadalupe Island but has not been sighted since soon after the turn of the century.

The spectacular Ivory-billed Woodpecker (*Campephilus principalis*) (Figure 60) is a species whose status is something of a mystery. At best, it is a gravely endangered bird; very probably the North American race is already extinct but occasional unconfirmed reports to the contrary are received. The status of the Cuban Ivory-billed Woodpecker (*C. p. bairdii*) is similarly doubtful; perhaps a few pairs remain. By the end of the century, if not sooner, the species will almost certainly be gone.

60 Ivory-billed Woodpeckers (*Campephilus principalis*). This species, known from Cuba and the southern states of North America, is gravely endangered and will probably be extinct by the end of the century.

RYUKYU KINGFISHER

Halcyon miyakoensis

PLATE XXXVII

Halcyon miyakoensis
(Kuroda, 1919)

Halcyon miyakoensis Kuroda, 1919 (Miyako-jima, Ryukyu Islands or Nansei-shotō)

Length 24 cm ($9\frac{1}{2}$ in)

Description
Type specimen (unsexed): lores blackish; small patch over eyes bluish white; stripe passing under eye then extending and widening to side of nape greenish blue; rest of head and neck deep cinnamon chestnut; back and scapulars dark bluish green; lower back and upper tail coverts bright cobalt blue; primaries dull brownish black; wing coverts and tail ultramarine tinged with green; underparts (including under wing coverts and axillaries) uniform cinnamon, a little darker on chest; colour of bill unknown (horny sheaths are not present and the species remained undescribed until long after the specimen's collection); feet dark red; claws dark brown.

Measurements
Type specimen: wing 105 mm; tail 80 mm; culmen 38 mm (probably longer when horny sheath is added); tarsus 17 mm.

The so-called Ryukyu (or Riukiu) Kingfisher is either a very enigmatic extinct species or an invalid one. It is represented by just a single specimen taken in 1887 and deposited in the musem at Tokyo, but was not scientifically described until more than 30 years after the date of collection in the Ryukyu Islands (or Nansei-shotō) south of Japan.

In these circumstances several points can be noted. First, a bird known from a unique type specimen should be regarded with some degree of caution – it might merely be a hybrid or 'sport'. Secondly, the provenance of this particular individual is somewhat doubtful, its description so long after collection rendering error a distinct possibility. The museum label gives the collection locality as Miyako shima – one of the Ryukyu Islands – but specimen labels are notoriously unreliable and with no additional examples forthcoming from this place there is room for uncertainty.

There is also doubt over an important aspect of the bird's appearance in life – the mandibles of the type lack their horny covering so beak colour cannot be determined. The bird's nearest extant relative, the Micronesian Kingfisher (*Halcyon cinnamomina*) found on Guam, Ponapé (or Pohnpei) and the Palau Islands, has a beak mostly black in colour but with an area of pale yellow at the base of the lower mandible. This species is, in fact, very similar to the Ryukyu bird; it is therefore quite reasonable to consider *miyakoensis* as simply a race of *cinnamomina*. The most significant morphological differences between the two lie in the colour of the feet – olive brown in *cinnamomina*, red in *miyakoensis* – and *miyakoensis*'s lack of a black nape band.

Notwithstanding these difficulties over provenance and interpretation, the Ryukyu Kingfisher has acquired a place in ornithological literature as a distinct species and there is, of course, no reason why a *Halcyon* kingfisher should not establish itself on the Ryukyu Islands and develop characteristics rendering it specifically distinct. Nothing is known of the bird in life.

XXXVII Ryukyu Kingfisher. Oil painting by Errol Fuller.

PERCHING BIRDS

ORDER
Passeriformes

EXTINCT SPECIES

Stephen Island Wren (*Xenicus lyalli*)
Bay Thrush (?*Turdus ulietensis*)
Grand Cayman Thrush (*Turdus ravidus*)
Kittlitz's Thrush (*Zoothera terrestris*)
Piopio (*Turnagra capensis*)
Four-coloured Flowerpecker (*Dicaeum quadricolor*)
Lord Howe Island White-eye (*Zosterops strenua*)
Kioea (*Chaetoptila angustipluma*)
Hawaii O'o (*Moho nobilis*)
Oahu O'o (*Moho apicalis*)
Molokai O'o (*Moho bishopi*)
Ula-ai-Hawane (*Ciridops anna*)
Koa 'Finch' (*Rhodacanthis palmeri*)
Kona Grosbeak 'Finch' (*Chloridops kona*)
Greater Amakihi (*Hemignathus sagittirostris*)
Mamo (*Drepanis pacifica*)
Black Mamo (*Drepanis funerea*)
Bonin Islands Grosbeak (*Chaunoproctus ferreirostris*)
Kusaie Island Starling (*Aplonis corvina*)
Mysterious Starling (*Aplonis mavornata*)
Norfolk and Lord Howe Starling (*Aplonis fusca*)
Bourbon Crested Starling (*Fregilupus varius*)
Rodrigues Starling (*Fregilupus rodericanus*)
Huia (*Heteralocha acutirostris*)

The Passeriformes are by far the largest and most highly developed of all bird orders. This enormous assembly of families contains well over half the world's species and, as might be expected where so many kinds are involved, their interrelationships are very complex and not in all cases clearly understood. The arrangement of the various families given here is that generally accepted but there are other, probably equally valid sequences of classification. The exact number of recognized species depends, of course, on personal taste or opinion but it is probably safe to say that there are rather more than 5,000. Most are small but there is considerable variation in size, ranging from tiny wren-sized birds to the larger crows or the Australian lyrebirds. The characteristic binding all together is the 'perching' position that the feet are able to assume – three toes at the front and one behind, never reversible, all joined at the same level. From this passerines derive their common name – perching birds.

Although a considerable number of passerine species and races have disappeared since 1600 (for extinct subspecies see Table 18), and many more are today threatened with extinction (see Table 21), when set against the great number and variety of forms, this loss seems quite small, particularly if compared with losses sustained or threatened in some other bird orders. From just over 300 species in the parrot order, nine have vanished, and the proportion of losses among rails is even greater, with eleven extinct from around 130 species. Yet from the thousands of passerine species less than thirty can with certainty be listed extinct. The tally will undoubtedly increase rapidly during the next few decades and the status of a number of species remains enigmatic.

The first three passerine families, the broadbills (Eurylaimidae), woodcreepers (Dendrocolaptidae) and ovenbirds (Furnariidae), contain almost 300 species, but from these only one race, the Trinidad Straight-billed Woodcreeper (*Xiphorhynchus picus altirostris*), is regarded by King (1981) as seriously threatened, and not one is extinct. In fact, it is not until the ninth family, the tyrant flycatchers (Tyrannidae), is reached that any extinct forms can be selected and not until the thirteenth, the New Zealand wrens (Acanthisittidae), that the toll of extinction becomes more dramatic. Some of the first families in the sequence do, however, contain rare or mysterious birds (see Table 19).

Formicariidae

In the fourth family, the antbirds (Formicariidae), for instance, are two closely related Colombian species, the Moustached Antpitta (*Grallaria alleni*) and the Brown-banded Antpitta (*G. milleri*), both named from very few specimens. *Grallaria milleri* is known from just the type series of seven examples collected in 1911 and *alleni* is represented by just the type alone, taken in the same year. The status of each is unknown – either may be extinct or exist in some numbers.

Rhinocryptidae

Two similarly mysterious species are included in the family of tapaculos (Rhinocryptidae). Stresemann's Bristlefront (*Merulaxis stresemanni*) and the Brazilia Tapaculo (*Scytalopus novacapitalis*) are known from two and three specimens, respectively.

Cotingidae

The cotingas (Cotingidae) have an exclusively South and Central American distribution and appear to have some natural relationship with the four families that follow them in the systematic list – the manakins (Pipridae), tyrant flycatchers (Tyrannidae), sharpbills (Oxyruncidae) and plantcutters (Phytotomidae). Largely because of difficulties presented by the kind of country they inhabit, a number of cotinga species are but poorly known, some having what seem very restricted ranges, although these restrictions may be apparent rather than real. More than one species has been discovered surprisingly recently but another, the Kinglet Calyptura (*Calyptura cristata*), could be extinct. In view of the recent finds in the family and the difficulties involved in fieldwork, its seeming absence for a period of very many years may not be significant. The species is

Table 18

SUBSPECIES OF PERCHING BIRD THAT ARE PROBABLY EXTINCT

Tyrannidae
Grenada Euler's Flycatcher *Empidonax euleri johnstonei*
Acanthisittidae
North Island Bush Wren *Xenicus longipes stokesi*
Stead's Bush Wren *Xenicus longipes variabilis*
Campephagidae
Cebu Black Greybird *Coracina coerulescens altera*
Cebu Barred Greybird *Coracina striata cebuensis*
Norfolk Island Long-tailed triller *Lalage leucopyga leucopyga*
Pycnonotidae
Cebu Slaty-crowned Bulbul *Hypsipetes siquijorensis monticola*
Cinclidae
Cyprus Dipper *Cinclus cinclus olympicus*
Troglodytidae
San Benedicto Rock Wren *Salpinctes obsoletus exsul*
Guadalupe Bewick's Wren *Thryomanes bewickii brevicauda*
San Clemente Bewick's Wren *Thryomanes bewickii leucophrys*
Martinique House Wren *Troglodytes aedon martinicensis*
Daito Wren *Troglodytes troglodytes orii*
Muscicapidae
(Turdinae)
Cebu Black Shama *Copsychus niger cebuensis*
Isle of Pines Solitaire *Myadestes elisabeth retrusus*
Lanai Thrush *Myadestes obscurus lanaiensis*
Oahu Thrush *Myadestes obscurus oahensis*
Muriel's Chat *Saxicola dacotiae murielae*

Muscicapidae (Turdinae) (cont'd)
Maré Island Thrush *Turdus poliocephalus mareensis*
Lord Howe Island Thrush *Turdus poliocephalus vinitinctus*
Lifu Island Thrush *Turdus poliocephalus pritzbueri*
Yakushima Seven Islands Thrush *Turdus celaenops yakushimensis*
(Timaliinae)
Burma Jerdon's Babbler *Moupinia altirostris altirostris*
(Sylvinae)
Huatrine Island Long-billed Reed Warbler *Acrocephalus caffra garretti*
Raiatea Long-billed Reed Warbler *Acrocephalus caffra musae*
Astrolabe Nightingale Reed Warbler *Acrocephaus luscinia astrolabii*
Laysan Millerbird *Acrocephalus familiaris familiaris*
Daito Japanese Bush Warbler *Cettia diphone restricta*
Chatham Islands Fernbird *Bowdleria punctata rufescens*
(Malurinae)
Lord Howe Island Grey Warbler *Gerygone igata insularis*
(Rhipidurinae)
Lord Howe Island Fantail *Rhipidura fuligonosa cervina*
(Monarchinae)
Tongatabu Flycatcher *Pomarea nigra tabuensis*
Maupiti Flycatcher *Pomarea nigra pomarea*
Paridae
Daito Varied Tit *Parus varius orii*
Dicaeidae
Cebu Orange-bellied Flowerpecker *Dicaeum trigonostigma pallida*
Zosteropidae
Cebu Everett's White-eye *Zosterops everetti everetti*

Zosteropidae (cont'd)
Seychelles Chestnut-flanked White-eye *Zosterops mayottensis semiflava*
Lord Howe Island Grey-backed White-eye *Zosterops lateralis tephropleura*
Meliphagidae
Chatham Islands Bellbird *Anthornis melanura melanocephala*
Emberizidae
(Emberizinae)
Santa Barbara Song Sparrow *Melospiza melodia graminea*
Guadalupe Rufous-sided Towhee *Pipilo erythrophthalmus consobrinus*
St Kitts Puerto Rican Bullfinch *Loxigilla portoricensis grandis*
Drepanididae
Hawaii Akialoa *Hemignathus obscurus obscurus*
Oahu Akialoa *Hemignathus obscurus ellisianus*
Lanii Akialoa *Hemignathus obscurus lanaiensis*
Oahu Nukupu'u *Hemignathus lucidus lucidus*
Laysan Honeycreeper *Himatione sanguinea freethi*
Lanai Creeper *Paroreomyza montana montana*
Icteridae
Grand Cayman Jamaican Oriole *Icterus leucopteryx bairdi*
Slender-billed Grackle *Quiscalus mexicanus palustris*
Fringillidae
McGregor's House Finch *Carpodacus mexicanus mcgregori*
Oriolidae
Cebu Dark-throated Oriole *Oriolus xanthonotus assimilis*
Callaeidae
South Island Kokako *Callaeas cinerea cinerea*

Table 19

SOME PERCHING BIRDS KNOWN ONLY FROM A FEW MUSEUM SPECIMENS

Formicariidae	
Moustached Antpitta	*Grallaria alleni*
Brown-branded Antpitta	*Grallaria milleri*
Rhinocryptidae	
Stresemann's Bristlefront	*Merulaxis stresemanni*
Brasilia Tapaculo	*Scytalopus novacapitalis*
Cotingidae	
Kinglet Calyptura	*Calyptura cristata*
Philepittidae	
Small-billed Wattled Sunbird	*Neodrepanis hypoxantha*
Muscicapidae	
(Turdinae)	
Rufous-headed Robin	*Erithacus ruficeps*
(Sylvinae)	
Long-legged Warbler	*Trichocichla rufa*
(Muscicapinae)	
Fanovana Newtonia	*Newtonia fanovanae*
Dicaeidae	
Obscure Berrypecker	*Melanocharis arfakiana*
Emberizidae	
(Emberizinae)	
Tumaco Seedeater	*Sporophila insulata*
Olive-headed Brush Finch	*Atlapetes flaviceps*
(Cardinalinae)	
Townsend's Finch	*Spiza townsendi*
(Thraupinae)	
Cherry-throated Tanager	*Nemosia rourei*
Azure-rumped Tanager	*Tangara cabanisi*
Turquoise Dacnis	*Dacnis hartlaubi*
Fringillidae	
São Tomé Grosbeak Weaver	*Neospiza concolor*
Ptilonorhynchidae	
Yellow-fronted Gardener Bower Bird	*Amblyornis flavifrons*

represented in museums by several skins, all collected during the nineteenth century. It was found at only a very few localities, all of them in the state of Rio de Janeiro, and has not even been seen during this century.

Two cotinga species in the genus *Ampelion* are hardly known at all. The Bay-vented Cotinga (*A. sclateri*) was described in 1874 on the basis of two specimens, a male and female collected in central Peru. For the next 98 years, little further was learned of it, just three additional examples being taken in 1921. In 1972, the species was found again; since this time it has not proved difficult to locate. Although a specimen of another bird in the genus, the White-cheeked Cotinga (*A. stresemanni*), had been lodged in the Philadelphia Academy of Sciences during the 1930s, the species was not formally described and named until 1954. The description was based upon two females collected east of Lima, Peru; males were not identified until 1966. An even more recently described species is the Grey-winged Cotinga (*Tijuca condita*).

Tyrannidae

In the related family of tyrant flycatchers (Tyrannidae) one otherwise quite plentiful species, Euler's Flycatcher (*Empidonax euleri*), appears to have lost a race. The species has a wide distribution in South America but a subspecies, *johnstonei*, of which there may be no more than three records, inhabited Grenada at least until early years of this century.

Acanthisittidae

The thirteenth family of passerines, the New Zealand wrens (Acanthisittidae), contains just four species but losses among these have been heavy. One, the Rock Wren (*Xenicus gilviventris*) (Figure 61), is a rare inhabitant of New Zealand's South Island. Another, the Bush Wren (*X. longipes*) (Figure 61), is critically endangered and may even now be extinct. If it survives still, its last stronghold is likely to be in the rugged territory known as Fiordland. Once a widespread inhabitant of forested country, two of three subspecies recognized are almost certainly extinct. The North Island Bush Wren (*X. l. stokesi*) has throughout the period of European colonization been excessively rare and was last sighted in the 1950s. Stead's Bush Wren (*X. l. variabilis*) from Stewart and surrounding islands, vanished from Stewart Island itself at about the same time. It survived for a while on an outlier, Big South Cape Island, but introduced rats eventually caused its extinction there and the race appears to have entirely vanished despite last-minute attempts to transfer birds to a safer island refuge. A third species, the extraordinary Stephen Island Wren (*Xenicus lyalli*), is extinct altogether. This bird is peculiar in a number of ways: it was both discovered and exterminated by a single animal – a cat belonging to a lighthouse keeper; it may have had the smallest natural range of any known bird, tiny Stephen Island in the Cook Strait; and it may have been the only passerine to have completely lost the power of flight.

Philepittidae

Like the New Zealand Wrens, the asites and false sunbirds (Philepittidae) are another family consisting of just four small species endemic to a land in which much of the original fauna is seriously threatened – in this case Madagascar. One of the false sunbirds, the Small-billed Wattled Sunbird (*Neodrepanis hypoxantha*), is a very mysterious bird indeed, identified from a few specimens (Greenway, 1958, lists seven; King, 1981, mentions eleven) taken between 1879 and 1925. It is very similar to the Wattled False

61 The three living species of New Zealand Wren, the sole surviving representatives of the family Acanthisittidae. Top: Rifleman (*Acanthisitta chloris*), male and female; middle: Rock Wren (*Xenicus gilviventris*); bottom: Bush Wren (*Xenicus longipes*). Chromolithograph after a painting by J.G. Keulemans from W.L. Buller's *History of the Birds of New Zealand*, Vol. 1 (London, 1887–8), Pl.12.

62 Superb Lyrebird (*Menura superba*).

Sunbird (*N. coruscans*) and was not actually recognized as distinct from this species until 1933, well after the most recently taken specimen was collected. Although the forests of Madagascar have been so drastically reduced, it seems likely that a few of these easily overlooked birds still live.

Menuridae

The most un-passerine-like of all passerines are the lyrebirds (Menuridae) (Figure 62). These wonderful creatures are shy, solitary birds that seldom fly and spend most of their time hidden in the underbrush of the forest of eastern Australia. Their large size and beautiful lyre-shaped tail feathers provided a great puzzle for early classifiers of birds and a layman might understandably assume lyrebirds to be in some way related to pheasants. Surprisingly, they are extraordinarily gifted mimics, perhaps the most able of all birds. They will imitate anything from a chainsaw to the sound of a human voice and can reproduce almost any noise that takes their fancy. There are two species – the Superb Lyrebird (*Menura superba*) and Prince Albert's Lyrebird (*M. alberta*). Although neither is perhaps seriously threatened, the Prince Albert's Lyrebird inhabits only a very limited area around the Queensland and New South Wales border.

Atrichornithidae

Also exclusively Australian are the lyrebird's closest relatives, but these are much more typically passerine in general appearance and size. They are the scrub-birds (Atrichornithidae) and this family also is made up of just two species, both of which are rare. The Noisy Scrub-bird (*Atrichornis clamosus*) was for many years from 1889 onwards considered an extinct species. In 1961, a year in which more than one lost Australian species was rediscovered, it was located again at Two People's Bay, Western Australia. This population is the only one located during this century and, although apparently stable, consisted in the 1970s of just under 100 individuals. The rediscovery was fortuitous but very timely. Part of Two People's Bay had been earmarked for development as a town and had the birds not been found when they were, their home grounds would no longer exist as an environment suitable for rare creatures. The Rufous Scrub-bird (*A. rufescens*) still persists in numbers at several localities, and the decline of this species appears to have been halted.

Campephagidae

From among the cuckoo-shrikes (Campephagidae), three races of three different and still-surviving species are probably extinct. The Long-tailed Triller (*Lalage leucopyga*) is distributed across islands in the New Hebrides (now Vanuatu) and Solomon groups, also occurring in New Caledonia and the Loyalties (Îsles Loyauté). This range was extended considerably southwards by the nominate race, which until recently inhabited Norfolk Island but probably is no longer extant. Several species of cuckoo-shrike inhabit the Philippines and two of these appear to have lost races, both from the island of Cebu. The subspecies *altera* of the Philippine Black Greybird (*Coracina coerulescens*) and the race *cebuensis* of the Barred Greybird (*C. striata*) both disappeared soon after the turn of the century.

Pycnonotidae

Among the bulbuls (Pycnonotidae) is another extinct race that vanished at about the same time and also came from Cebu. This is *Hypsipetes sequijorensis monticola*, a subspecies of the Slaty-crowned Bulbul. Although

Cebu is a comparatively large island, its birds have been subjected to intense pressure because of the scale of deforestation that has occurred there. Even by the first decades of the twentieth century little remained of the original forest cover and birds unable to adapt rapidly to changing conditions had little chance of survival.

Cinclidae

The dippers (Cinclidae) are the only passerines that can be described as truly aquatic. There are just five species, all of which are very similar in size, build and habit and all of which live in and around fast-flowing mountain streams. Only one, the Rufous-throated Dipper (*Cinclus schulzi*), can be considered rare as a species but its status is not clear. This bird is an inhabitant of northwestern Argentina. The other four species are all quite widespread but a race of the Common Dipper (*C. cinclus olympicus*) may have disappeared from Cyprus since World War II; in reality, the subspecies may not be separable from others.

Troglodytidae

The true wrens (Troglodytidae) are an almost exclusively New World family of about 60 species, yet the only one to inhabit Europe and Asia, the Common or Winter Wren (*Troglodytes troglodytes*), is by far the most celebrated and has acquired a position in folklore and legend second, perhaps, to no other bird. Several races of this common species have only small populations and the Daito Wren (*T. t. orii*), a subspecies from the Borodino (Daitō) Islands, south of Japan midway between the Ryukyus (Nansei-shotō) and the Bonins (Ogasawara-shotō), is probably extinct.

Various exclusively New World species have also lost races. The House Wren (*Troglodytes aedon*) was believed to have lost two Caribbean subspecies, *martinicensis* from Martinique and *guadeloupensis* from Guadeloupe, but the ostensibly extinct population on Guadeloupe was rediscovered in 1969 after having been missed from the time of World War I. The form from Martinique became extinct after 1886, when it was collected for the last time. Then, the wrens on the island seemed quite common but they have never since been recorded. Another subspecies, *mesoleucus* from St Lucia, is at the verge of extinction or is, perhaps, already lost. On the other hand, the St Vincent race *musicus* has recovered from seemingly imminent eradication and may even be no longer threatened. Other races have been described but none is thought to be seriously at risk.

Bewick's Wren (*Thryomanes bewickii*) occurs across a large part of North America and Mexico and is also found on islands off the coast of California. A race from Guadalupe Island some 322 km (200 miles) southwest of San Diego (not Guadeloupe of the West Indies), *brevicauda*, was last collected in 1892 and last seen in 1897. It is likely that combined effects of severe habitat destruction caused by introduced goats, together with the depredations of cats, also introduced to the island, brought about the destruction of this bird soon after the last record. Another race, *leucophrys* from San Clemente Island, may have vanished several decades later.

Vanished, too, seems to be a subspecies of the Rock Wren (*Salpinctes obsoletus*), which once inhabited San Benedicto Island some 550 km (340 miles) off the west coast of Mexico, and was given the name *exsul*.

Muscicapidae

The vast assembly of birds brought together under the heading of the Muscicapidae has, of necessity, been split into several subfamilies. It might be expected that such a huge group would contain a large number of extinct and threatened forms; the Muscicapidae certainly does but a very significant proportion of them come from the subfamily Turdinae, which contains the thrushes and chats.

This subfamily is represented throughout the world, being absent from only the highest latitudes and some Pacific islands – most notably New Zealand although species have been introduced since European colonization. In the Turdinae are some of the world's most accomplished songsters including the Song Thrush (*Turdus philomelos*), the Blackbird (*T. merula*) and the Nightingale (*Erithacus megarhynchos*).

Despite the losses sustained, many species of thrush have shown themselves to be among the world's most adaptable birds and have adjusted rapidly to changing conditions. Those that did not and now seem quite extinct are the mysterious Bay Thrush (*Turdus ulietanus*), of which no specimen exists, the Grand Cayman Thrush (*T. ravidus*) and Kittlitz's Bonin Island Thrush (*Zoothera terrestris*).

In addition to these full species, several subspecies have been lost. Races of the Hawaiian Islands Thrush or Omao (*Myadestes obscurus*) have been described from various islands in the group (for reasons for change of name from *Phaeornis* see Pratt, 1981). Only one, the nominate race from Hawaii itself, remains reasonably common. Of the other races, two are apparently extinct and two others seriously threatened. The Oahu Thrush (*M. o. oahensis*) presumably became extinct during the first half of the nineteenth century. Individuals were collected by Andrew Bloxham in 1825 during the voyage of HMS *Blonde*, which carried back to Honolulu the embalmed bodies of King Liholiho and Queen Kamamalu, who had both died of measles during a visit to London. Bloxham recorded the birds to be common but the situation clearly changed rapidly and today even Bloxham's examples no longer exist. The subspecies *lanaiensis* survived for much longer. It was reasonably common on Lanai Island until the building of Lanai City in the 1920s. Then, the bird population quickly declined and by 1931 this race of thrush was on the verge of extinction. It has not been seen since.

Molokai has become notorious as something of a disaster area for native flora and fauna and little remains of its original forests. The race that was once common upon it, *rutha*, was considered extinct from 1936 until 1963, but in the latter year two individuals were seen. It has not been observed since and may by now no longer exist. The subspecies inhabiting the island of Kauai, *myadestinus*, is restricted to the Alaka'i Swamp Forest Reserve and is seriously endangered. A closely related but smaller thrush also lives on Kauai and shares the same Alaka'i Swamp with the larger species. This is the Small Kauai Thrush (*Myadestes palmeri*), which has been excessively rare since the time of its discovery during the 1890s and has sometimes been considered extinct. Its numbers are now reported as very low.

Various reasons have been advanced to account for the decline of these species, among them introduced avian diseases, introduced predators, destruction of natural forests together with habitat deterioration and the rapid increase in the human population.

In the West Indies the related Isle of Pines Solitaire (*Myadestes elisabeth retrusus*) has not been seen on its island home (now called Isla de la Juventud) since 1934. The nominate race survives still in Cuba.

Two races of the Canary Islands Chat have been recognized, *Saxicola dacotiae dacotiae* from the island of Fuerteventura and *S. d. murielae* from Allegranza Island. The Allegranza form, hardly separable from the nominate race, was plentiful in 1913 when D.A. Bannerman visited the island, discovered its chat and named it for his wife. According to King (1981), it has not come to notice since.

The Island Thrush (*Turdus poliocephalus*) occurs on many Pacific islands, both large and small; on many of these, distinct subspecies have developed – R. Howard and A. Moore (1980) recognize no less than forty-nine. Owing chiefly to the limited size of their home islands, many of these races undoubtedly have small populations and may therefore be considered threatened to some degree but only one – the nominate race from Norfolk Island – is recognized by W.B. King (1981) as presently endangered. Three other races, however, are believed extinct. The Lord Howe Island subspecies, *vinitinctus*, vanished in the years following World War I, a presumed victim of rats that came ashore from the wreck of the *Mokambo* at Ned's Beach in 1918; two populations from the Loyalty (Loyauté) group, *mareensis* from Maré and *pritzbueri* from Lifou, probably were lost during World War II. Among the islands south of Japan occurs a related species – the Seven Islands Thrush (*T. calaenops*); a race, *yakushimensis*, apparently confined to Yaku-shima, seems to have remained uncollected since 1904 and is, very likely, extinct.

Also included among the Turdinae is the genus *Copsychus*. The Black Shama (*C. niger*) is confined to the Philippines, with two subspecies recognized. The nominate race is probably in no danger but the race *cebuensis* – from Cebu, which has lost so many of its birds – is now either very rare or absent altogether. In 1956 an expedition in search of it saw but a single individual, which was collected. There are no more recent records. Another species in the genus, the Seychelles Magpie Robin (*C. sechellarum*), has now only a tiny surviving population – perhaps of no more than thirty individuals. It is restricted to Fregate Island where its existence until very recently was threatened by cats but these predators are now supposed to have been eliminated. The species remains, however, one of the world's most vulnerable.

The babblers (Timaliinae) are another large subfamily incorporated in the Muscicapidae group, but surprisingly few are in any way threatened. One subspecies that may have been lost is the nominate race of Jerdon's Babbler (*Moupinia altirostris*). This form was recorded from southern Burma but has not been located since World War II. Two other races inhabit parts of northern Burma and also the Indian subcontinent.

The Old World warblers (Sylvinae), like the babblers in the Muscicapidae, also include a large number of forms, but rather more of these are endangered and some species have lost races. Losses have been most severe within the genus *Acrocephalus*, at least three species having each dropped races. The Long-billed Reed Warbler (*A. caffra*) from the Society Islands has lost an apparently distinct population named *garretti*, once an inhabitant of Huatrine Island. The nominate race from Moorea may be endangered and a possibly distinct form may have formerly inhabited Raiatea. Although no specimen of the Raiatea race exists this presumed population has been named *A. c. musae*. An illustration by George Forster is thought to depict it (see Holyoak and Thibault, 1978). No members of the species inhabit Raiatea today.

The Nightingale Reed Warbler (*Acrocephalus luscinia*) is recorded from various Pacific islands and several races are recognized. One rather mysterious form, *A. l. astrolabii*, is known from two examples, probably collected by the French vessels *L'Astrolabe* and *La Zelée* during the 1830s at an unknown locality in the Pacific. Wherever they came from the

63 Laysan Millerbird (*Acrocephalus familiaris familiaris*), photographed at its nest in May 1902 by Walter K. Fisher. This bird, one of two races of Millerbird indigenous to the Hawaiian Islands, was once common on Laysan Island but became extinct in the 1920s. Courtesy of Denver Museum of Natural History.

64 Fernbirds from New Zealand and its offshore islands: *Bowdleria punctata punctata* (above) and the now-extinct race from Mangere in the Chatham group, *B. p. rufescens* (below). Engraving after a drawing by J.G. Keulemans from W.L. Buller's *Manual of the Birds of New Zealand* (Wellington, 1882).

population of which these birds were members is now almost certainly extinct.

From the Hawaiian Islands two races of a species known as the Millerbird (*Acrocephalus familiaris*) have been described, the peculiar vernacular name deriving from a great liking for a particular kind of moth called 'millers'. Only one now survives, the Nihoa Millerbird (*A. f. kingi*), which lives on the small island of Nihoa with a population of just over 200 individuals. This is considered by some as a species in its own right but most authorities regard it as conspecific with the Laysan Millerbird (*A. f. familiaris*) (Figure 63), which failed to survive the years between 1913 and 1923. When first collected in 1891 this was a common bird but severe habitat deterioration turned Laysan Island temporarily into what Wetmore (1925) described as a 'barren waste of sand', and the Millerbird was not able to cling to existence until better times came.

One very mysterious bird included in the Sylvinae is the Fijian Long-legged Warbler (*Trichocichla rufa*), described from three specimens taken in 1890, then collected again in 1894, but following this not recorded for 73 years. After the Whitney South Seas Expedition's failure to find the species during the 1920s, it was listed as extinct, but not only was it rediscovered, an additional race (*cluniei*) was identified in 1974.

A dozen races of the Japanese Bush Warbler (*Cettia diphone*) are recognized. One of these, *restricta*, from the Borodino (Daitō) Islands is either excessively rare or extinct and another, *panafidinica*, can perhaps be similarly categorized.

In few areas have birds suffered such drastic declines as they have in the islands south of Japan, but one with a comparable record is New Zealand and its islands – and, from here, a race belonging to another species in the subfamily Sylvinae has vanished. The rather aberrant Fernbird (*Bowdleria punctata*) (Figure 64) is found on both main islands and on some of New Zealand's smaller offshore isles. In 1868 a distinct form was found on Mangere, one of the Chatham group, and given the name *rufescens*. It occurred also on nearby Pitt Island but, subject to the depredations of introduced cats and the excessive zeal of collectors, the birds were gone by the turn of the century. Another subspecies, *wilsoni*, is restricted to Codfish Island in the Stewart group where the population has declined to around 100 individuals.

Although the Fernbird is often listed with the Old World warblers, some taxonomists place it in the next subfamily, the Australian wrens (Malurinae). Birds from three quite distinct groups go by the name of 'wren' – the true wrens (Troglodytidae) of the Americas and the Palaearctic, the New Zealand wrens (Acanthisittidae) and the Australian wrens. The relationship of all of these to other passerines, and indeed to one another, is uncertain but the Malurinae seem to show affinity to other Muscicapidae subfamilies rather than to other 'wrens'.

In the Malurinae are several little-known and rare species. Perhaps the strangest are the two Emu-wrens, which in form are as curious as any bird in the world (Figure 65). These tiny creatures have tails quite unlike their relatives – the feathers, unwebbed, resembling in structure those of the Emu (*Dromaius novaehollandiae*). The birds thus appear almost like 'made-up' creatures. Populations of the Southern Emu-wren (*Stipiturus malachurus*) are widely distributed but declining in number. A race of the second species, the Mallee Emu-wren (*S. ruficeps mallee*), has a restricted range and is rare.

Although there are no species within the subfamily that are actually

65 The endangered Southern Emu-wren (*Stipiturus malachurus*).

extinct, more than one have, until recently, been assumed to be so, and some, because of very limited distribution, may be considered endangered. Among the least known is Campbell's Fairy Wren (*Malurus campbelli*), discovered as recently as 1980 when two birds were caught in mist nets set upon the slopes of Mount Bosavi, Papua New Guinea. These were examined, photographed and released but a year later another was taken and subsequently two more were caught. At present, nothing further can be said about the species' status, but it is possible that this wren will prove much more widespread than the very localized trapping records indicate. A closely related New Guinea species, the Broad-billed Fairy Wren (*M. grayi*), has been known since 1861 but is only occasionally encountered. Whether it is actually rare or only apparently so is not clear.

The Purple-crowned Fairy Wren (*Malurus coronotus*) from northern parts of Australia is likely to be rather more seriously endangered. Numbers have been drastically reduced in parts of its range and this beautiful and unusually coloured little bird may soon be threatened with extinction. Similarly, the future of the Red-winged Fairy Wren (*M. elegans*) from the southwestern corner of Australia and that of the Blue-breasted Fairy Wren (*M. pulcherrimus*) cannot be regarded as secure.

It is among the birds known as 'grasswrens' that the most enigmatic species in the subfamily occur. One, the Grey Grasswren (*Amytornis barbatus*), lives in dense brakes of lignum on floodwaters in the inhospitable interior of Australia. It is known with certainty from only two localities and the story of its discovery is rather unusual. In September 1942 an ornithologist, N. Favaloro, was passing part of the Bulloo River in an automobile when he startled a grasswren from cover that he could not identify. Unable to stop at the time, he did not return to the remote scene of this encounter for 25 years. Then, in 1967, he came across five individuals perched upon canes in a clump of lignum. Eight years later a population of similar birds was found living several hundred miles to the west of the Bulloo River site. In neither of these localities do the birds seem abundant but they may, of course, occur in other places. On the other hand, they may not.

The Carpentarian Grasswren (*Amytornis dorotheae*) has been seen by only a very few observers. The range, as far as can be determined, is rather limited but although birds of the species are exceptionally difficult to find, this may not necessarily indicate great rarity. The Thick-billed Grasswren (*A. textilis*), on the other hand, ranges over a wide area but appears to be declining almost everywhere.

The Eyrean Wren (*Amytornis goyderi*) has all the romance of a once-lost species attached to its name. Like the Takahe (*Notornis mantelli*) of New Zealand (see page 71), it was known from just a very few examples taken during the nineteenth century but this species went missing for an even longer period. Originally discovered during the 1870s, these birds were not again located until 1961, although possible sightings were occasionally recorded during the intervening years. Even the 1961 observation was not universally accepted as genuine. In 1972 the species was seen in the Simpson Desert near to the spot indicated on maps as the meeting place of Northern Territory, South Australia and Queensland. In 1976 specimens were collected, the first since the bird's discovery a little over a century earlier, and the existence of the Eyrean Grasswren was proved beyond doubt. It is now realized that this species occurs over a large area and it is no longer considered rare or, in view of the desolate nature of its home grounds, endangered.

Australia is remarkably rich in rediscovered species and yet another grasswren, the Black Grasswren (*Amytornis housei*), was found, lost and found again. This species first came to attention in 1901 and, although occasionally searched for, was not seen again until 1968. The known range of the species is limited but within this very remote area the birds may not be scarce.

Also sometimes included in the Malurinae are birds of the genus *Gerygone*, which are distributed widely across Australia, New Guinea and the South Pacific. The New Zealand Grey Warbler (*G. igata*) occurs throughout New Zealand and is common, but the species had also colonized Lord Howe Island almost 1,610 km (1,000 miles) to the northwest. This isolated population was described under the name *insularis* but no longer exists, presumably having been exterminated after the invasion of Black Rats (*Rattus rattus*) to the island in 1918 after the grounding of the SS *Makambo*.

Another member of the original avifauna of Lord Howe was *Rhipidura fuliginosa cervina*, which is one of the fantail flycatchers (Rhipidurinae). Like *Gerygone igata, Rhipidura fuliginosa* is a common species but it had an even wider range. Just as the Grey Warbler vanished from Lord Howe, so, too, did the fantail population that once lived there. This loss was also sustained soon after 1918 and presumably can be attributed to the same cause.

Among the Old World flycatchers (Muscicapinae) is placed a species that may exist in smaller numbers than any other kind of bird. This is the Chatham Islands Robin (*Petroica traversi*). The surviving population is closely monitored by the New Zealand Wildlife Service, which recorded in 1977 that only three pairs and a lone male were left alive. By 1983 the total population had increased to eleven. Once widespread over the whole Chatham group, the Black Robin disappeared rapidly from all but Little

Mangare Island, where it clung to existence in a small patch of low scrub forest. By 1976 this habitat had deteriorated to such an extent that it became necessary for the entire colony to be trapped and relocated on Mangare Island itself. Obviously, this species may become extinct at any time. In the same subfamily is *Newtonia fanovanae*, a bird known from only a single specimen collected in the Fanovana forest of eastern Madagascar during 1931.

Related to the Old World Flycatchers are the monarch flycatchers (Monarchinae), and within this group the genus *Pomarea* contains extinct and threatened kinds. Several distinct forms of the Society Islands Flycatcher (*P. nigra*) seem to have existed once but only the nominate race survives today and this itself is very rare. Restricted to the island of Tahiti it has been seen only occasionally during this century and several searches have failed to locate it. Recently, a small colony consisting of several dozen pairs was located in the forested mountain valleys behind Papeete.

That other races of this species once existed is indicated by a unique specimen collected in 1823 on Maupiti, a mountainous, forested island towards the west of the Society group. Naturalists aboard the French exploring vessel *La Coquille* took this individual and upon it was eventually founded a new species, but this is today recognized as just a race of the Society Islands Flycatcher (*Pomarea nigra pomarea*). As might be expected for a bird so long vanished, a survey (Thibault, 1974) indicates that the subspecies is indeed extinct. Another race, *P. n. tabuensis*, was named in 1929 by G.M. Mathews on the basis of a detailed description given by J.R. Forster, who saw the birds on the island of Tongatapu in 1773 during Captain Cook's second circumnavigation. Forster's descriptions can be relied upon. On the Society Islands he had collected specimens of *P. nigra* and when Cook's expedition called at Tongatapu, the main island of the Tonga group, Forster saw similar birds. Although he compared the two forms, no specimens of the Tongatapu birds seem to have been taken. No one since seems ever to have found it.

Two other species in the genus are endangered. Several races of the Marquesas Flycatcher (*Pomarea mendozae*) have been described. All have only small populations and one, *nukuhivea*, may even be extinct. Rarotonga, in the Cook Islands, has its own endemic species *Pomarea dimidiata*, the Rarotonga Flycatcher, which has a very small population threatened by habitat destruction.

The last of the Muscicapidae subfamilies, the whistlers (Pachycephalinae), are renowned for their fluty, melodious singing. They occupy most of the Australasian region, New Guinea, Indonesia, the Philippines and various Pacific islands. One species, the New Zealand Thrush or Piopio (*Turnagra capensis*), is almost certainly extinct. Although showing some resemblance to the thrushes, any similarity is superficial, the vernacular name owing much to the nostalgia of early European colonists in New Zealand.

Paridae

Three families of tits follow the Muscicapidae and from one of these, the typical tits and chicadees (Paridae), a race has been lost – yet another casualty from the fauna of the islands south of Japan, in this case the Borodinos (Daitō). The Varied Tit (*Parus varius*) inhabits Korea, Taiwan, Japan and associated islands, with a number of races. Borodino birds, named *orii*, are either exceptionally rare or have become extinct since their description in 1923.

Dicaeidae

The flowerpeckers (Dicaeidae) are spread throughout Australasian and Oriental regions, finding their greatest diversity in New Guinea. As their pretty name suggests, they are small, delicate creatures that can often be seen pecking or flitting around bushes containing flowers or buds, searching for food. One species in the family, the Four-coloured Flowerpecker (*Dicaeum quadricolor*), is entirely extinct and another, the Obscure Berrypecker (*Melanocharis arfakiana*), probably an inhabitant of western New Guinea, is known from just two specimens. The Orange-bellied Flowerpecker (*Dicaeum trigonostigma*) is a very widespread species. Several races have been described from the Philippines, one of which, *pallida*, seems to have vanished from Cebu along with the Four-coloured Flowerpecker and much else of the original avifauna.

Zosteropidae

The family of white-eyes (Zosteropidae), like the flowerpeckers, are found in the Oriental, Pacific and Australasian regions, but these birds also inhabit much of Africa. Although upwards of 80 species are recognized, these are quite similar in appearance. One species, the Lord Howe Island White-eye (*Zosterops strenua*) is extinct and so, too, may be the Lord Howe race (*tephropleura*) of the widely distributed Grey-backed White-eye (*Z. lateralis*). Two more races of otherwise extant white-eye species also now seem to have disappeared. From Cebu, the nominate race of Everett's White-eye (*Z. everetti*) was probably last seen during the early years of this century. On the island of Marianne in the Seychelles, the race *semiflava* of the Chestnut-sided White-eye (*Z. mayottensis*) became extinct during the 1880s, primarily as a result of the destruction of suitable habitat. The nominate race of this species still survives on Mayotte Island.

Meliphagidae

The honeyeaters (Meliphagidae) attain their greatest proliferation in the Australian region but are also distributed across much of the Pacific area generally. It is among the group that colonized the Hawaiian Islands that the greatest losses have been sustained. These include some very notable species, most particularly those of the genus *Moho*. Four species make up this exclusively Hawaiian genus but only one, *Moho braccatus*, the Kauai O'o, survives, and even this is a very rare bird. Once considered extinct, it was rediscovered in 1960 and lives in the same Alaka'i Swamp area inhabited by the two Kauai thrushes. The size of population is unknown but it is certainly very small and has been established as low as two individuals by some commentators. Perhaps more than this survive. The three extinct *Moho* species are the Hawaiian O'o (*M. nobilis*), the Molokai O'o (*M. bishopi*) and the Oahu O'o (*M. apicalis*). Another honeyeater from Hawaii, the Kioea (*Chaetoptila angustipluma*), is also extinct, having disappeared during the middle of the last century.

One of the subspecies of the New Zealand Bellbird (*Anthornis melanura*) is no longer extant. This otherwise quite common bird was once represented on the Chathams by the rather large race *melanocephala*, which was observed by several eminent New Zealand ornithologists during the nineteenth century. By 1900 it was notable chiefly for its great rarity and was subjected to heavy pressure from collectors. Within a very few years there seemed none left to collect and search of the islands in 1924 failed to find evidence of survivors.

Although the Bellbird is not threatened as a species, another interesting

66 Stitchbirds (*Notiomystis cincta*), male (front) and female (behind). This species has become extinct on the New Zealand mainland but a small population still survives on Little Barrier Island. Chromolithograph after a painting by J.G. Keulemans from W.L. Buller's *History of the Birds of New Zealand*, Vol. 1 (London, 1887–8), Pl.11.

member of the Meliphagidae and also a New Zealand form may be. The Stitchbird (*Notiomystis cincta*) (Figure 66) is an attractive yellow and black bird that rapidly became rare as European settlement progressed in New Zealand. At the beginning of the century it seemed very unlikely that Stitchbirds would be able to survive and, in fact, the species failed to do so on the mainland. Fortunately, a few individuals lingered on one of New Zealand's offshore islands – Little Barrier – and these, extraordinarily, have since flourished. Although the birds are now common on this island, the total population is comparatively small and, although numbers may be stable, the Stitchbird remains a vulnerable species.

Emberizidae

Like the Muscicapidae, the Emberizidae is a large family that, for convenience, is split into several subfamilies. The group contains the buntings (Emberizinae), the cardinal grosbeaks (Cardinalinae), tanagers (Thraupinae) and two species that are each placed in separate subfamilies – the Swallow Tanager (*Tersina viridis*) in Tersininae and the Plush-capped Finch (*Catamblyrhynchus diadema*) in Catamblyrhynchinae. Within the group are some extinct and some very rare forms.

Two extinct forms are races of generally plentiful North American species. A great many subspecies of the Song Sparrow (*Melospiza melodia*) have been described, only one of which, *amaka*, is thought to be seriously at risk. The population named *graminea* from Santa Barbara Island, California, is thought, however, to have become extinct in the 1960s. A race of the Rufous-sided Towhee (*Pipilo erythrophthalmus consobrinus*) has been lost from the same general area. This race once inhabited Guadalupe Island off the west coast of Mexico where several endemic forms failed to survive. The most recent specimens date from 1885 and living birds were last seen in 1897.

Two subspecies of the Puerto Rican Bullfinch (*Loxigilla portoricensis*) have been recognized, the nominate race from Puerto Rico itself, which is rare, and the race *grandis* from St Kitts (St Christopher). The larger St Kitts birds are believed to have been exterminated by Green Monkeys (*Cercopithecus aethiops*) introduced to the island, and died out at some time after 1880.

The Cherry-throated Tanager (*Nemosia rourei*) is known only from the type collected on the bank of the Rio Paraíba do Sul, Brazil, in 1870 and subsequently deposited in the Berlin Museum. Nothing is known of the bird's habits and the species may now be extinct. Slightly better known is the Azure-rumped Tanager (*Tangara cabanisi*), which until recently was known from just three museum specimens – one taken before 1866, a second in 1937 and a third in 1943. During the 1970s it was seen several times, one observation being of a flock of sixteen birds in the cloud forest of southern Chiapas, Mexico.

A problematical form known as *Spiza townsendi* could conceivably constitute a valid and, if so, extinct species. Audubon (1832–9) described *Eberiza townsendi* from a unique specimen taken at New Garden, Chester County, Pennsylvania, but this mystery bird is usually thought to bear some kind of affinity to the Dickcissel (*Spiza americana*). Whether this example is a freak, a hybrid or a member of an otherwise unrecorded species has not been determined.

The most celebrated of all of these enigmatic forms is one of Darwin's famous finches. When, in 1835, the great naturalist collected from the Galápagos Islands a variety of finch-like birds, he could hardly have realized that from these little dead creatures would be assembled an idea destined 24 years later to send shock waves around the world. Darwin's Finches, as they became known, are among the cornerstones upon which the concept of evolution was founded. Among the various species collected were seven particularly large specimens, four females and three males, but no individuals identical to these could be found by those who followed Darwin.

With the passage of time, it was assumed that Darwin had collected from a population that vanished soon after his visit to the Galápagos. The seven large individuals were designated as the nominate race of the species *Geospiza magnirostris*, with considerable smaller birds, which later collectors did find, being attributed to a subspecies (*G. m. strenua*) that inhabits several of the Galápagos Islands. Then, in 1957, a single individual reported to be identical to Darwin's original seven examples was taken on Floreana (Bowman, 1961). The status of the form remains obscure, however, for Darwin's Large Ground Finch has not been located since. Perhaps it survives still, perhaps not.

Drepanididae

Of all the known bird families that have evolved in the recent geological past, perhaps none has been more drastically pruned than the Hawaiian honeycreepers (Drepanididae). Many are endangered, and at least six full species and an additional six subspecies seem now altogether extinct (see Table 20). A considerable amount of time has elapsed since these vanished birds were last located but hopes for the survival of some may not be completely groundless for several other forms have reappeared since their supposed extinction. Those parts of the Hawaiian Islands to which many honeycreepers are restricted – the dripping mountain forests – are not the most convenient places for the viewing and identification of rare birds. The Oahu Akepa (*Loxops coccinea rufa*), for instance, may have been sighted in 1977 after having been missed throughout this century and *Melamprosops phaeosoma*, the Po'o'uli, was not even described until 1974.

The family provides a striking and indeed classic example of adaptive radiation. From a supposed single ancestral species these creatures developed into an extraordinary diversity of forms, showing particular variation in bill size and shape. Presumably, the principal factors allowing them to vary to such an astonishing degree were, first, the great remoteness of the Hawaiian Islands from any of the major landmasses and, second, the considerable distances separating each of the larger islands one from another. The ancestral honeycreepers simply fanned out to fill the many vacant ecological niches, which under more normal circumstances would be filled by members of other bird groups. Naturally, those that began to plug a particular gap grew in some ways to resemble birds that filled a similar position in more accessible parts of the world. Thus, some honeycreepers have short, chunky parrot- or finch-like beaks, whereas others have long, slender, sickle-shaped bills. Some have beaks in which the upper and lower mandibles are quite different in length, others show an altogether more normal development. Although comparatively little known, the family has undergone much taxonomic revision, the most recent, which is followed here, being an unpublished analysis by Pratt completed in 1979.

All of the six extinct species, the Koa 'Finch' (*Rhodacanthis palmeri*), the Kona Grosbeak 'Finch' (*Chloridops kona*), the Greater Amakihi (*Hemignathus sagittirostris*), the Ula-ai-Hawane (*Ciridops anna*), the Mamo

Table 20

STATUS OF THE HAWAIIAN HONEYCREEPERS (DREPANIDIDAE)

Scientific name	Common name	Status
Telespyza cantans		
T. c. cantans (*Psittirostra cantans cantans*)	Laysan Finch	Not endangered
T. c. ultima (*Psittirostra cantans ultima*)	Nihoa Finch	Not endangered
Rhodacanthis palmeri (*Psittirostra palmeri*)	Koa finch	Extinct
Loxioides bailleui (*Psittirostra bailleui*)	Palila	Endangered
Chloridops kona (*Psittirostra kona*)	Grosbeak Finch	Extinct
Psittirostra psittacea	O'u	Endangered
Pseudonestor xanthophrys	Maui Parrotbill	Endangered
Oreomystis bairdi (*Loxops* or *Paroreomyza maculata bairdi*)	Kauai Creeper	Endangered
Oreomystis mana (*Loxops* or *Paroreomyza maculata mana*)	Hawaii Creeper	Endangered
Loxops coccinea		
L. c. coccinea	Hawaii Akepa	Endangered
L. c. ochracea	Maui Akepa	Endangered
L. c. rufa	Oahu Akepa	Endangered, possibly extinct
Loxops caeruleirostris (*Loxops coccinea caeruleirostris*)	Kauai Akepa	Endangered
Hemignathus obscurus		
H. o. obscurus	Hawaii Akialoa	Extinct
H. o. lanaiensis	Lanai Akialoa	Extinct
H. o. ellisianus	Oahu Akialoa	Extinct
H. o. procerus	Kauai Akialoa	Endangered, possibly extinct
Hemignathus lucidus		
H. l. lucidus	Oahu Nukupu'u	Extinct
H. l. hanapepe	Kauai Nukupu'u	Endangered, possibly extinct
H. l. affinus	Maui Nukupu'u	Endangered
Hemignathus munroi (*Hemignathus wilsoni*)	Akiapola'au	Endangered
Hemignathus virens		
H. v. virens (*Loxops virens virens*)	Hawaii Amakihi	Not endangered
H. v. wilsoni (*Loxops virens wilsoni*)	Maui Amakihi	Not Endangered
H. v. chloris (*Loxops virens chloris*)	Oahu Amakihi	Endangered
Hemignathus stejnegeri (*Loxops virens stejnegeri*)	Kauai Amakihi	Not endangered
Hemignathus parvus (*Loxops parva*)	Lesser Amakihi	Not endangered
Hemignathus sagittirostris (*Loxops sagittirostris*)	Greater Amakihi	Extinct
Ciridops anna	Ula-ai-Hawane	Extinct
Drepanis pacifica	Mamo	Extinct
Drepanis funerea	Black Mamo	Extinct
Drepanis coccinea (*Vestiaria coccinea*)	Iiwi	Not endangered
Palmeria dolei	Crested Honeycreeper	Endangered
Himatione sanguinea		
H. s. sanguinea	Apapane	Not endangered
H. s. freethi	Laysan Honeycreeper	Extinct
Paroreomyza maculata (*Loxops maculata maculata*)	Oahu Creeper	Endangered
Paroreomyza flammea (*Loxops maculata flammea*)	Molokai Creeper	Endangered
Paroreomyza montana		
P. m. montana (*Loxops maculata montana*)	Lanai Creeper	Extinct
P. m. newtoni (*Loxops maculata newtoni*)	Maui Creeper	Endangered
Melamprosops phaeosoma	Po'o'uli	Endangered

67 Laysan Honeycreeper (*Himatione sanguinea freethi*) photographed by Donald Dickey in 1923, just a few days before the race became extinct. Courtesy of Denver Museum of Natural History.

(*Drepanis pacifica*) and the Black Mamo (*D. funerea*) disappeared during the 1890s or the first years of the twentieth century. The six subspecies that are no longer extant vanished over a rather longer period of time.

The first subspecies to become extinct was probably *Hemignathus obscurus ellisianus*. This was the Oahu race of the Akialoa, which was never located with certainty after 1837 although two individuals may have been seen during the 1890s. The species as a whole has suffered very heavily. Two other races are lost – *H. o. obscurus* from Hawaii and *lanaiensis* from Lanai Island. Both were missed from around the turn of the century onwards. Just one race remains extant, *procerus* from Kauai, but even about this there is doubt. If it still exists, this bird shares the Alaka'i Swamp with several other endangered forms, but it has not been seen since 1965 when just one individual was observed; before 1960 it was considered extinct. Known to Hawaiians as the Akialoa, the species is peculiar in that its long, downcurved bill has an upper mandible considerably longer (sometimes by as much as a quarter of the length) than the lower one.

The closely related Nukupu'u (*Hemignathus lucidus*) shows this feature even more markedly than the Akialoa and in this case the lower mandible reaches only about half the length of the upper. The nominate race from the island of Oahu is probably extinct, having become so in the years between 1860 when it was still considered quite common and the 1890s, when experienced collectors failed to find it. The two other recognized races are very rare, *hanapepe* from Kauai having been seen on only a handful of occasions during this century and *affinis* of Maui rediscovered in 1967 following its apparent absence after 1896.

The species *Himatione sanguinea* is divided into two races. The nominate race, known as the Apapane, inhabits all of the main Hawaiian Islands showing remarkably little variation and is probably the most common of the honeycreepers. Despite the success that *H. s. sanguinea* has enjoyed, the second race, *freethi*, is extinct. Known as the Laysan Honeycreeper (Figures 67 and 68), it was restricted to the island of that name. Walter Rothschild's collector Henry Palmer first collected it in June of 1891. In his diary for 17 June he remarked: 'I caught a little red honeycreeper in the net, and when I took it out the little thing began to sing in my hand. I answered it with a whistle, which it returned and continued to do so for some minutes, not being in the least frightened.'

Rabbits (*Oryctolagus cuniculus*), introduced to Laysan in a forlorn attempt to establish a meat canning industry, destroyed the natural vegetation and by the 1920s this tiny island had become virtually a desert. In 1923, three Laysan Honeycreepers were left alive. One of these, a male, was filmed in full song just days before a gale and dust storm swept all away. For 3 days clouds of swirling sand raged around the island covering everything. When the weather cleared the last honeycreepers were gone.

The Lanai Creeper, the nominate race of the species *Paroreomyza montana*, is extinct also. It is thought to have vanished during the 1930s, a victim of the increase in human population and the great development of the island this century. A second subspecies, *newtoni*, is found on Maui and is considered endangered.

68 Nest and eggs of the Laysan Honeycreeper, photographed by Walter K. Fisher in May 1902. Courtesy of Denver Museum of Natural History.

Icteridae

The family of New World blackbirds (Icteridae) consists of the birds known as New World orioles and blackbirds, cowbirds, grackles, troupials and oropendolas. Two species have each lost races. A subspecies, *bairdi*, of the Jamaican Oriole (*Icterus leucopteryx*), known also as the

Banana Bird, seems no longer extant. An inhabitant of Grand Cayman Island, it has perhaps not been reliably recorded since World War II.

A form that is probably best regarded as a subspecies of the Great-tailed Grackle (*Quiscalus mexicanus*) has come to be known as the Slender-billed Grackle. Originally described as *Scaphidurus palustris* by W. Swainson (1827), it is still often given full species status. Specimens are said to have been taken in marshes around Mexico City, but this locality has been questioned (Dickerman, 1965) and it is now believed that the form was restricted to the marshes of the upper Rio Lerma. The birds have not been found since the first decade of the twentieth century and they are generally presumed extinct.

Fringillidae

Numbered among the finches (Fringillidae) is *Carpodacus mexicanus*, the House Finch, a species which has differentiated into several races in North and Central America. One of these races, *mcgregori* from the island of San Benito, is listed by King (1981) as extinct, not having been recorded since 1938.

Also usually placed within the finch family are the Bonin Islands Grosbeak (*Chaunoproctus ferreirostris*), which is extinct, and the São Tomé Grosbeak Weaver (*Neospiza concolor*), which may very well be. *Neospiza concolor* has not been seen on its island home since 1888 and was not found in 1949 by an Oxford University Expedition that searched for it. Whether it survives or not can only be guessed at. Because it is so little known, the species' correct taxonomic position can be speculated upon and some authors place it in another family (Ploceidae).

Ploceidae

The family of weavers and sparrows (Ploceidae), to which the São Tomé Grosbeak may belong, contains an interesting group of birds known as fodys. Three species are endangered – the Rodrigues Fody (*Foudia flavicans*), the Mauritius Fody (*F. rubra*) and the Seychelles Fody (*F. sechellarum*) – but the most puzzling form is one known as *F. bruante*, which is sometimes listed as an extinct bird. It seems likely, however, that this supposed species, based upon an illustration in E.L. Daubenton's *Planches Enluminées* (1765–81), has been confused with the Madagascan Fody (*F. madagascariensis*). Certainly, it can be regarded as no more than hypothetical.

Sturnidae

Several extinctions have occurred within the family of starlings (Sturnidae). Two Mascarene species, the Huppe or Bourbon Crested Starling (*Fregilupus varius*) and the Rodrigues Starling (*F. rodericanus*), are extinct. So, too, are three members of the genus *Aplonis* – the Kusaie Starling (*A. corvina*), the Mysterious Starling (*A. mavornata*), represented by just a single skin, and the Norfolk and Lord Howe Starling (*A. fusca*).

Oriolidae

As is evident from lists of extinct passerines, the Philippine island of Cebu, like the islands of Norfolk and Lord Howe, is one of the great theatres for bird extinctions. From here has vanished a race of the Dark-throated Oriole (*Oriolus xanthonotus*). Other races of this species are not endangered, but the Cebu birds, named *assimilis*, were last reliably recorded as far back as 1906.

69 Saddlebacks (*Philesturnus carunculatus*). This New Zealand species was once common on the mainland but the North and South island races now survive only on island sanctuaries offshore. Engraving after a drawing by J.G. Keulemans from W.L. Buller's *Manual of the Birds of New Zealand* (Wellington, 1882).

70 Kokakos (*Callaeas cinerea*), close relatives of Saddlebacks, may now survive only on New Zealand's North Island in small and probably dwindling numbers. Engraving after a drawing by J.G. Keulemans from W.L. Buller's *Manual of the Birds of New Zealand* (Wellington, 1882).

Callaeidae

The New Zealand wattle birds (Callaeidae) are a family containing just three species, all endemic to New Zealand – the Huia (*Heteralocha acutirostris*), the Saddleback (*Philesturnus carunculatus*) and the Kokako (*Callaeas cinerea*). The three are believed to be descended from crow-like ancestral stock which invaded the islands long ago. They have been considered closest to the starlings (Sturnidae) by some, to the crows (Corvidae) by others, and attention has also been drawn to connections with the birds of paradise (Paradiseidae), bower birds (Ptilonorhynchidae) and the butcher birds (Cractidae). Most recent authorities believe it more likely, however, that their closest affinities lie with the magpie larks (Grallinidae).

All three species are interesting and curious. The divergence in shape between the bills of male and female Huias is well known; no such marked difference has been recorded among the sexes of any other bird species. It is intriguing to find that the Saddleback (Figure 69), too, shows another odd characteristic, which is believed to be unique among birds. The young of the two races of this species are quite dissimilar to one another in the development rate of mature plumage. Individuals belonging to the subspecies *rufusater*, which formerly ranged across much of New Zealand's North Island, grow rapidly to resemble their parents but those of the nominate race from the South Island remain chocolate brown for perhaps 2 years and do not quickly assume the black head, tail and underparts so characteristic of the adult bird. The difference, completely unlooked for, was great enough to cause confusion and at one time immature South Island birds were thought of as a species distinct from their parents and were called 'Jackbirds'.

The Huia, which of the three species always had the most restricted range, is now almost certainly extinct, having vanished after 1907 (see page 229). The Saddleback and the Kokako are both rare birds. Populations of each have been divided into two races, both being represented by a subspecies on the South Island and another on the North. The Kokako (Figure 70), a beautiful grey bird that leaps, hops and crashes through the New Zealand bush uttering one of the most melodious of all bird songs, survives only in small and probably declining numbers. The blue-wattled North Island subspecies, *wilsoni*, is still recorded from a variety of localities but the nominate South Island race, carrying wattles bicoloured orange and blue, may now be extinct. There have been no reports since 1961 but during the first months of 1987 some individuals were rumoured to be still inhabiting Stewart Island off the extreme southern tip of mainland New Zealand. These sightings have not, however, yet been confirmed.

The Saddleback, originally widely distributed across the mainland, occurred also on several small offshore islands and it is because of some of those populations that the species has been preserved. Common during early years of settlement, numbers of this handsome brown and black bird rapidly declined until by the nineteenth century's close the species stood at the brink of extinction. From mainland New Zealand these birds seem to have vanished entirely but both the North and the South Island races managed to survive on small islands until the 1960s. During this decade individuals were transferred to several additional island sanctuaries and populations on these seem to have reached reasonably healthy levels. These are fragile birds, however, and are only able to exist for as long as their refuges are kept free from mammalian predators.

Table 21

SOME LITTLE-KNOWN, RARE OR ENDANGERED PERCHING BIRDS

Formicariidae
Narrow-billed Antwren *Formicovora iheringi*
Bicoloured Antpitta *Grallaria rufocinerea*
Black-hooded Antwren *Myrmotherula erythronotos*
Swainson's Fire-eye *Pyriglena atra*
Slender Antbird *Rhopornis ardesiaca*
Cotingidae
Dusky Purpletuft *Iodopleura fusca*
Buff-throated Purpletuft *Iodopleura pipra*
Bay-vented Cotinga *Ampelion sclateri*
White-cheeked Cotinga *Ampelion stresemanni*
Ridgeway's Cotinga *Cotinga ridgewayi*
Banded Cotinga *Cotinga maculata*
White-winged Cotinga *Xipholena atropurpurea*
Antonia's Cotinga *Carpodectes antoniae*
Grey-winged Cotinga *Tijuca condita*
Black and Gold Cotinga *Tijuca atra*
Scimitar-winged Piha *Chirocylla uropygialis*
Black-faced Cotinga *Conioptilon mcilhennyi*
Crimson Fruit Crow *Haematoderus milataris*
Long-wattled Umbrella-bird *Cephalopterus penduliger*
Pittidae
Koch's Pitta *Pitta kochi*
Gurney's Pitta *Pitta gurneyi*
Acanthisittidae
New Zealand Bush Wren *Xenicus longipes*
New Zealand Rock Wren *Xenicus gilviventris*
Manuridae
Prince Albert's Lyre Bird *Menura alberti*
Atrichornithidae
Noisy Scrub-bird *Atrichornis clamosus*
Rufous Scrub-bird *Atrichornis rufescens*
Alaudidae
Raza Lark *Calandrella razae*
Hirundinidae
White-eyed River Martin *Pseudochelidon sirintarae*
Motacillidae
Sokoke Pipit *Anthus sokokensis*
Campephagidae
Grauer's Cuckoo Shrike *Coracina graueri*
Réunion Cuckoo Shrike *Coracina newtoni*
Mauritius Cuckoo Shrike *Coracina typica*
Pycnonotidae
Olicaveous Bulbul *Hypsipetes borbonicus*
Laniidae
Black-capped Bush Shrike *Malaconotus alius*
Kupe Mountain Bush Shrike *Telophorus kepeensis*
Vangidae
Bernier's Vanga *Oriolia bernieri*
Van Dam's Vanga *Xenopirostris damii*
Pollen's Vanga *Xenopirostris polleni*
Cinclidae
Rufous-throated Dipper *Cinclus schulzi*
Troglodytidae
Apolinar's Marsh Wren *Cistothorus apolinari*
Zapata Wren *Ferminia cerverai*
Mimidae
White-breasted Thrasher *Ramphocinclus brachyurus*
Muscicapidae
(Turdinae)
Usambara Robin Chat *Dryocichloides montana*
Iringa Robin Chat *Dryocichloides lowei*
Seychelles Magpie Robin *Copsychus sechellarum*
Dappled Mountain Robin *Modulatrix (Phyllastrephus) orostruthus*
Hawaiian Thrush *Myadestes (Phaeornis) obscurus*
Muscicapidae (Turdinae) (cont'd)
Small Kauai Thrush *Myadestes (Phaeornis) palmeri*
Taita Olive Thrush *Turdus helleri*
Ashy Ground Thrush *Zoothera cinerea*
(Orthonychinae)
Western Whipbird *Psophodes nigrogularis*
(Timaliinae)
Spiny Babbler *Turdoides nipalensis*
(Picarthartinae)
White-necked Bald Crow *Picathartes gymnocephalus*
Grey-necked Bald Crow *Picathartes oreas*
(Sylvinae)
Millerbird *Acrocephalus familiaris*
Rodrigues Brush Warbler *Brebrornis rodericanus*
Seychelles Brush Warbler *Brebrornis sechellensis*
Japanese Marsh Warbler *Megalurus pryeri*
Aldabra Brush Warbler *Nesillas aldabranus*
(Malurinae)
Eyrean Grasswren *Amytornis goyderi*
Grey Grasswren *Amytornis barbatus*
Carpentarian Grasswren *Amytornis dorotheae*
Black Grasswren *Amytornis housei*
Thick-billed Grasswren *Amytornis textilis*
Campbell's Fairy Wren *Malurus campbelli*
Broad-billed Fairy Wren *Malurus greyi*
Purple-crowned Fairy Wren *Malurus coronatus*
Red-winged Fairy Wren *Malurus elegans*
Blue-breasted Fairy Wren *Malurus pulcherrimus*
Eastern Bristlebird *Dasyornis brachypterus*
Western Bristlebird *Dasyornis longirostris*
Silktail *Lamprolia victoriae* (Cont'd)

SOME LITTLE-KNOWN, RARE OR ENDANGERED PERCHING BIRDS (cont'd)

Muscicapidae (cont'd)
(Muscicapinae)
Chatham Islands Black Robin *Petroica traversi*
(Monarchinae)
Truk Monarch *Metabolus rugiensis*
Rarotonga Flycatcher *Pomarea dimidiata*
Marquesas Flycatcher *Pomarea mendozae*
Tahiti Flycatcher *Pomarea nigra*
Tinian Island Monarch *Monarcha takatsukasae*
Seychelles Black Flycatcher *Terpsiphone corvina*
Mascarene Paradise Flycatcher *Terpsiphone bourbonnensis*
Sittidae
Kabylian Nuthatch *Sitta ledanti*
Dicaeidae
Forty-spotted Pardalote *Pardalotus quadragintus*
Nectariniidae
Amani Sunbird *Anthreptes pallidigaster*
Zosteropidae
Ponapé Greater White-eye *Rukia longirostra*
Truk Greater White-eye *Rukia ruki*
White-breasted Silver-eye *Zosterops albogularis*
Seychelles White-eye *Zosterops modesta*
Meliphagidae
Bonin Islands Honeyeater *Apalopteron familiare*
Kauai O'o *Moho braccatus*

Meliphagidae (cont'd)
Black-eared Miner *Manorina melanotis*
Stitchbird *Notiomystis cincta*
Emberizidae
(Emberizinae)
Mangrove Finch *Camarhynchus heliobates*
Wilkin's Finch *Nesospiza wilkinsi*
Zapata Sparrow *Torreornis inexpectata*
Puerto Rican Bullfinch *Loxigilla portoricensis*
(Thraupinae)
Seven-coloured Tanager *Tangara fastuosa*
Parulidae
Kirtland's Warbler *Dendroica kirtlandii*
Bachman's Warbler *Vermivora bachmanii*
Semper's Warbler *Leucopeza semperi*
Icteridae
Yellow-shouldered Blackbird *Agelaius xanthomus*
Fringillidae
Red Siskin *Carduelis cucullata*
Yellow-throated Serin *Serinus flavigula*
Estrildidae
Black-faced Waxbill *Estrilda nigriloris*
Arabian Waxbill *Estrilda rufibarba*
Pink-billed Parrotfinch *Erythrura kleinschmidti*
Mindanao Parrot Finch *Erythura coloria*

Ploceidae
Rodrigues Fody *Foudia flavicans*
Mauritius Fody *Foudia rubra*
Seychelles Fody *Foudia sechellarum*
Clarke's Weaver *Ploceus golandi*
Golden-naped Weaver *Ploceus aureonucha*
Sturnidae
Ponapé Mountain Starling *Aplonis pelzelni*
Santo Mountain Starling *Aplonis santovestris*
Rothschild's Starling *Leucopsar rothschildi*
Callaeidae
Kokako *Callaeas cinerea*
Saddleback *Philesturnus carunculatus*
Corvidae
Marianas Crow *Corvus kubaryi*
Hawaiian Crow *Corvus tropicus*
Cuban Crow *Corvus nasicus*
Palm Crow *Corvus palmarum*
Flores Crow *Corvus florensis*
Formosan Blue Magpie *Urocissa caerulea*
Ceylon Blue Magpie *Urocissa ornata*
Lidth's Jay *Garrulus lidthi*
Sooty Jay *Perisoreus internigrans*
Dwarf Jay *Cyanolyca nana*
White-throated Jay *Cyanolyca mirabilis*
Silver-throated Jay *Cyanolyca argentigula*
Beautiful Jay *Cyanolyca pulchra*
Tufted Jay *Cyanocorax dickeyi*
White-tailed Jay *Cyanocorax mystacalis*
Azure-naped Jay *Cyanocorax heilprini*
Persian Ground Jay *Podoces pleskei*

Paradiseidae and Ptilonorhynchidae

Among the last of the passerine families are the birds of paradise (Paradiseidae) and the bower birds (Ptilonorhynchidae). Because of the wildness of the home grounds of many of these species, it is unlikely that any are presently threatened with extinction. This situation may, of course, change during the next few decades as the resources of New Guinea are developed. Although no recognized species in either family are extinct, there are a large number of problematical forms, the status of each of which remains obscure. Generally, these problematical forms, known sometimes as the 'rare' birds of paradise, are considered to be hybrids and many of them most certainly are. There are at least twenty-four distinct

71 An enigmatic bowerbird (family Ptilonorhynchidae), *Ptilonorhynchus rawnsleyi*. Hand-coloured lithograph by Joseph Wolf and Joseph Smit from D.G. Elliot's *Monograph of the Paradiseidae* (London, 1873).

kinds in the Paradiseidae alone and almost all are known from nothing more than a few ageing museum skins. The possibility exists that a few may have been 'good' species and if this is so, some may now be extinct. These enigmatic forms are discussed at more length in the last chapter entitled 'Hypothetical species and mystery birds'.

Among the bower birds are two forms, both of which were, until very recently, rather enigmatic, and one of which, remains so. *Ptilonorhynchus rawnsleyi* (Figure 71) was described from a single specimen collected a few miles from the city of Brisbane in July 1867. It is possible that this is the sole example known of an extinct species but far more likely that it is in fact a hybrid. The skin has vanished but pictures drawn from it exist and in plumage the bird stands very well between the Satin Bower Bird (*Ptilonorhynchus violaceus*) and the Regent Bower Bird (*Sericulus chrysocephalus*). The second enigmatic form is the Yellow-fronted Gardener Bower Bird (*Amblyornis flavifrons*). For many years this species was known from nothing more than three, perhaps four, trade skins, all taken in the nineteenth century and all coming without locality data. Several expeditions went in search of the home grounds of this bird but none met with any success until the American naturalist Jared Diamond met with the Yellow-fronted Gardener Bower Bird in the Gauttier (Foja) Mountains of western New Guinea during the late 1970s. Perhaps more clearly than any other, this discovery of a vanished bird indicates how passerine species may survive overlooked for generations.

STEPHEN ISLAND WREN

Xenicus lyalli

PLATE XXXVIII

Lying in the Cook Strait between the North and the South islands of New Zealand, rising some 305 m (1,000 ft) above sea level, is a rocky piece of land known as Stephen or Stephen's Island. Covering about 2.6 km^2 (1 sq. mile) in area, this otherwise unremarkable spot was the home of a little bird that caught the attention of naturalists for a few months in 1894 but vanished almost before the scientific community became acquainted with its existence.

There are perhaps three aspects concerning the Stephen Island Wren that render it particularly notable. First, it may have had the smallest natural range of any bird species; second, it may have been the only passerine truly incapable of flight; and, third, the manner of its finding and subsequent extinction was quite remarkable – it was both discovered and exterminated by a lighthouse keeper's cat. This single feline brought in some individuals that proved to belong to a hitherto undescribed species. After a while, the supply of birds apparently failed – at any rate the cat ceased to deliver its little victims into the lighthouse keeper's hands. Since no further examples have been seen or taken, it is assumed that Tibbles destroyed the entire population very shortly after finding it.

Specimens are today divided between museums in London, Christchurch (New Zealand), New York, Pittsburgh and Cambridge, Massachusetts. Initially, however, Sir Walter Buller (Figure 72) – celebrated biographer of New Zealand birds – obtained one and Walter Rothschild acquired rather more (nine) for his museum at Tring, Hertfordshire. Over the years there have been many instances of

Xenicus lyalli (Rothschild, 1894)

Traversia lyalli Rothschild, 1894 (Stephen Island, New Zealand)
Xenicus insularis Buller, 1895
Traversia insularis Buller, 1905

Length 10 cm (4 in)

Description
Male: top of head, neck, back, tail and wing coverts dark olive with brownish-black margins to feathers; narrow, yellowish superciliary streak; chin, throat, and breast olivaceous yellow, each feather edged with greyish brown; flanks and abdomen olivaceous brown, paler towards centre of each feather; upper mandible dark brown with horn-coloured tip; lower mandible, legs and feet pale brown.
Female: similar to male but rather duller with mottled appearance less conspicuous.

Measurements
Wing 46 mm; tail 21 mm; culmen 19 mm; tarsus 19 mm.

72 Sir Walter Buller, *c.* 1900. Engraving by W.A. Cox; frontispiece to W.L. Buller's *Supplement to the 'Birds of New Zealand'*, Vol. 1 (London, 1905).

unpleasantness arising from the naming of species new to science; the controversy surrounding *Astrapia mayeri* – the Ribbon-tailed Bird of Paradise – described from its tail feathers alone; or the squabble over another bird of paradise, Wilson's, *Diphyllodes respublica*, immediately spring to mind. The race to be first into print with a description of the Stephen Island Wren was given added piquancy by the extraordinary dislike Buller and Rothschild held for one another. Although Buller had supplied Rothschild with examples of rare New Zealand birds for many years, a smouldering resentment lay between them (see also page 163).

The pages of Rothschild's *Extinct Birds* (1907) are full of bitter remarks made at Buller's expense. The *Supplement* (1905) to the second edition of Buller's *A History of the Birds of New Zealand* (1888) is, for instance, described in the following terms (p. xxvi):

> these two volumes are rather disappointing. They contain very little that is new, and are mainly composed of quotations from other people's writings or letters. Buller's former great book on the Birds of New Zealand was a most important and creditable work, though not without shortcomings. Our knowledge of New Zealand Birds might have been brought up to date in his supplement, but we cannot say that this has been done properly, and errors are frequent.

Remarks (p.16) concerning differences between the robins of the Chathams (*Petroica traversi*) and those of Snares Island (*P. dannefaerdi*) typify exchanges between the two men:

> Buller's doubts about the distinctness of the latter [Snares Island Robin] might easily have been removed, if he had taken the trouble to compare them, for it does not require any genius to see the differences . . . I may add that Buller . . . has not quoted my description correctly, for in his rendering are several disturbing misprints, and in the fourth line from the bottom occurs a 'not' which ought not to be there, and which makes the sentence incomprehensible. Also the name itself is spelt incorrectly.

This outburst had been provoked by Buller's quite deliberately restrained comment in the second volume of the *Supplement* (p.125):

> I am very much in doubt about the propriety of separating this form, as a species . . . but I do not with to disallow, without further enquiry, a species set up by so careful a naturalist as Mr Walter Rothschild.

Whether because he valued Rothschild as a customer or for some other reason, Buller's tone generally retained a more respectful front, but he was not completely above scoring points off his rival (*Supplement*, Vol.2, p.111):

> I think that Mr Rothschild is under a misapprehension in supposing that he possesses all the known specimens [of the Stephen Island Wren] except that described by me in the 'Ibis'. Besides a pair in my son's collection, I purchased a specimen from Mr Henry Travers for Canon Tristram; and the former gentleman has since offered to sell me two more.

Where Buller is concerned, Rothschild's writings could degenerate into extraordinarily frantic outpourings. His justification for rushing into print over the Stephen Island Wren is just such a case (1907, p.24):

I received nine specimens of this new bird, and was not aware that any others had been taken at that time. As I was unable to attend the December meeting, 1894, of the British Ornithologists' Club, I asked Dr Hartert to exhibit the birds in my name. When he had done so and had read the description, the Chairman, Dr P.L. Sclater, said that the bird had also been received for illustration and description in the Ibis, from Sir Walter Buller, and he asked Dr Hartert if I would not withdraw my description. Dr Hartert said that this was unfortunate, but he had no authority to withdraw my description, and he and Dr Sharpe thought that the proceedings of the meeting should be printed without consideration of any manuscripts which might refer to the same bird. No doubt this was hard luck on Sir Walter Buller, but it would have been equally hard luck for me if he had forestalled me with the new bird. He had only one speciman, I had nine, of both sexes, and I had paid a high price for them, as types of a new bird. My type is in Tring, and, as everybody knows, available for study by any competent ornithologist, while Buller's type was not in any museum, and it was uncertain to whom he would sell it afterwards. I suppose it is now in the Carnegie Museum, Pittsburgh, to which Buller's 'third collection', 625 specimens, was sold for a thousand pounds, as Buller himself tells us in his Supplement II, p.167, under the heading of *Glaucopis wilsoni*. On the same page Sir Walter Buller also tells us that his 'second collection' was sold to me, but he makes a mistake about the price, as I certainly did not pay a thousand pounds for it.

I mentioned these unimportant details, because Buller rather bitterly and severely complained about my describing the Stephens' Island Wren, on p.111 of his supplement. I may only add that of course my name, being published in December, 1894, has the priority over his, which was not published before April, 1895.

Buller, in the *Supplement* (Vol.2, p.110), had confined himself largely to relevant facts:

My description . . . appeared in the 'Ibis', 1895, pp.236–237, accompanied by a beautiful plate from the pencil of Mr Keulemans. Unfortunately, the preparation of this plate delayed the publication of my paper three months, and in the meantime Mr Rothschild received the bird from Mr Henry Travers. Mr Rothschild's description of *Traversia lyalli* appeared in the same number of the 'Ibis', but later on, pp.268–269. It had, however, previously been announced in the 'Bulletin of the British Ornithologists' Club', issued on December 29th, 1894 . . . I have in my possession a letter from one of the Editors, Dr Sclater (dated April 7th, 1896), in which he informs me that he told Mr Rothschild I had already described the bird, and tried to dissuade him from doing what he considered an unfair thing . . . Any technical construction of the rules of nomenclature notwithstanding, I think I am justified, under the circumstances, in adhering to the specific name originally bestowed by me, leaving ornithologists, with a knowledge of the facts, to adopt it or not as they may think fit.

Rothschild may well have behaved rather badly over the Stephen Island Wren; certainly, he showed none of the generosity of spirit that might have been expected had, say, a Wallace or a Darwin been involved – but since this seems so uncharacteristic of him, it can only be assumed that his behaviour was fired by existing grievances.

Buller, undoubtedly advantaged by being the man on the spot,

XXXVIII Stephen Island Wrens. Hand-coloured lithograph by J.G. Keulemans from W.L. Buller's *Supplement to the Birds of New Zealand*, Vol. 2 (London, 1905), Pl.10.

ultimately paid the penalty for that very circumstance. He was thousands of miles from the great centres of the scientific world at a time when such distances represented weeks of travel. Rothschild won this particular skirmish in the feud precisely because he was in a position to manipulate events in London.

To avoid confusion, the editors of *Ibis* (1895) felt obliged to publish the following explanation:

> *Note on Xenicus insularis* – The bird described and figured under this name in the last number of the 'Ibis' . . . is identical with Mr Rothschild's *Traversia lyalli*. There can be no question that the latter name has precedence in point of date of publication. Sir Walter Buller's description, together with a specimen of the bird for illustration, were received by the Editors in this country and were in their hands before Mr Rothschild's communication was made to the 'B.O.C.'.

Discussion over which man merited the privilege of being first to describe this bird may be of general interest but is otherwise irrelevant. Rothschild's *Traversia lyalli* takes precedence over Buller's *Xenicus insularis*, although the genus *Traversia* has fallen into disuse. Stephen Island Wrens bore a close relationship to birds of three species inhabiting the main island of New Zealand; generic separation from all of these is not appropriate.

These birds make up the family Acanthisittidae or New Zealand Wrens; a group bearing no close relationship to the typical wrens. Stephen Island Wrens differed in several respects from other members of their genus, most particularly in the striking patterning of their plumage and their stouter, more robust bills. In size they showed a tendency often developed in isolated island forms, being somewhat larger than their mainland relatives.

Little is known of the species' habits. Apparently, living birds were observed only upon two occasions, both times by the lighthouse keeper Lyall, and both times during the evening. It seems likely that they were nocturnal; certainly, these tiny creatures were active during the twilight. Disturbed from holes in the rocks, they were seen to run very fast – like mice, and were never seen in the air. Probably, most of their time was spent skulking among the stones, for what evidence there is suggests that they did not fly at all. Weak flight – at best – is indicated by their short, rounded wings and soft plumage; whether or not they could actually rise from the ground must now be left an open question.

The last word on the Stephen Island Wren may be left to an anonymous correspondent of the Canterbury *Press* (quoted by Buller, 1905):

> And we certainly think that it would be as well if the Marine Department, in sending lighthouse keepers to isolated islands where interesting specimens of native forms are known or believed to exist, were to see that they are not allowed to take any cats with them, even if mouse-traps have to be furnished at the cost of the State.

BAY THRUSH

?Turdus ulietensis

PLATE XXXIX

During the early years of exploration several islands of the South Pacific acquired notoriety for the ferocity of their human inhabitants and some were for many years quite deliberately avoided by European voyagers. Even well into the twentieth century a number of islands continued to carry reputations sinister enough to cause them to be shunned by all but the most intrepid travellers. Others, due to the more receptive nature of the natives, had quickly become favourite ports of call for those whose business took them regularly to the South Seas. From such friendly havens ships could be reprovisioned; on their beaches vessels could be careened and repaired and crews, released temporarily from their confined quarters, could enjoy the comforts of life ashore. When the whaling fleets were in, a carnival atmosphere may certainly have prevailed but the seamen were a law unto themselves and, if circumstances permitted, their visits often soon turned into drunken and promiscuous riots.

Of all the popular resorts used by Europeans perhaps none received more favour than the Society Islands – Tahiti, Moorea, Raiatea, Huahine and Bora-bora. The significance of this as far as the bird population was concerned lay in the likelihood that each new arrival would bring with it an invasion of rats. Although other mammals were introduced later and some, including the Polynesian rat (*Rattus exulens*) had already arrived, it

?*Turdus ulietensis* (Gmelin, 1789)

Turdus ulietensis Gmelin, 1789 (Ulietea, now Raiatea, Society Islands); *ex* Latham, 'Bay Thrush'
Turdus badius Forster, 1844
Merula ulietensis Ramsay, 1879

Length 21 cm (8½ in)

Description
Adult(?) (from Latham): upperparts greyish brown, head marked with darker brown, but rest with feathers edged rufous although this edging is browner on wings and tail; underparts ochraceous; iris dark yellow; bill reddish pearl; legs and feet black.

was probably the coming of the Black Rat (*R. rattus*) and Brown Rat (*R. norvegicus*) that signalled the early decline of certain birds. Several species from the Society Islands vanished during the period immediately following European contact and are known today only through specimens collected or illustrations produced by those naturalists who travelled with Captain Cook on his epic voyages around the world. A shorebird (*Prosobonia leucoptera*), a rail (*Rallus pacificus*), a parrot (*Cyanoramphus ulietanus*) and a passerine (*Turdus ulietensis*) are among very early casualties from the list of Pacific birds, and there can be little doubt that rats played an important part in the extinction of some, if not all, of these species.

Little is known of these birds but perhaps the most enigmatic is *Turdus ulietensis*, a form so puzzling that J.C. Greenway (1958) called it by the rather romantic vernacular name of 'Mysterious Bird of Ulieta'. First described by John Latham under the heading of 'Bay Thrush', this bird was given its scientific name by J.F. Gmelin in 1789. Many birds had originally been listed by Latham in his *General Synopsis of Birds* (1781–5) but only given by him English names, thus providing Gmelin with the opportunity, just a few years later, to found Latin titles upon the earlier descriptions.

The specimen from which Latham made his description was brought back from the South Pacific by naturalists who took part in Cook's second voyage, aboard HMS *Resolution*. Latham apparently saw the example in the collection of Sir Joseph Banks but what subsequently became of it is unknown; probably it rotted away during the nineteenth century along with others of Cook's types. Undoubtedly, the methods of preservation used in preparing these early natural history specimens were somewhat crude. All that remains today of the encounter with this species is a drawing by George Forster (plate 146 in the British Museum (Natural History) collection) marked 'Raiatea♀, June 1st, 1774' and George and Johann Reinhold Forster's own writings.

The 'mysterious' bird has caused a certain amount of confusion among ornithologists. Although the Forsters were competent naturalists, it is not known whether the vanished species really was a thrush and as the only specimen taken no longer exists it seems unlikely that this doubt will ever be resolved. George Forster himself described the bird as a thrush only with some reluctance. J.C. Greenway (1958) remarks that the scutellated tarsi and bifid tongue described by Forster are not suggestive of the thrush family, and another puzzling feature is a bill notched near the tip, noted by Latham. In 1890, R.B. Sharpe tried to match Forster's picture with a unique skin, then of unknown origin, in the British Museum collection. But this skin, a starling named *Aplonis mavornata*, has no connection with Latham's Bay Thrush. Neither E. Stresemann (1949) nor Ernst Mayr (in Greenway, 1958) were able to correlate the type of *mavornata* with the Forster illustration or description and since they wrote the full provenance of the specimen has become known.

All that is known of the habits of the Bay Thrush comes from Forster; no other naturalist ever recorded seeing the species. It had a soft, fluting voice and lived among thickets chiefly in the valleys of its Society Island home.

In conclusion, it can simply be said that the Bay Thrush seems to have been a valid species, but of uncertain affinities, which vanished from Raiatea (called Ulietea in Cook's time) at a date some time after the visit of the *Resolution* in 1774.

XXXIX (Left): Bay Thrush. Watercolour by George Forster; Pl.146, Forster Portfolio, British Museum (Natural History). Courtesy of the Trustees.
XXXX (Right): Grand Cayman Thrush. Oil Painting by Errol Fuller.

GRAND CAYMAN THRUSH
Turdus ravidus
PLATE XXXX

Turdus ravidus (Cory, 1886)

Mimocichla ravida Cory, 1886 (Grand Cayman, West Indies)

Length 28 cm (11 in)

Description
Adult: General plumage colour slate grey but white on abdomen, under tail coverts and tips of tail feathers; bill, feet and naked eye-ring coral red.

A large and handsome grey thrush once lived on the comparatively isolated West Indian island of Grand Cayman. It was first described by C.B. Cory in the *Auk* of 1886 and by 1911 – when a professional bird catcher took the last specimens ever collected – it was to be found only in remote areas of woodland. By the outbreak of World War II the last recorded sighting – in the extreme east of the island – had been logged.

The reasons for this rather abrupt disappearance are not understood but presumably destruction of woodland was the crucial factor, for the species was an inhabitant of dense forest. Little is known about its habits. The song was not particularly fine, tending to weakness and hesitancy (one writer describes it as a subdued warbling) though it was noted to be pleasant enough and quite lengthy. Apart from this information, there is virtually nothing on record of the species.

A close relative is the Red-legged Thrush (*Turdus* (*Mimocichla*) *plumbeus*), a bird found on many West Indian islands. *Turdus ravidus* had barely passed the point at which it could be regarded as a separate species.

KITTLITZ'S THRUSH
Zoothera terrestris
PLATE XXXXI

Zoothera terrestris (Kittlitz, 1830)

Turdus terrestris Kittlitz, 1830 (Boninsima = Peel Island, now Chichi-jima, Ogasawara group)
Geocichla terrestris Bonaparte, 1850
Cichlopasser terrestris Bonaparte, 1854
Zoothera terrestris Gray, 1869

In June 1829 F.H. von Kittlitz landed on Peel Island (Chichi-jima), the largest in the Bonin group (Ogasawara-shotō) – one of the island clusters that string out into the Pacific, south-east of Japan. During the previous year the island had been visited by HMS *Blossom* under the command of a Captain Beechey and several bird species new to science were recorded. Despite their observations, Beechey and his crew seem to have overlooked a rather handsome ground thrush and it was left to von Kittlitz to make this discovery. This is particularly surprising since von Kittlitz made his find at the very landing place that Beechey had called Port Lloyd and here, in fact, von Kittlitz reported seeing the bird regularly.

Four specimens seem to have been collected during the visit and these –

Zoothera terrestris (cont'd)
Length 23 cm (9 in)

Description

Leiden specimen (description of Seebohm and Sharpe, 1902): 'General colour of upperparts olive brown, shading to chestnut brown on rump, upper tail coverts and tail; inside web of each feather much darker, approaching black on the back; lores dark brown; eye stripe very obscure; lesser wing coverts brown, darkest on inside web; median coverts dark brown with large olive-brown tips, greater coverts nearly black, broadly tipped and narrowly margined towards base with olive brown; primary coverts black with broad olive-brown patch on outer webs; tertials dark brown on inner web, olive brown on outer web; secondaries brown, margined with olive brown on the outer webs; primaries brown with basal half of outer webs and a spot where the emargination begins olive brown; tail feathers chestnut brown; ear coverts brown; underparts olive brown, shading into white on the chin, throat and centre of belly; under tail coverts dark brown, with irregular diamond-shaped white tips; axillaries brown; under wing coverts brown; bill (possibly faded to) horn-colour; legs, feet and claws pale.'

Measurements

Leiden specimen: wing 99 mm; tail 66 mm; culmen 22 mm; tarsus 28 mm.

according to J.C. Greenway (1958) – are now divided between the museums at Frankfurt, Leningrad, Leiden and Vienna. After a short stop, von Kittlitz departed from the island with his skins; the ground thrush was never seen again by an ornithologist.

Peel Island was uninhabited (apart from two castaways) when Beechey arrived in the *Blossom* but in 1830, 2 years after von Kittlitz landed, the situation changed dramatically. First, a small settlement was made by a mixed party of British and American sailors together with several Polynesians; then whalers began to use the Bonins increasingly as a convenient stopping place for effecting repairs on their vessels and it can be assumed that rats escaped onto the islands through this agency, bringing destruction to unwary birds.

However it was that the fragile balance of life on the islands was disrupted, the endemic bird life suffered rapidly. Two American naval expeditions visited the islands in the 1850s and reported that birds were rare. By 1889, when the collector P. Holst worked over the islands on behalf of Henry Seebohm (then compiling his *Monograph of the Turdidae*), there was no sign of the thrush.

It is not known whether *Zoothera terrestris* occurred on other islands of the group or whether it was restricted to Peel. All that can be said is that Peel was the only one on which it was ever found.

XXXXI Kittlitz's Thrush. Hand-coloured lithograph by J.G. Keulemans from H. Seebohm's *Monograph of the Turdidae* (London, 1902), Pl.33.

PIOPIO

Turnagra capensis

PLATE XXXXII

Turnagra capensis
(Sparrman, 1787)

Tanagra capensis Sparrman, 1787
Turdus crassirostris Gmelin, 1789
Lanius crassirostris Cuvier, 1817
Campephaga ferruginea Vieillot, 1817
Tanagra macularia Quoy and Gaimard, 1830
Keropia crassirostris Gray, 1840
Loxia turdus Forster, 1844
Otagon turdus Bonaparte, 1850
Otagon tanagra Schlegel, 1865
Turnagra hectori Buller, 1869
Turnagra tanagra Gray, 1869
Turnagra turdus Gray, 1869
Keropia tanagra Finsch, 1870

Length 28 cm (11 in)

Description
Turnagra capensis capensis (Adult): upperparts olive brown but forehead striped with rufous, upper tail coverts and tail bright rufous, and wing coverts tipped with rufous; throat and sides of neck olive brown striped rufous; rest of underparts olive brown broadly streaked with yellowish white; abdomen and flanks tinged with yellow; bill and feet dark brown; iris yellow. Sexes alike.
Turnagra capensis tanagra (Adult): upperparts olive brown but tail feathers (apart from the two central ones) and coverts bright rufous; throat white; breast olivaceous grey; abdomen yellowish white; sides olive brown washed with yellow; bill and feet dark brown; iris yellow.

Measurements
Turnagra capensis capensis: wing 128–132 mm; tail 128–130 mm; culmen 20–23 mm; tarsus 32–38 mm.

April 2nd 1773. Dusky Bay
A large Wattle-bird [*Callaeas cinerea*] was killed by my son [George] . . . and a . . . bird with a ferruginous long tail, a short strong bill and . . . brown with white strokes on the belly and breast. We returned after having enjoyed the pleasure of a fine day, a thing that we seldom met with in New Zealand.

Johann Reinhold Forster made this note in his *Resolution Journal* and with it registered the first observation of the Piopio. Here at Dusky Bay and again at their landfall in Queen Charlotte Sound (Figure 73), where James Cook was so entranced by the early morning chorus of New Zealand's birds, the naturalists participating in Cook's second great exploratory voyage noticed this now-extinct species.

In his evident dislike of the new land, Forster proved the rule rather than the exception among the early visiting artists and naturalists. Sixty years later – still in the infancy of New Zealand's history – the young Charles Darwin gave expression to a similar position in his *Narrative* (1839):

> The meridian of the Antipodes has . . . been passed; and now every league, it made us happy to think, was one league nearer to England . . . Only the other day I looked forward to this airy barrier as a definite point in our voyage homewards; but now I find it, and all such resting-places for the imagination, are like shadows, which a man moving onwards cannot catch . . . In the afternoon we stood out of the Bay of Islands, on our course to Sydney. I believe we were all glad to leave New Zealand. It is not a pleasamt place.

Just a little after Darwin, William Strutt – one of the major painters of the nineteenth-century Australasian scene – was rather more specific in voicing his objections, mentioning an 'indescribable feeling of loneliness', and, concerning his own particular home, adding, 'The awful stillness of this spot, walled in by the mighty forest and shut out from all human intercourse beyond one's own immediate circle, began to feel oppressive and to tell upon me.'

Even today, these intimations of a powerful sense of isolation and a strangely alien land may not be completely unfamiliar to those who visit New Zealand. Yet now the country is no more than a few hours' air travel from the world's major centres; how much more severe must it have seemed to those travellers or settlers carrying with them the knowledge that they were many weeks' sailing distance from all they held most dear!

That this dreadful sense of separation and loneliness was felt by many – and not just by those who left written records – can be demonstrated in all sorts of ways but perhaps never more clearly than by the early colonial attitude to birds. Not content with the strange and wonderful avifauna of the 'Land of Birds', European settlers quickly imported the more familiar species of home to join them in exile. Nor was this all. Quite understandably (and as took place in other new lands) any bird showing the remotest resemblance to one from home was immediately named after the European species, no matter how vague the actual relationship.

Thus Forster's Wattle Bird became the Wattled Crow; a small dark bird

Turnagra capensis (cont'd)
Turnagra capensis tanagra: wing 128–133 mm; tail 120–126 mm; culmen 20–23 mm; tarsus 32–38 mm.

of the Chathams – discovered somewhat later – was named the Black Robin; the tiny creatures making up the family Acanthisittidae were called wrens; and the Piopio of the Maori became the Thrush.

It is still often referred to as the New Zealand Thrush, but the Piopio bears no apparent relationship to the family Turdidae; in fact, it seems quite distinct from all other passerine groups, although affinity with the New Zealand wattle birds (Callaeidae), the cat and bower birds (Ptilonorhynchidae) and to various other families has been suggested; most often it is included among the whistlers (Pachycephalinae).

Even the scientific name *Turnagra capensis* is inappropriate and came about through simple error. Anders Sparrman, who named the species, took duplicate specimens from Cook's expedition at the Cape of Good Hope and, as seems commonly to happen, labels on these became mixed so leaving the describer to make the mistake of assuming that the Piopios had come from South Africa!

Two well-differentiated races are recognizable *T. c. capensis* from the South Island characterized by a striped undersurface, and *T. c. tanagra* (something of a nomenclatural mouthful) found on the North. A third subspecies, described by J.H. Fleming (1915) as *T. c. minor*, is only doubtfully distinct. It was identified from four specimens taken on tiny Stephen Island showing shorter measurements for wing and tail than is usually the case with birds from the mainland.

This was a tame and confiding species, common during early days of settlement over much of the South Island and southern districts of the North. In fact, the species seemed able to inhabit a variety of differing landscapes occurring in coastal areas and also the high alpine country; always, however, a marked preference was shown for the forest and scrub

73 Captain Cook's landfall in Queen Charlotte Sound, Cook Strait, New Zealand in 1773. Watercolour by John Webber (?1750–93).

XXXXII Piopios: North Island (front) and South Island (back) forms. Chromolithograph after a painting by J.G. Keulemans from W.L. Buller's *History of the Birds of New Zealand*, Vol. 1 (London, 1887–8), Pl.4.

of the valleys usually quite close to water.

The species remained so undisturbed by the presence of humans that in some ways it seems surprising that it could not adapt to changing circumstances, but it was certainly this adventurous and guileless manner that brought about its downfall. Like other original members of New Zealand's avifauna, Piopios were almost grotesquely unwary – even friendly – birds. They could become very tame, hopping around doorsteps and sills seeking scraps in much the same way as the Robin does in Britain. Several accounts detail the terrible inroads made by introduced dogs and cats on colonies of these birds close to settlements and camps. Even when foraging in the forests they were vulnerable because they spent a large proportion of their time grubbing on the forest floor.

By the 1880s – little more than a century after the last of Cook's exploratory voyages – the North Island Piopios were all but gone. The last indubitably reliable record is of two individuals encountered in the eastern Urewera ranges; one of them is now preserved in the War Memorial Museum, Auckland.

South Island birds, always apparently the more widespread and plentiful, seem to have fared slightly better although by 1880 it was evident that they, too, were fast disappearing. By the century's turn they were virtually gone with just isolated individuals or pairs lingering here and there. Two were reported from Lake Hauroko in 1949 (they were seen in 1947) and two more were alleged to be living at Caswell Sound; no good observations appear to have occurred since these.

Despite vanishing before the onslaught of introduced mammalian predators, the Piopio was by no means a bird lacking in aggression. An instance is on record of a captive Piopio killing two parakeets and attempting to eat the head of one, while another account tells of an aviary inhabitant stealing eggs from Australian ring-doves.

Nor was the flight of the Piopio lacking in power; Walter Buller (1888) described it in the following terms:

> It haunts the undergrowth of the forest, darting from tree to tree, and occasionally descending to the ground, but rarely performing any long passage on the wing. It is very nimble in its movements; and when attempting on one occasion to catch one of these birds with an almost invisible horsehair noose, it repeatedly darted right through the snare, and defeated every effort to entrap it.

The choice of food was quite catholic though parrot heads, presumably, did not form part of the regular diet. All kinds of insects and other invertebrates were consumed and so, too, were fruits, buds, seeds and foliage.

Piopios nested in manuka, tutu or coprosma and usually the nest was built some 244 cm (8 ft) from the ground although any elevation between 91 and 366 cm (3–12 ft) might be chosen. Supple twigs and mosses were interwoven to form a firm and tidy cup-shaped structure lined with fine grasses or the soft down from tree ferns. The clutch consisted of two white eggs spotted or blotched, particularly at the larger end, with both light and dark brown; these eggs measured about 33 mm × 25 mm. They were only ever found in December but T.H. Potts (1882) believed that Piopios bred twice a year.

During much of the day Piopios remained silent but in the early morning they gave a variety of calls – one of them the short, sharp and rapidly repeated whistling cry from which the Maori derived their name for the bird. According to Potts (1873), the dawn was welcomed with a long-drawn, plaintive whistle. Close to water, the birds piped three times but the regular song, delivered with spread tail and slightly drooping wings, was similar to the song of a lark, interrupted sometimes with rattles and rasps. Almost inevitably, when considering New Zealand birds, one turns to Sir Walter Buller (1888) for the finest and most evocative description and so it is with the song of the Piopio:

> It was when I obtained a caged Piopio that I first became acquainted with its superior vocal powers. In 1866 I purchased one for a guinea from a settler in Wellington, in whose possession it had been for a whole year. Although an adult bird when taken, it appeared to have become perfectly reconciled to confinement; but on being placed in a new cage it made strenuous assaults on the wire bars, and persevered till the feathers surrounding its beak were rubbed off and a raw wound exposed . . . On being removed, however, to a spacious compartment of the aviary, it immediately became reconciled to its condition, made no further efforts to escape, and for a period of fifteen months (when it came to an untimely end) it continued to exhibit the contentment and sprightliness of a bird in a state of nature.

XXXXIII Four-coloured Flowerpecker. Oil painting by Mike Latter.

> I observed that this bird was always most lively during or immediately preceding a shower of rain. He often astonished me with the power and variety of his notes. Commencing sometimes with the loud strains of the Thrush, he would suddenly change his song to a low flute-note of exquisite sweetness; and then abruptly stopping, would give vent to a loud rasping cry, as if mimicking a pair of Australian Magpies confined in the same aviary. During the early morning he emitted at intervals a short flute-note, and when alarmed or startled uttered a sharp repeated whistle.

Anyone wishing to learn more of the 'untimely' fate of this little creature may read on:

> During the day the Piopio was unceasingly active and lively; at night he slept on a perch, resting on one leg, and with the plumage puffed out into the form of a perfectly round ball, the circular outline broken only by the projecting extremities of the wings and tail. Every sound seemed to attract his notice, and he betrayed an inquisitiveness of disposition which in the end proved fatal; for having inserted his prying head through an open chink in the partition, it was seized and torn off by a vicious Sparrow-Hawk in the adjoining compartment of the aviary.

OTHER PUBLISHED ILLUSTRATIONS OF THE *PIOPIO*

Buller, W.L. 1872. *A History of the Birds of New Zealand* (first edn), pl. 14. Artist, J.G. Keulemans.

Fleming, C.A. and Lodge, G.E. 1983. *George Edward Lodge, Unpublished Bird Paintings*, pl. 89.

FOUR-COLOURED FLOWERPECKER

Dicaeum quadricolor

PLATE XXXXIII

Dicaeum quadricolor
(Tweeddale, 1877)

Prionochilus quadricolor Tweeddale, 1877

Length 9 cm ($3\frac{1}{2}$ in)

Description
Male: most of head, neck and mantle deep blue black but cheeks white; most of back orange red with black bases to the feathers giving a mottled appearance; lower back and rump olive yellow; upper tail coverts blue black; scapulas black; wing coverts and quills black-edged with some admixture of blue; tail black;

In 1877, at the very start of the year, the ornithological collector A.H. Everett arrived in Manila with the objective of forming a collection of Philippine birds – one as large and representative as possible. By April he had put together a cache of more than 350 specimens and was ready to move on to the nearby island of Cebu. On arrival at this new collecting locality he received something of a setback – he was stricken with an attack of fever and prevented from working for much of his stay. His activities were not curtailed completely, however, for during the first few days of May he carried away from Cebu the skins of more than 280 birds. In itself this is perhaps unremarkable, such zealous collecting being hardly unusual at the time, but among the haul were several forms new to science and one that was destined to become extinct within just a few years.

The Four-coloured Flowerpecker, a small and quite typical member of the Dicaeidae (a family containing more than 50 species), is certainly the most intriguing of Everett's finds and the specimens – of both sexes – that he brought away from the Philippines are among the very few of this species now existing.

The Marquis of Tweeddale, who first described Everett's birds, named this species *Prionochilus quadricolour* but it has more recently – and probably quite properly – been assigned to the large genus *Dicaeum*. It is also often

Dicaeum quadricolor (cont'd)
underparts mostly pale ashy grey but with white on throat, centre of abdomen, under tail coverts, axillaries and under wing coverts; bill black; iris brown; feet black.
Female: similar to male but upperparts more olive brown.

Measurements
Wing 52 mm; tail 27 mm; culmen 12 mm; tarsus 15 mm.

listed under the name 'Four-coloured Flowerpecker', but it remains a creature about which virtually nothing is known. W.B. King, author of *Endangered Birds of the World: the ICBP Bird Red Data Book* (1981), suggests that the species became extinct during the early years of this century.

Along with at least two other large Philippine islands – Panay and Bohol – Cebu has been deforested and is now virtually barren. It is doubtful if any original forest exists there. Unless the Four-coloured Flowerpecker occurs on a neighbouring island there is no hope of its having survived.

A race, *pallidus*, of another flowerpecker species, the Orange-breasted Flowerpecker (*Dicaeum trigonostigma*), vanished at about the same time as the Four-coloured; this species survives, however, elsewhere in the Philippines.

LORD HOWE ISLAND WHITE-EYE

Zosterops strenua

PLATE XXXXIV

Zosterops strenua (Gould, 1855)

Zosterops strenuus Gould, 1855 (Lord Howe Island)
Nezozosterops strenua Mathews, 1913

Length 13 cm (5 in)

Description
Adult: head and rest of upperparts bright olive green with a band of grey across shoulders and a ring of white around each eye showing a black line beneath; wings and tail slaty brown margined with olive green; chin and throat yellow; flanks pale vinaceous brown; abdomen almost white; under wing coverts and axillaries white; under tail coverts pale yellow; bill and feet bluish black; iris light brown. Sexes alike.

Measurements
Male: wing 71 mm; tail 54 mm; culmen 11 mm; tarsus 22 mm.
Female: wing 67 mm; tail 52 mm; culmen 11 mm; tarsus 22 mm.

John Gould, 'The Bird Man', introducing this little bird in his *Supplement to the Birds of Australia* (1869) wrote: 'The present new species is the largest member yet discovered of a group of birds comprising numerous species'. In terms of species the group is indeed well endowed – there are not very far short of a hundred – and these are all remarkably similar in appearance. They are 7–13 cm (3–5 in) long; almost all show a distinctive white ring around the eye; and they display a limited, though beautiful, range of colours in the plumage – usually olive green to greyish brown on the upperparts, from yellow to grey or white beneath. Although not particularly strong flyers, they have proved adept exponents in the art of colonization and in addition to a presence in Africa and Australasia, they have spread through many islands of the Pacific. For the most part they seem quite vigorous species and there is even evidence that the process of range expansion continues to the present day.

As Gould suggested, the white-eye he described was a particularly large one and on account of this comparatively handsome size he gave it the vernacular name of 'Robust Zosterops'. As a description of size his name might appear justifiable, but, in terms of resilience, time has shown it to be quite inappropriate – it was in fact a very ironic choice. Only a single skin – collected in September 1852 by the British naturalist John MacGillivray who sailed aboard HMS *Herald* under command of Captain Denham – was available to Gould when he wrote the *Supplement* but the species proved common enough; within 60 years it had gone, presumably exterminated by Black Rats (*Rattus rattus*) accidentally introduced to Lord Howe Island in 1918 after the grounding of SS *Makambo*.

Gould's 'Robust Zosterops' was endemic to Lord Howe Island in the Tasman Sea, which it shared, surprisingly, with another white-eyed species (*Z. tephropleura*). It is curious that tiny Lord Howe, 11 km (7 miles) long by 1.6 km (1 mile) wide, should provide a home for two species of *Zosterops* when the entire continent of Africa contains only four.

Zosterops strenua bred during November and December. A cup-shaped nest, 10 cm (4 in) wide by 5 cm (2 in) deep, was made in the fork of a twig using fine rootlets and grasses as building material and lined with skeleton

XXXXIV (Opposite): Lord Howe Island White-eye. Hand-coloured lithograph by J. Gould and H.C. Richter from Gould's *Supplement to the Birds of Australia* (London, 1869), Pl.48.
XXXXV (Right): Kioea. Hand-coloured lithograph by J.G. Keulemans from W. Rothschild's *Avifauna of Laysan and the Neighbouring Islands* (London, 1893–1900), Pl.71. Courtesy of The Hon. Miriam Rothschild.

leaves and other soft things. Two or three blue eggs made up the clutch, each measuring around 18 mm × 13 mm.

Locally, these white-eyes were known as 'Big Grinnels', a name distinguishing them from the other white-eye of Lord Howe. In the days of their abundance they were regarded as something of a pest; they were destructive to fruit and crops and also were alleged to suck the eggs of other birds. Within just a few years of the rat invasion of 1918, however, all were gone. A search by Michael Sharland in 1928 and another search and enquiry in 1936 revealed no trace of them.

KIOEA

Chaetoptila angustipluma

PLATE XXXXV

Chaetoptila angustipluma (Peale, 1848)

Entomiza angustipluma Peale, 1848 (Hawaii)

Length 32 cm (13 in)

Description
Adult: top of head and neck blackish brown striped with (cont'd)

The honeyeaters (Meliphagidae) are an essentially Australian passerine family that has spread north and east through New Guinea to islands of the Pacific. They make up a quite diverse group that has radiated to fill a variety of vacant niches. Thus there are honeyeaters that in some ways resemble warblers; others appear more similar to thrushes; and others are almost jay-like. The honeyeaters known to have ventured farthest from the original Australian homeland were the colonists of the Hawaiian Islands. All these birds – the o'os and the Kioea – are now extinct or at the verge of extinction. Although the four species of o'o were all quite similar, the Kioea was very distinct from these, showing a greater resemblance to other more geographically distant honeyeaters.

Chaetoptila angustipluma (cont'd) greenish yellow with a greyish-white stripe above the eyes; lores, sides of face and ear coverts dull black with a patch of feathers immediately below the eye mottled with greyish white; chin and throat dull white tinged with yellow but feather shafts and whiskers black; wing coverts and back greenish brown tinged with ochraceous colour on rump, the feathers of the mantle showing a white shaft streak widening to a tear shaped spot towards the tip; tail deep brown with outer margins yellow making the whole appear greenish; breast and abdomen dull white striped longtitudinally with darkish brown; flanks strongly tinged ochraceous; bill, legs and feet dark brown, almost black.

Measurements
Wing 140 mm; tail 153 mm; culmen 34 mm; tarsus 45 mm.

In all probability the Kioea has been extinct for many years; certainly, it has not been encountered by anyone able to recognize it since 1859. At about this time several specimens were taken by J. Mills, a local shopkeeper, naturalist and ardent collector – possibly in the woods above Hilo, his favourite collecting ground. Just four examples represent the species in museums, none collected earlier than 1840, none more recently than the Mills specimens.

The Kioea was first noticed by Charles Pickering and Titian R. Peale during the visit to Hawaii of the United States Exploring Expedition in 1838–42, and, as might be expected since the species was encountered by Europeans only during such a very short space of time, their brief description of the birds in life is the only reliable one that has come down to us. Peale (1848) wrote: 'It is very active and graceful in its motions, frequents the woody districts, and is disposed to be musical, having most of the habits of a Meliphaga; they are generally found about those trees which are in flower.'

Apart from the fact that its call was described as a loud chuck nothing further is recorded of the bird in life. It is not even known with certainty exactly what parts of the island the Kioea inhabited, for locality data of those specimens that still exist are not very complete. The birds are supposed to have been inhabitants of the high plateau between the mountains and the edges of the forest but whether they were restricted to this habitat cannot be said.

Similarly, it was for many years uncertain whether or not the species was restricted to Hawaii. During the nineteenth century S.B. Dole (1879) recorded that the Kioea also occurred on Molokai but there was no corroboration of this. Now, however, according to H.D. Pratt (1979), it seems that fossils of the Kioea have been found an another of the Hawaiian Islands – Oahu.

ANOTHER PUBLISHED ILLUSTRATION OF THE *KIOEA*

Wilson, S.B. and Evans, A.H. 1890–9. *Aves Hawaiiensis*, pl. 42. Artist, F.W. Frohawk.

HAWAII O'O

Moho nobilis

FRONTISPIECE

The Hawaii O'o had the very dubious privilege of being chosen as a 'royal' bird by the ancient kings and princes of the island. This conferred upon the birds the honour of supplying for the royal robes and capes their beautiful yellow feathers. From the curious tuft growing beneath the wings, choice plumes were plucked and then woven into a bed of coarse netting to make up the wonderful and intricate garments now so prized in ethnological collections. In past times, the kings' bird collectors must have caught many thousands of individuals belonging to this species – chiefly by trapping them with bird-lime. Sometimes, the bird's own pugnacity contributed to its undoing for trappers are said to have used the Iiwi (*Drepanis* (*Vestiaria*) *coccinea*) as bait to entice the large and bold O'o to attack and thus ensnare itself.

After plucking, the bird catchers were supposed to let their o'o go but

Moho nobilis (Merrem, 1786)

Gracula nobilis Merrem, 1786 (Hawaii)
Acrulocercus niger Cabanis, 1847
Moho niger Gray, 1847

Length 32 cm (12½ in)

Description
Adult: general colour of plumage glossy black although this shades to dull brown on the abdomen; axillary tufts and under tail coverts bright yellow; varying amounts of white on two outer pairs of tail feathers; central pair of tail feathers elongated and twisted at tips; iris dark reddish brown; bill, feet and legs black. Sexes similar but female considerably smaller and middle pair of retrices not so twisted.
Immature: similar to adults but lacking the yellow axillary tufts.

Measurements
Male: wing 150 mm; tail 190 mm; culmen 33 mm; tarsus 38 mm.
Female: wing 110 mm; tail 150 mm; culmen 30 mm; tarsus 35 mm.

---

whether they were reliable in this is doubtful, for even the small amount of meat involved must have seemed very desirable; according to S.B. Wilson and A.H. Evans (1890–9), o'os fried in their own fat were considered a great delicacy. If, after surrender of the treasure, o'os were released, they probably had every chance of survival, for removal of the largely ornamental feathers would not leave any significant areas that were exposed to the cold.

Of the four species of o'o to inhabit the Hawaiian Islands, the one from Hawaii itself was the largest. The last legitimate record of a Hawaii O'o is possibly that of a bird heard on the slopes of Mauna Loa about 1934 but reports of alleged sightings still occasionally come in.

During the earliest period of contact between Europeans and native Hawaiians, the species came to notice. Examples were brought back from the Pacific by those who sailed with Captain Cook on his ill-fated third and final voyage of discovery. Specimens in the Leverian Museum in London were the first to be described in publications of the day and John Latham (1782) based his 'Yellow tufted Bee-eater' upon these. Although the species was noticed by several visitors to the Hawaiian Islands during the years following Latham's account, it was not until the visit of Titian R. Peale with the United States Exploring Expedition of 1838–42 that details of the bird's habits became available.

Although such large numbers were trapped, Europeans found the Hawaii O'o to be very shy and wary. It lived by taking nectar from the flowers of ochias, lehuas, lobelias and other plants by means of a long tubular tongue. Also, the O'o was seen to eat bananas and took insects and their larvae. The call was harsh and quite distinctive and it was from the double syllable expressed that the natives derived the name of O'o.

Wilson and Evans (1890–9) compared its dipping flight to that of the Magpie (*Pica pica*), but George C. Munro (1944) said that the wings flapped quickly to produce a continuous buzzing unlike the sound made by any other forest bird. A preference for frequenting topmost branches of the trees, perhaps 30.5 m (100 ft) or so from the ground, was noticed. Nest and eggs seem never to have been found, possibly because they were placed very high in the trees, but R.C.L. Perkins caught some young birds recently out of the nest.

In its heyday, the Hawaii O'o must have been a common bird over almost all of the island. It seems to have been able to live at almost any altitude up to 1,220 m (4,000 ft) and is even supposed to have been observed above 1,830 m (6,000 ft).

It is unlikely that feather hunting was the primary cause of extinction because the plumes had been in demand for generations with no effect on numbers – certainly none that can be detected from this distance in time. Far more likely is that the Hawaii O'o responded in an adverse way to the changes in the environment following European contact. Like other endemic Hawaiian species it seems to have been unable to tolerate alteration of any kind. Perhaps, though, hunting may have been responsible for a final dip towards extinction as the bird was still common in parts of Hawaii until the 1890s although disappearing fast from its strongholds. H.W. Henshaw (1903) recorded that even as late as 1898 hunters took a thousand O'os in the woods north of the Wailuku.

2
1

OAHU O'O
Moho apicalis
PLATE XXXXVI

Moho apicalis (Gould, 1860)

Moho apicalis Gould, 1860

Length 30 cm (12 in)

Description
Adult: general plumage colour sooty black but tail brown with outer feathers tipped with large patches of white; two central tail feathers lack white tipping and are narrower than others tapering to very fine filamentous points; sides and under tail coverts yellow; under wing coverts white; bill and legs black. Sexes similar but female generally smaller.

Measurements
Male: wing 114 mm; tail 165 mm; culmen 38 mm; tarsus 38 mm.
Female: wing 105 mm; tail 145 mm; culmen 35 mm; tarsus 34 mm.

Only one of the four known species making up the genus *Moho* is likely to survive today. This is *Moho braccatus*, the O'o of the Hawaiian island of Kauai but there is considerable doubt over whether even this species remains extant. For many years it was believed to be extinct but was rediscovered in the Alaka'i Swamp in 1960. If the Kauai O'o still clings to existence in the dripping rainforest, it can certainly be counted as one of the most critically endangered species in the world.

A great intolerance to any kind of disturbance of their habitat – shown by so many endemic Hawaiian species – seems to have affected the *Moho* species. They were intensively hunted by native Hawaiians for their yellow feathers yet this seems not to have brought about a plunge to extinction. The more insidious dangers of encroachment into their territory by introduced cattle, by plants foreign to the islands and by such predators as rats helped to produce the disastrous depletion of the original Hawaiian avifauna. To these difficulties can be added competition from introduced birds and, probably, a great vulnerability to the diseases they carried with them. Given the record of Hawaiian birds in general and the o'os in particular, it will be remarkable if *Moho braccatus* manages to survive and flourish.

While the O'o of Kauai is the last of the o'os, the earliest species to vanish was probably *Moho apicalis*. In 1837, from the hills behind Honolulu, a Herr Deppe collected a series of o'os belonging to a species that seems to have been confined to the island of Oahu. Not since then has this bird been recorded although at least two specimens exist of which the provenance is unknown; individuals may have been taken more recently than the birds procured by Deppe. By the late 1880s and 1890s, the period of the great rush of interest in Hawaiian birds, the Oahu O'o seems to have gone. Certainly, none could be found by such experienced field collectors as R.C.L. Perkins, who were searching for all rare Hawaiian Islands' birds.

Information about the Oahu O'o is almost totally lacking; all that remains are the skins – three collected by Deppe now in the museums of Berlin, Vienna and New York, another specimen in New York taken in 1824 during the voyage of HMS *Blonde*, a skin in the British Museum (Natural History) once belonging to the great book producer John Gould but of which the full provenance is unknown and another in Paris of similarly unknown origin, and, finally, an example in Cambridge, Massachusetts taken by J.K. Townsend, a companion of Deppe.

MOLOKAI O'O
Moho bishopi
PLATE XXXXVII

It is to Walter Rothschild that we owe much of our knowledge of rare and curious species of birds. His intense interest in these creatures was one of the driving forces of his life and nowhere has his influence been felt more powerfully than in the study of the vanished and vanishing endemic avifauna of the Hawaiian Islands.

Rothschild (Figure 74) was one of the great instigators behind the flood

XXXXVI Oahu O'os. Hand-coloured lithograph by J.G. Keulemans from W. Rothschild's *Avifauna of Laysan and the Neighbouring Islands* (London, 1893–1900), Pl.73. Courtesy of The Hon. Miriam Rothschild.

Moho bishopi (Rothschild, 1893)

Acrulocercus bishopi Rothschild, 1893 (Molokai, Hawaiian Islands)

Length 29 cm (11½ in)

Description
Adult: head deep black with slightly metallic gloss; plumage of crown slightly curled; upper parts black with brownish tinge on back and light shafts visible on feathers of the neck; underparts brownish black; under tail coverts, axillary tufts and a tuft of feathers directed backwards from near to the ear coverts golden yellow; tail black with narrow white fringes to the tips, two middle tail feathers long and pointed; bill and feet black. Sexes similar but female rather smaller.

Measurements
Male: wing 102 mm; tail 109 mm; culmen 29 mm; tarsus 33 mm. *Female:* wing 100 mm; tail 107 mm; culmen 28 mm; tarsus 31 mm.

---

of interest in the birdlife of these remote islands that took place towards the end of the nineteenth century and it was collectors in his pay who contributed so greatly to our awareness of many species. Indeed, some species were discovered and observed almost solely at Rothschild's expense. Had he never caused such a spurt of activity our knowledge of native Hawaiian birds and in particular the remarkable variation of the honeycreepers (Drepanididae) would have been so much the poorer. And yet his own work was sometimes curiously erratic; Rothschild was after all mostly a cabinet naturalist and made many of the mistakes to which such workers are prone. This is particularly apparent in an inference he made about the o'os.

It is well known that many of the Hawaiian honeycreepers give off a peculiar and quite powerful odour noticeable even in museum skins. Rothschild believed that birds in the genus *Moho* had a similar but even stronger smell than the honeycreepers, were this true it would be very surprising since the affinities of the o'os lie with a completely different family – the Meliphagidae or honeyeaters. Clearly, the mistake came about because o'o skins had been packed for despatch to Rothschild along with those of various species of honeycreeper and during the long trip these latter had passed on their pungent smell. When Rothschild unwrapped his birds at Tring in Hertfordshire, comparatively odourless birds had been infected with the honeycreeper scent.

74 The Hon. Walter Rothschild (*c.* 1900), the naturalist who took a special interest in rare and curious species of birds, particularly the vanished and vanishing native birds of the remote Hawaiian Islands. Courtesy of The Hon. Miriam Rothschild.

Although the name of Rothschild is closely associated with all Hawaiian birds, it has a particularly close association with the Molokai O'o. Specimens had been taken during the 1880s, but it was Rothschild himself who first described the species from individuals collected by his agent Henry Palmer at Kaluaaha in 1892. Later, R.C.L. Perkins located these O'os and watched them visiting the flowers of ohias, lobelias and bananas. Although individuals were reported from the Wailau trail in 1915, the last authenticated sighting of the species seems to have occurred in 1904 when George C. Munro saw a group of birds. This particular o'o species may not have been restricted to Molokai and possibly inhabited the neighbouring island of Maui. H.W. Henshaw recorded seeing a single adult male in the forest north-east of Olinda at an altitude of 1,375 m (4,500 ft) in June 1901. He was an experienced observer and was certain that he had seen an o'o, but no such bird was recorded from Maui either before or since Henshaw's experience.

ANOTHER PUBLISHED ILLUSTRATION OF THE *MOLOKAI O'O*

Wilson, S.B. and Evans, A.H. 1890–9. *Aves Hawaiiensis*, pl. 41. Artist, F.W. Frohawk.

ULA-AI-HAWANE

Ciridops anna

PLATE XXXXVIII

Ciridops anna (Dole, 1879)

Fringilla anna Dole, 1879 (Hawaii)

Length 13 cm (5 in)

Description
Adult: crown black shading into silvery grey and white on nape becoming tinged with brown on back; tail, flight feathers, scapulars and breast black; wing coverts, back, rump, and middle of abdomen bright red; lower abdomen and under tail coverts brownish buff; iris dark hazel; bill and feet pinkish brown.

Measurements
Wing 76 mm; tail 48 mm; culmen 10 mm; tarsus 17 mm.

In 1937, on the Kahua ditch trail, close to a spot he had passed many years before in company with Walter Rothschild's collector Henry C. Palmer, the old Hawaiian ornithologist George C. Munro glimpsed a bird with a striking black crown and grey neck. It could not easily be associated with any of the more familiar Hawaiian species and on no field trip since his youth had Munro encountered anything quite like this. Long ago, 45 years earlier, quite near to this very place a native had procured a specimen of just such a bird and brought it to Palmer. No one had seen the Ula-ai-Hawane since. Perhaps, thought Munro, after all these years he had found it again. But perhaps he was mistaken – how could he be sure after so brief a view? What is certain is that Munro had a hand in the last record of the Ula-ai-Hawane, but whether in 1892 or 1937 cannot be said.

Very little is known of *Ciridops anna*. This small bird species is represented in museums by just five specimens – the first taken by J. Mills, a Hilo shopkeeper and, apparently, a painter of some talent – about 1859, the last being the bird brought to Palmer from the headwaters of the Awini River on Mount Kohala on 20 Febuary 1892.

As the species was known to the natives by the name of Ula-ai-Hawane – 'the red bird that feeds upon the Hawane palm' – it may be assumed that at an early period it was once fairly widely distributed in the mountain forests of Hawaii; some very localized Hawaiian birds seem to have no native name at all. By the time of European interest in Hawaiian birdlife, the species had become much more restricted in range although it was recorded from the Hilo and Kona districts and also from the Kohala Mountains. According to R.C.L. Perkins (1903), it was wild and shy and only ever found close to the Hawane palm.

Nothing is known of the voice, nest, or eggs of the species.

XXXXVII Molokai O'os. Hand-coloured lithograph by J.G. Keulemans from W. Rothschild's *Avifauna of Laysan and the Neighbouring Islands* (London, 1893–1900), Pl.74. Courtesy of The Hon. Miriam Rothschild.

KOA 'FINCH'

Rhodacanthis palmeri

PLATE XXXXIX

Rhodacanthis palmeri (Rothschild, 1892)

Rhodacanthis palmeri Rothschild, 1892 (Kona, Hawaii)
Rhodacanthis flaviceps Rothschild, 1892
Psittirostra palmeri Amadon 1950
Psittirostra flaviceps Amadon, 1950

Length 20 cm (8 in)

September 30, 1891. In the heat of the day I shot another of the great finches. It has a little yellow over its bill. It was feeding in a koa top. Palmer killed another at the same time, a much smaller bird with golden head and neck and light yellow breast. It was also in the koa. The tree has seed pods on it and it seems they are feeding on the seeds.

This extract comes from the field notes of George C. Munro who accompanied Walter Rothschild's agent Henry C. Palmer for 15 months collecting in the Hawaiian Islands. It introduces one of the mysteries of Hawaiian ornithology: were there two distinct species of Koa 'Finch', a greater and a lesser, or was there just a single kind?

During the late summer and autumn of 1891, Palmer collected a series of very large 'finches' all of which eventually arrived at Rothschild's museum

Rhodacanthis palmeri (cont'd)
Description
Male: head and throat rich scarlet orange, sometimes more yellowish but crown with a golden sheen that lost its lustre soon after death; upper breast dull reddish orange shading into dull orange yellow on abdomen and pale yellow on under tail coverts; back greenish olive brown washed with orange, particularly on rump; wings and tail blackish brown, feathers margined yellow on outer vanes; axillaries and under wing coverts greyish olive washed with orange; bill bluish grey; legs and feet almost black.
Female: upperparts olive brownish green showing yellow on forehead, yellow green on rump and upper tail coverts; throat and sides of body yellowish olive green; breast and middle of abdomen white washed with green.

Measurements
Wing 92–110 mm; tail 55–72 mm; culmen 17–21 mm; tarsus 24 mm.

at Tring in Hertfordshire. Neither Palmer nor Munro appear to have felt any doubt that the birds collected were other than examples of a single species; but from the series he received Rothschild described two – the Greater 'Finch' (*Rhodacanthis palmeri*), larger with a more orangey head, and the Lesser 'Finch' (*R. flaviceps*), smaller with a yellower head. Eight individuals were assigned to *flaviceps* – two males, six females.

For five more years birds referable to *palmeri* were observed or collected by field workers but the smaller ones were never seen again. Then, the Greater Koa 'Finch', too, was gone – there is no reliable record for this form later than 1896. With the birds themselves no more, argument over whether there once existed two quite separable forms continues and it is now unlikely to be resolved to the satisfaction of all parties. In an authoritative recent work on the endemic avifauna of the Hawaiian Islands, H.D. Pratt (1979) tentatively concluded that only a single species is justifiable. In this he goes against the opinion of Amadon (1950), Wilson and Evans (1890–9) and, of course, Rothschild (1892) himself.

Pratt's position, backed up by Munro and the feeling of R.C.L. Perkins (1903), can be easily summarized. The smaller 'finches' were all obtained in the same locality, on the same days and from the same koa trees as the larger ones. The collectors believed they were taking individuals belonging to a single species. The females assigned to *flaviceps* are identical in colour to those of *palmeri*. The skins of several males referred to *palmeri* but marked 'juv.' or 'imm.' show head colouring identical to the supposed males of *flaviceps*. Although the basis on which these determinations of age were made remains unknown, head colour of the males cannot be considered a diagnostic character for *flaviceps* or *palmeri*. This, then, leaves just size discrepancy as a puzzling factor but Pratt indicates that this also lends little support for Rothschild with the only clear differential lying in the length of the tail, which naturally may be significantly affected by feather wear. Of four *flaviceps* specimens examined by Pratt, two females and a male were indeed in a bad state of wear. Pratt cautiously suggests that perhaps the specimens assigned to *flaviceps* are simply the smallest and most heavily worn first-year individuals of *palmeri*. Others have considered that the *flaviceps* birds might all be hybrids, perhaps between *R. palmeri* and *Psittirostra psittacea*.

XXXXVIII (Left): Ula-ai-Hawane. Hand-coloured lithograph by F.W. Frohawk from S.B. Wilson and A.H. Evans's *Aves Hawaiiensis* (London, 1890–9), Pl.11.
XXXXIX (Right): Koa 'Finches'. Hand-coloured lithograph by J.G. Keulemans from W. Rothschild's *Avifauna of Laysan and the Neighbouring Islands* (London, 1893–1900), Pl.68. Courtesy of The Hon. Miriam Rothschild.

The last record of observations of the living Koa 'Finch' was given by Perkins (1903):

> Although spending most of its time in the tops of the loftiest Koa trees, *Rhodacanthis* occasionally visits the lesser trees . . . chiefly for the sake of the caterpillars that feed upon them . . . it sometimes devours large quantities of gaudily coloured species as well as the more sombre brown or green looper caterpillars. Its chief food however is the green pod of the Koa tree, which it swallows in large-sized pieces and its blue beak is often stained with the green juice and fragments of the pods . . . The song, if such it can be called . . . is entirely different from that of any other native bird. It consists of four, five, or even six whistled notes, of which the latter ones are much prolonged . . . Although the notes are not loud, they are very clear, and are very easily imitated, and the bird responds most freely to an imitation. Were it not for this fact *Rhodacanthis*, when keeping to the leafy crowns of tall Koa trees, as it often does, would be most difficult to get sight of . . . The green-plumaged young, which greatly resemble the female, are fed partly on large fragments of Koa pods, such as their parents eat, both sexes being assiduous in feeding them . . . I have seen the male bird come down to the ground for building material and carry this to the top of one of the tallest Koa trees, and in this situation, in the locality frequented by the bird, certain largish nests which became visible later, when the trees were stripped by caterpillars, I have no doubt were built by *Rhodacanthis*.

By the 1890s the species appears to have existed in only a very limited area but it was by no means uncommon in the correct locality. Its ultimate decline must have been quite abrupt for its home, at elevations above 1,220 m (4,000 ft) in the koa forest of the Kona District of Hawaii, has been searched many times since 1896. A report that a Koa 'Finch' was heard in 1937 may be accurate but there has been no further record.

ANOTHER PUBLISHED ILLUSTRATION OF THE *KOA 'FINCH'*

Wilson, S.B. and Evans, A.H. 1890–9. *Aves Hawaiiensis*, pl. 36. Artist, F.W. Frohawk.

KONA GROSBEAK 'FINCH'

Chloridops kona

PLATE L

These large 'finches' with exceptionally heavy beaks were observed and collected in the Kona District of Hawaii for a period of about 7 years during the late 1880s and early 1900s. A single specimen was obtained by S.B. Wilson on 21 June 1887, Henry Palmer took a series for Walter Rothschild in September 1891 and R.C.L. Perkins found birds in 1892 and again in 1894; then, like the Koa 'Finches' from the same area, these birds vanished never to reappear.

At the time of discovery, the Kona 'Finch' appears to have enjoyed only a very limited distribution. It perhaps occupied an area no more than 10 km^2 (4 sq. miles) on the slopes of the volcano Mauna Loa, although a record of Palmer indicates that a colony may have existed 16 km (10 miles) or so to the south. Like the Koa 'Finch', this species seems never to have

Chloridops kona (Wilson, 1888)

Chloridops kona Wilson, 1888 (Kona, Hawaii)
Psittirostra kona Amadon, 1950

Length 18 cm (7 in)

Description
Adult: general colour of upperparts bright olive green; underparts similar but throat is more golden green, the abdomen is whitish, the under wing coverts brownish buff; lores dusky black; bill dull flesh colour; iris dark hazel; legs and feet dark brown almost blackish. Sexes alike.

Measurements
Wing 90 mm; tail 52 mm; culmen 22 mm; tarsus 24 mm.

acquired a Hawaiian name, obviously an indication of excessive rarity or extremely local distribution.

Kona 'Finches' were most often heard in naio trees, where the sound of hard seeds being split by the birds' heavy beaks attracted the attention of field ornithologists. The germ from the centre of these seeds formed the staple diet but sometimes the nut itself was swallowed, a food supply that was supplemented with leaves and caterpillars.

By those few who saw them, it is reported that they rarely left the area of recent clinker lava flows, such areas being covered with medium-sized trees but little undergrowth. The only calls recorded are a squeak and a light, sweet song – sometimes quite long and rendered with a variety of notes. Perkins (1893) gives the best account of the species in life:

> It is a dull, sluggish, solitary bird, and very silent – its whole existence may be summed up in the words 'to eat'. Its food consists of the seeds of the fruit of the aaka, and as these are very minute, its whole time seems to be taken up in cracking the extremely hard shells of this fruit . . . Only once did I see it display any real activity, when a male and female were in active pursuit of another amongst the sandal trees. Its beak is nearly always very dirty, with a brown substance adherent to it, which must be derived from the sandal-nuts.

G.C. Munro (1944) expressed the hope that the species might survive in the rough lava flows of Hawaii at high elevations but this hope now appears forlorn.

GREATER AMAKIHI

Hemignathus sagittirostris

PLATE LI

Hemignathus sagittirostris (Rothschild, 1892)

Viridonia sagittirostris Rothschild, 1892 (Lower Hilo slopes of Mauna Kea, Hawaii)
Loxops sagittirostris Amadon, 1950

Length 15 cm (6 in)

Description
Adult: upperparts bright olive green showing a more yellowy tinge on forehead, sides of head and upper tail coverts; throat, breast and abdomen yellowish olive green; wings blackish brown, primaries margined with yellowish green; undersurface of wing ash-coloured; bill black or blackish brown becoming bluish or grey towards base; (cont'd)

This is one of the species discovered by Europeans during the surge of interest in Hawaiian birdlife that took place in the 1880s and 1890s, then encountered by one or two collectors for a few years, only to vanish soon after the turn of the century, never to reappear. Apparently completely unknown to native Hawaiians, it was restricted to just a very small area of the island of Hawaii – the dense, dripping forest to either side of the Wailuku River at elevations of 300–900 m (1,000–3,000 ft).

The field collector Henry Palmer was the first to find it when, during 1892, he obtained four skins for Walter Rothschild's Zoological Museum at Tring, Hertfordshire. R.C.L. Perkins located more birds belonging to this species in December 1895; in 1900 H.W. Henshaw collected a series now divided between the Museum of Comparative Zoology, Harvard University, and the Bernice P. Bishop Museum, Honolulu, and a year later A.M. Woolcott found individuals in the same general area. None has been located since.

The best account of the species in life is given by Perkins (1903) who, while considering the birds very rare, was fortunate enough to see twelve individuals, either singly or in pairs, one morning. Perkins was able to locate the species chiefly because its voice differed from that of its near ally, the Amakihi (*Hemignathus virens*). Although its song was in many respects similar, the Greater Amakihi would introduce two or three additional whistles at the end of its trill.

Hemignathus sagittirostris (cont'd) iris hazel with reddish tinge; legs and feet black. Sexes alike.

Measurements
Wing 83 mm; tail 53 mm; culmen 23 mm; tarsus 24 mm.

Almost always the birds were seen creeping along the branches of ohia trees or in the foliage that grew upon them. They seemed particularly fond of certain kinds of cricket that could be found upon these trees but caterpillars, spiders and beetles also formed part of the diet and Perkins observed an individual which was feeding on nectar from the flower of the ohia.

That the Greater Amakihi will almost certainly never be seen again is confirmed by D. Amadon who reported in 1950 that its forest home had gone – the area is now covered with sugar-cane.

L (Left): Kona Grosbeak 'Finch'. Hand-coloured lithograph by J.G. Keulemans from W. Rothschild's *Avifauna of Laysan and the Neighbouring Islands* (London, 1893–1900), Pl.70. Courtesy of The Hon. Miriam Rothschild.
LI (Right): Greater Amakihi. Hand-coloured lithograph by J.G. Keulemans from W. Rothschild's *Avifauna of Laysan and the Neighbouring Islands* (London, 1893–1900), Pl.54. Courtesy of The Hon. Miriam Rothschild.

MAMO
Drepanis pacifica
PLATE LII

The rump of the Mamo, the upper and lower tail coverts and the thighs are coloured a deep yellow. To adorn their royal war cloaks with this rich hue, generations of Hawaiians trapped Mamos and took their plumes. While larger cloaks are chiefly composed of the yellow feathers of the Hawaii O'o (*Moho nobilis*) (see page 208), here and there in diamond-shaped patterns are crocus-yellow feathers of the Mamo. A few smaller cloaks are made entirely from Mamo feathers but it seems to have been impractical to create the larger robes from them alone. However, for the most famous cloak of all, that of Kamehameha I, it has been estimated that 80,000 Mamos would have been sacrificed before work could be brought to completion. It is said that this robe was manufactured through the reigns of eight monarchs before it came to Kamehameha.

The Kings of Hawaii are supposed to have issued an edict guaranteeing the safety of their 'royal' birds. When the birdcatchers trapped one, they were forbidden to kill it but were required to turn their prize loose once the yellow feathers had been plucked. How strictly the system was applied is not known. It would have been virtually impossible to enforce, but although the Hawaiian O'o may have been able to survive under these conditions a plucked Mamo must have made a very sorry sight.

The Mamo was endemic to the island of Hawaii and was once widely

Drepanis pacifica (Gmelin, 1789)

Certhia pacifica Gmelin, 1789
Melithreptus pacificus Vieillot, 1817
Drepanis pacifica Temminck, 1820
Vestiaria hoho Lesson, 1840

Length 20 cm (8 in)

Description
Adult: generally glossy black but lower part of body, rump, upper and lower tail coverts, thighs and anterior margins of the wing rich yellow, and larger primary wing coverts and under wing coverts white; four central tail feathers black, the rest brown showing a patch of dull white near the tip; bill blackish brown, the upper mandible longer than the lower; legs probably deep brown. Sexes similar.

Measurements
Wing 104 mm; tail 64 mm; culmen 44 mm; tarsus 34 mm.

distributed there. It was first brought to attention by the naturalists who accompanied Captain Cook and was originally described by John Latham in 1782 as the 'Great Hook-billed Creeper' from specimens in the Leverian Museum in London. Despite its early description and the fact that individuals probably survived into the early 1900s, little is known of it. Mamos probably fed chiefly on nectar and are supposed to have been particularly fond of that from arborescent lobelias but they may also have eaten insects. According to H.W. Henshaw (1902), the flight of the Mamo was similar to that of the Cuckoo 'not rapid but smooth and well sustained'. Natives described its call as a single long, rather mournful note.

Although hunting perhaps may have contributed to the bird's downfall, it seems likely that habitat deterioration was equally to blame. No Mamo has been taken since 16 April 1892. Henry Palmer, collecting for Walter Rothschild, lay injured, having been kicked by a horse, and sent his assistant Wolstenholme and an old native bird-catcher called Ahulan into the woods on Mauna Loa above Hilo. Wolstenholme (see Rothschild, 1893–1900) noted the circumstances surrounding the taking of the Mamo:

> I heard a call from the other side of the gulch, and thought it was a native calling, but immediately afterwards a bird flew across, and I saw in a moment it was the bird we were after. I was going to follow it up to shoot it, but Ahulan begged me not to shoot as it would scare the other away, which I had heard calling a little way off. Ahulan fixed the snares and bird-lime on a haha, which growed out on a treefern, and which has flowers somewhat

LII Mamo. Hand-coloured lithograph by J.G. Keulemans from W. Rothschild's *Avifauna of Laysan and the Neighbouring Islands* (London, 1893–1900), Pl.61. Courtesy of The Hon. Miriam Rothschild.

75 Hawaiian man in feather cape and helmet. Engraving after a painting by John Webber.

like those of a fuschia. Ahulan fulfilled his promise and caught the Mamo! He is a beauty, and takes sugar and water eagerly and roosts on a stick in the tent. I now feel as proud as if someone had sent me two bottles of whisky up.

Palmer was altogether more business-like. When the party returned 5 days or so later with the living bird, he killed and skinned it. Neither is the last record of an encounter with living Mamos a particularly happy one. In a letter to Rothschild, H.W. Henshaw described an experience he had had in 1898:

About a year ago last July I found what, no doubt, was a family of mamos in the woods above Kaumana. I am sure that I saw at least three individuals, possibly four or five. They were flitting silently from the top of one tall Ohia-tree to another, apparently feeding upon insects. The locality was a thick tangle, and a momentary glimpse of a slim, trim body as it threaded its way through the leafy tree-tops was all I could obtain. After about two or three hours I succeeded in getting a shot at one bird in the very top of a tall Ohia-tree. It was desperately wounded, and clung for a time to the branch, head downwards, when I saw the rich yellow rump most plainly. Finally, it fell six or eight feet, recovered itself, flew round to the other side of the tree, where it was joined by a second bird, perhaps a parent or its mate, and in a moment was lost to view.

BLACK MAMO

Drepanis funerea

PLATE LIII

Drepanis funerea (Newton, 1893)

Drepanis funerea Newton, 1893 (Molokai, Hawaiian Islands)

Length 20 cm (8 in)

Description
Adult: lustreless black plumage apart from a buffy white shade on outer webs of primaries; bill black but with a patch of yellow at base of maxilla; iris pale yellowish brown; legs and feet black. Sexes similar although beak of male is perhaps longer and female may be generally smaller.

Measurements
Wing 102 mm; tail 70 mm; culmen 64 mm; tarsi 32 mm.

The Black Mamo, sometimes called Perkins's Mamo after its discoverer, was found on the island of Molokai at a height of 1,525 m (5,000 ft) on 18 June 1893. During the next few years, Black Mamos were located on several occasions but none has been seen since 1907. The chance that this species has survived is very slim indeed for the environment that it enjoyed on Molokai has been largely destroyed.

What is known of the Black Mamo comes almost exclusively from the account of R.C.L. Perkins (1903). He believed that in most respects, including the voice, this species closely resembled *Drepanis pacifica* (see page 218), its counterpart on Hawaii. However, Perkins recorded it only in the underbrush, rarely seeing these birds more than 3.6 m (12 ft) from the ground and never observing them to land high up in a large tree. He described the species as a 'true honey-sucker', by which was meant that none of the birds dissected contained animal food. During inspection of wet moss the tongues of these birds darted in and out with such rapidity that they appeared like 'liquid streaks'. So tame were these Mamos that their discoverer was able to watch them at very close quarters working their way from one large flower to another. Like the Hawaiian Mamo, these on Molokai favoured arborescent lobelias. Perkins wrote:

I saw three adult males of this bird in one low bush passing from flower to flower and spending only a few seconds over each. These were very tame and I was able to watch their movements in this and neighbouring bushes for at least an hour. Even those flowers which were at a height of no more than a foot from the ground were carefully explored. The crown of the head of each of these birds was plentifully encrusted . . . with the sticky white or purplish-white pollen of the lobelias and gave them a singular appearance . . . they will sit quietly preening their feathers, when they have a very comical appearance, much stretching of the neck being necessary to enable them to reach the fore parts of the body with the tip of their long beaks.

As is the case with the Mamo itself, the record of man's last encounter with the Black Mamo is rather unpleasant. Alanson Bryan (1908) reported killing three individuals during June 1907 at Moanui. Accounting for the first of these he wrote:

To my joy I found the mangled remains hanging in the tree in a thick bunch of leaves, six feet or more beyond where it had been sitting. It was, as I feared, badly mutilated. However, it was made into a very fair cabinet skin.

BONIN ISLANDS GROSBEAK

Chaunoproctus ferreirostris

PLATE LIV

Chaunoproctus ferreirostris (Vigors, 1828)

Coccothraustes ferreirostris Vigors, 1828 (Bonin Shima = Peel Island, now Chichi-jima, Ogasawara group)
Chaunoproctus ferreirostris Bonaparte, 1851

Length 20 cm (8 in)

Description
Male: forehead, stripe above eye, cheeks, throat and upper breast reddish orange; lower breast paler shading to white on belly; rest of plumage olive brown, striped darker brown on back; beak feet and legs probably blackish brown.
Female: all mid brown except yellowish forehead and flank feathers tipped darker brown.

One of the great curiosities of the extinction of bird species is that certain comparatively tiny areas of the earth's surface have suffered the most appalling losses, while the avifaunas of major landmasses remain, by and large, intact. In great measure this is due to the catastrophic effect of the coming of modern people on the fragile bird populations of certain island groups. One might immediately think of the terrible depletion of forms on the Hawaiian Islands, New Zealand, the Mascarenes or the two tiny islands of the Tasman Sea – Lord Howe and Norfolk. Between them these small areas account for a very considerable percentage of those species and subspecies that have become extinct during the last few hundred years. Another of the world's blackspots are the islands south of Japan – Ryukyu Islands (Nansei-shotō), the Borodinos (Daitō-jima) and the Bonins (Ogasawara-shotō). From these inconspicuous places have vanished not only several full species but also a series of distinct races of birds that otherwise remain widespread.

One of the most striking forms endemic to Ogasawara is the Bonin Islands Grosbeak (*Chaunoproctus ferreirostris*). Known from little more than the skins collected in the first half of the nineteenth century, the affinities of the species remain obscure. It is assumed to be a member of the Fringillidae and, on zoogeographical grounds, J.L. Peters (1931–86) placed it close to the *Carpodacus* species. Apart from the knowledge that can be gleaned from the specimens, there is very little on record.

They were certainly inhabitants of Peel Island (Chichi-jima), where two examples were procured in June 1827 by men of HMS *Blossom*. Peel is just over 14.5 km (9 miles) long by 6.4 km (4 miles) wide and was the only one of the group on which a landing was attempted. Although two shipwrecked sailors were found here, the state of the island led Captain Beechey of the *Blossom* to believe that it had had no permanent human population. Birds were very tame and could be approached with ease. A year later, F.H. von Kittlitz, a naturalist whose name is connected with other extinct birds (e.g. Kittlitz's Thrush, see page 198), landed on Peel and collected a few additional specimens. His notes, published in 1832, provide the only record of the species' habits. According to von Kittlitz,

LIII Black Mamo. Hand-coloured lithograph by J.G. Keulemans from W. Rothschild's *Avifauna of Laysan and the Neighbouring Islands* (London, 1893–1900), Pl.62. Courtesy of The Hon. Miriam Rothschild.

the birds inhabited the forest, being found either singly or in pairs. They were most often seen on the ground, a trait that may have made them particularly vulnerable to the predation of introduced mammals. Presumably, these birds searched low down for the small fruits and buds of which von Kittlitz found evidence in their gullets. The call was a 'soft, pure, high piping note, given sometimes shorter, sometimes longer'.

Following von Kittlitz's report the Bonin Islands Grosbeak was never recorded again. It is known that in 1830 a party of American and British sailors landed together with a group of Polynesians and colonized Peel. Whether they noticed the Grosbeak is now known but certainly their presence along with a large number of introduced mammals must have wrought great changes on the environment. During the 1850s at least two American naval expeditions visited Peel and the naturalist William Stimpson participated in the second of these. He collected from Peel in 1854 but was unable to find the Grosbeak. Rumours that the species survived until at least the 1890s were never confirmed.

ANOTHER PUBLISHED ILLUSTRATION OF THE *BONIN ISLANDS GROSBEAK*

Beechey, F.W. 1838. *The Zoology of Captain Beechey's Voyage to the Pacific*, pl. 8. Artist, Edward Lear.

LIV Bonin Islands Grosbeaks: male (left); female (right). Hand-coloured lithographs from C.L. Bonaparte and H. Schlegel's *Monographie des Loxiens* (Paris, 1850), Pl.37 and 38.

KUSAIE ISLAND STARLING

Aplonis corvina

PLATE LV

Two species vanished during the nineteenth century from Kusaie (or Kosrae) Island in the Pacific Ocean, a rail (*Porzana monasa*) and a large, glossy black, long-tailed starling (*Aplonis corvina*). Both have in common the fact that they are known from only a very few specimens collected during early years of regular European contact with the island. Just two examples of the rail exist – both now in the Leningrad Museum where, too, are kept the only surviving skins of the starling, which may have been also formerly represented in the collection of the Frankfurt Museum.

Aplonis corvina (Kittlitz, 1833)

Lamprothornis corvina Kittlitz, 1833 (Kusaie (Kosrae), Caroline Islands)
Lamprocorax corvinus Hartlaub, 1854
Calornis corvina Grey, 1859
Sturnoides corvina Finsch and Hartlaub, 1867
Kittlitzia corvina Hartert, 1891
Aplonis corvina Reichenow, 1914

Length 26 cm (10 in)

Description
Adult: plumage glossy black; iris red; bill black (according to von Kittlitz's illustration).

Kusaie is one of the Carolines but lies to the east of the group's main concentration. It is volcanic in origin and covers an area of about 105 km² (40 sq. miles). During the nineteenth century, the island was much favoured by whalers who used it extensively for beaching their vessels prior to cleaning or repair. As a result, rats were repeatedly introduced to Kusaie, which at one time was reported to be completely overrun with them. There seems little doubt that these mammals were in some degree responsible for the extinction of both the starling and the rail.

Aplonis corvina was first described by F.H. von Kittlitz in 1833 on the basis of specimens that he had collected on his travels through the Pacific in the 1820s. According to this authority, the bird was solitary and even at the time of discovery seemed rare. Quite when the species vanished is not certain, for it was an inhabitant of the island's mountainous forests and these were probably only occasionally penetrated by nineteenth-century Europeans. The species may have been gone by 1880 when Otto Finsch (1881) was unable to find it. There now seems little likelihood that the starling is other than extinct for the island has been thoroughly searched several times since 1880 and Kusaie is not really very extensive.

For a starling the species was surprisingly large, reaching 25 cm (10 in) in length, but its most singular feature was, perhaps, its long, curved bill. A closely related, but smaller species, *A. pelzelni*, occurs in the mountainous forests of Ponapé (or Pohnpei), the largest island of the Caroline group, but this form is itself now rare and threatened with extinction.

MYSTERIOUS STARLING

Aplonis mavornata

PLATE LV

Aplonis mavornata (Buller, 1888)

Aplonis mavornata Buller, 1888
Aplonis inornata Sharpe, 1890

Length 18 cm (7 in)

Description
Specimen in BM(NH): brown above and below with slight bronzy gloss on head and a few hair-like whitish shafts on the breast; iris yellow.

Measurements
Wing 105 mm; tail 65 mm; culmen 20 mm; tarsus 25 mm.

A unique skin of hitherto unknown origin has existed in the British Museum (Natural History) for many years. As long ago as 1890 the British ornithologist Richard Bowdler Sharpe wrote of it in the following terms: 'The specimen is very old and remained for many years unnoticed in the collection of mounted birds.'

Two years before Sharpe wrote this, the New Zealand writer Walter Buller (1888) had named the form *Aplonis mavornata* but Sharpe himself referred to it as *Aplonis inornata*, a name he found inscribed upon the bird's stand (as indicated above, it was once a stuffed bird, but has now been reduced to a skin). Sharpe's feeling was that a published description with priority over Buller's could very well be in existence and, if so, might one day come to light – perhaps in some particularly obscure work; but even for the very complete *Catalogue of Birds in the British Museum* (Vol. 13, 1890) he was unable to trace any definite record relating to the specimen's provenance and until 1986 this remained a mystery.

In the absence of information to the contrary, this brown and rather drab bird was considered on zoogeographical grounds to have come from a Pacific island, and it was felt probable that it was brought back from one of Captain Cook's three voyages around the world. Over the years, attempts were made to correlate the enigmatic skin with other little-known species – most particularly with *Turdus ulietensis*, Latham's obscure Bay Thrush (see page 196). These efforts to tidy up ornithological history met with no success; Erwin Stresemann (1949) compared the skin with a painting by George Forster of *Turdus ulietensis* (no specimen of this species exists today) and found several outstanding discrepancies between the two. In 1951, the type of *A. mavornata* was examined by Ernst Mayr, a specialist in the birds of Pacific islands, who declared that it belonged to no known species but bore marks of affinity to *A. tabuensis*, a widespread and generally successful starling that has diversified into a number of races across the South Pacific.

And so the matter rested until December 1986 when *Notornis*, the journal of the Ornithological Society of New Zealand, carried an article by Storrs L. Olson, an expert on extinct and fossil birds at the Smithsonian Institution in Washington DC. Olson had studied an old manuscript (No. M8S BLO) that for many years had lain undisturbed in the British Museum (Natural History) and managed, almost with complete certainty, to correlate *Aplonis mavornata* with a bird listed in it.

The manuscript was that of Andrew Bloxham, who had sailed from England as ship's naturalist aboard HMS *Blonde* under the command of George Anson, Lord Byron, on 28 September 1824.

The main purpose of the voyage was rather unhappy. It was to carry back from London to Honolulu the bodies of Liholiho and Kamamalu, King and Queen of Hawaii, both of whom had sadly died of the measles while visiting England. During May 1825 the *Blonde* arrived in the Hawaiian Islands and discharged its doleful commission; then, in July, it headed south, eventually arriving home by way of South America and the island of St Helena.

While still in the Pacific, though, the *Blonde* touched at the Cook Islands and stopped briefly at the island of Mauke (which Bloxham called Mauti). Here, during the afternoon of 9 August 1825, Bloxham shot three birds – a pigeon, a kingfisher and a starling.

The visit to the island was apparently short – no more than 2 hours or so. By about 4.00 in the afternoon (the time is noted quite particularly in the manuscript) the shore party had returned to their ship and the excursion was over. Bloxham had previously collected birds in the Hawaiian Islands and was to collect more in South America; these were presented to the Admiralty on the expedition's return and in due course ended up at the British Museum where many of them can still be traced.

As might be expected, an account of the voyage was published (see Byron, 1827) but this was actually compiled from notes by a Mrs Maria Graham. Poor Mrs Graham's work has long been noted as being inadequate and her account of the landing at Mauke is cursory. It is almost certainly because of this deficiency that the Mysterious Starling has remained so long a puzzle. Amazingly, no ornithological collector after Bloxham took birds from Mauke until the early 1970s when D.T. Holyoak visited the island; it goes without saying that he saw no sign of the starling. Who can say what caused it to vanish in the intervening years?

As the type of *Aplonis mavornata* corresponds exactly with Bloxham's description, and as his collection was deposited at the British Museum, it appears virtually certain that the provenance of the Mysterious Starling has finally been determined. Ironically, all that is known of the living bird is that it was, in Bloxham's words, 'killed hopping about a tree'.

ANOTHER PUBLISHED ILLUSTRATION OF THE *MYSTERIOUS STARLING*

Greenway, J.C. 1958. *Extinct and Vanishing Birds of the World*, frontispiece. Artist, D.M. Henry.

NORFOLK & LORD HOWE STARLING

Aplonis fusca

PLATE LV

Aplonis fusca (Gould, 1836)

Aplonis fuscus Gould, 1836 (Murrumbidgee River (error) = Norfolk Island)

Length 20 cm ($7\frac{3}{4}$ in)

Description

Aplonis fusca fusca (Male): generally dark greyish in colour but head glossed with bottle green, which spreads to throat, breast and more dully to scapulas, back rump, upper tail coverts and wing coverts; underparts dark ashy grey but under tail coverts dull white; bill black; iris reddish orange; feet black.

Aplonis fusca fusca (Female): similar to male but a little lighter in colour with greenish gloss less evident; throat and breast shading to brown on sides of body and flanks; centre of lower breast washed with pale ochraceous; lower abdomen and under tail coverts yellowish white.

Aplonis fusca hulliana: similar to *A. f. fusca* but overall appearance is browner, less grey.

Measurements

Male: wing 103 mm; tail 68 mm; culmen 13 mm; tarsus 25 mm.
Female: wing 98 mm; tail 63 mm; culmen 13 mm; tarsus 25 mm.

The islands of Norfolk and Lord Howe represent something of a disaster area for certain kinds of birds. Lying in the Tasman Sea, far removed from any major landmasses, they were colonized mostly from Australia or New Zealand – by a variety of bird forms, being just large enough to provide a selection of 'niches' for these colonizing birds to fill as they developed in isolation. When humans started regularly to visit Norfolk and Lord Howe – and either accidentally or purposefully to introduce terrestrial mammals – the equilibrium maintained by the fragile avifaunas was shattered.

From Lord Howe were lost several insular races of otherwise extant species – a pigeon (*Columba vitiensis godmanae*), a parakeet (*Cyanoramphus novaezelandiae subflavescens*), an owl (*Ninox novaeseelandiae albaria*), a thrush (*Turdus poliocephalus vinitinctus*), a warbler (*Gerygone igata insularis*), a fantail (*Rhipidura fuliginosa cervina*) and a white-eye (*Zosterops lateralis tephropleura*); in addition, a full species of white-eye (*Zosterops strenua*) and a species of swamphen (*Porphyrio albus*), both endemic to the island, became extinct (see pages 205 and 84). Norfolk lost its island races of a pigeon (*Hemiphaga novaeseelandiae spadicea*), a parrot (*Nestor meridionalis productus*) and a triller (*Lalage leucopyga leucopyga*).

As well as these extinct forms, there once lived upon both islands a starling that can no longer be found on either. This was *Aplonis fusca*, which ornithologists split into two subspecies – *hulliana* of Lord Howe and the nominate race from Norfolk Island.

The grounding of the SS *Makambo* at Ned's Beach during June of 1918 proved disastrous for many of Lord Howe's birds. The pigeon, the swamphen and the parakeet were already gone, but Black Rats (*Rattus rattus*) escaping from the distressed ship speedily accounted for the warbler, the thrush, the fantail, two forms of white-eye and the starling. Previous to this, Lord Howe seems to have been protected from rats by arrangements for landing on the island. Visiting ships found it necessary to anchor offshore, the actual landing being effected by means of small boats; as far as can be determined it was solely this circumstance that rendered the tiny area of land rat-free for so long. Then the *Makambo* drifted onto the beach! For 9 days the ship lay grounded until finally it was refloated; during this period – so it is believed – the rats left the vessel and escaped onto the island.

Within 10 years the vulnerable bird populations had vanished; but well before this (1921) a Lord Howe resident, Alan McCulloch, felt moved to write (see Hindwood, 1940):

> Within two years this paradise of birds has become a wilderness, and the quietness of death reigns where all was melody. One cannot see how the happy conditions are to be restored. The very few birds remaining are unable to breed, being either destroyed upon their nests or driven from them by rats, and their eggs eaten.

Before the calamity, the starlings were plentiful and because of their liking for fruit and other crops something of a pest.

Reasons for the species' disappearance from Norfolk Island cannot be so neatly classified but its evident and general fragility is obviously to be

LV (Left): Three extinct starlings: Kusaie Island Starling (top); Norfolk and Lord Howe Starling (middle): Mysterious Starling (bottom). Oil painting by Mike Latter.
LVI (Right): Bourbon Crested Starling. Chromolithograph after a painting by E. de Maes from Baron E. Sélys Longchamp's *Collections Zoologique*, Fasc. 31, *Oiseaux* (Brussels, 1910), Pl.1.

held responsible. It presumably vanished at much the same time as its Lord Howe counterpart.

The nest was built in a rather loose manner from small twigs and grasses positioned in a tree hollow (Hindwood, 1940). Three to five bluish-green eggs were laid and these were speckled and blotched with pale red, particularly towards the larger end; average measurements were 27 mm × 19 mm.

By local people this starling was called the *cudgimaruk* (a name derived from the call) or the Red-eye.

BOURBON CRESTED STARLING

Fregilupus varius

PLATE LVI

Fregilupus varius (Boddaert, 1783)

Upupa varia Boddaert, 1783
Upupa capensis Gmelin, 1789
Upupa madagascariensis Shaw, 1812
Fregilupus capensis Lesson, 1831
Fregilupus borbonica Vinson, 1868

Length 30 cm (12 in)

Description
Adult: most of head, elevated crest, and hind neck light ash grey, feathers with white shafts; lores and eyebrows, cheeks, throat and underparts white;

Until the middle years of the nineteenth century there lived on the Mascarene island of Réunion (once named Bourbon) a beautiful but rather aberrant crested starling with a long, downcurved beak. Although usually called the Bourbon Crested Starling (*Fregilupus varius*) it was known to islanders and early naturalists as the *huppe* or *calandre* but surprisingly little information regarding habits and so forth can be attached to whatever name is used.

The species certainly survived into the 1830s. Jules Verreaux shot a crested starling in 1832 and Julien Desjardins received four living birds – sent by a friend on Réunion – at his home in Mauritius during May 1835; two of these escaped into the forests several months after their arrival and seem to have been shot sometime later. The last specimens taken were obtained before 1840. However, numerous rumours that individuals survived hidden deep in the interior of the island persisted for several decades after this.

That the final decline of the species to extinction may have been quite rapid is indicated by a letter written in 1911 by one De Cortimoy (quoted by Hachisuka, 1953):

Fregilupus varius (cont'd)
upperparts (except for head and neck) brown, lower back, rump and upper tail coverts washed with rufous; under wing coverts and axilliaries white; primary coverts white for terminal half.

Measurements
Wing 147 mm; tail 114 mm; culmen 41 mm; tarsus 39 mm.

> When I was a child, I have known this bird to inhabit the forest and live in flocks; their song was a clear note and they had a beautiful plumage. The bird was very tame and, being young, I killed dozens of them. When I returned to the Island after ten years in Paris, I found no further trace of them. I used to keep them in a cage without any trouble. They eat bananas, potatoes, cabbage etc.

Other evidence confims that these beautiful starlings were very trusting and tame. Long after the birds had vanished, older inhabitants of the island recalled how in years gone by they could easily be knocked down with sticks. Whether extinction was directly due to such persecution by the islanders is difficult to say but wherever birds show any tendency towards tameness they will certainly suffer for it at the hand of man. Another agent of destruction may well have been the rats that were introduced to the island accidentally, while other mammals put there quite deliberately might also have taken a toll.

Considering the early date of extinction, and our almost total ignorance of the bird in life, a surprisingly large number of specimens have been preserved and are still held in European and North American museums, including some pickled examples. A few stuffed *huppes* are even retained in the Mascarenes.

RODRIGUES STARLING

Fregilupus rodericanus

FIGURE 76

Fregilupus rodericanus (Günther and Newton, 1879)

Necropsar rodericanus Günther and Newton, 1879 (Rodrigues)

Length 25 cm (10 in)

Description
Appearance in life unknown.

In any investigation of the species known today as *Fregilupus rodericanus*, three possibly quite unconnected strands of evidence need to be considered. First, there are the bones that were found on the Mascarene island of Rodrigues by the Revd H.H. Slater during the Transit of Venus Expedition in 1874. These were described by A. Günther and E. Newton in 1879 and establish that a starling-like bird was in recent times an inhabitant of Rodrigues. The two authors wished to refer the skeletal remains to the genus *Fregilupus* because it seemed clear that the bird from which they originated was quite nearly related to the extinct Bourbon Crested Starling (*F. varius*). In deference, however, to what they termed 'present ornithological practice', they kept the two generically separated, with the Rodrigues birds placed in the genus *Necropsar*; later writers usually refer the species back to *Fregilupus*.

Although there may be no connection, an early and rather enigmatic account of an otherwise unidentifiable bird has come to be associated with the skeletal remains. This account is contained in the anonymous manuscript known as the *Relation de l'Île Rodrigue* documenting a visit to the island thought to have taken place about the year 1730:

> A little bird is found which is not very common, for it is not found on the mainland. One sees it on the *Islet au Mât*, which is to the south of the main island, and I believe it keeps to that islet on account of the birds of prey which are on the Mainland, and also to feed with more facility on the eggs of the fishing birds which feed there, for they feed on nothing else but eggs, or some turtles dead of hunger, which they well know how to tear out of their shells. These birds are a little larger than a blackbird, and have white

76 The specimen of a mystery starling in the Merseyside County Museum, Liverpool. Called *Necropsar leguati*, its provenance is unknown but circumstantial evidence suggests that it came from Rodrigues and that it could be the skin of the extinct Rodrigues Starling (*Fregilupus rodericanus*). Chromolithograph after a painting by J.G. Keulemans from W. Rothschild's *Extinct Birds* (London, 1907), Pl.2. Courtesy of The Hon. Miriam Rothschild.

> plumage, part of the wings and tail black, the beak yellow, as well as the feet, and make a wonderful warbling. I say a warbling since they have many and altogether different notes. We brought up some with cooked meat, cut up very small, which they eat in preference to seeds.

Whether this account of the 'Turtle Bird' is justifiably matched with the bones is debatable; possibly there is a connection for there is no reason to suppose that a starling could not turn to a diet of meat and eggs.

The link with the third strand in the story is perhaps more tenuous. Several authors have suggested that a unique bird skin now in the Merseyside County Museum, Liverpool, is a specimen of *Fregilupus rodericanus*. This skin has also been used as a basis upon which to found a new species, *Necropsar leguati*. H.O. Forbes (1898) described its discovery:

> In a cabinet in the Derby Museum, where it had reposed for nearly fifty years, among species of Bulbuls, *Hypsipetes*, there was discovered in the early part of this year, during an examination of our unmounted ornithological collection . . . a flat skin in perfect preservation. That it had been left undetermined for so long a period is probably due to the fact of its being taken for an albino of some species of the above-named genus or perhaps for a white starling.
>
> At first sight the bird is, in general appearance, not dissimilar from those with which it was associated, but a closer examination soon proved that it differs greatly from them in the form of the bill and in the external character of its legs, which are sturnine. Its wings, however, are quite unlike those of any starling, and while the bill persisted in associating itself in my mind with that of the Crested Starling of Réunion [*F. varius*] the form of the wings and the absence of a crest long prevented me from including it in that alliance.
>
> As to its history, I regret that I can supply no more than the meagre information afforded by its label, that it was 'purchased', probably by its former noble owner, Lord Derby, on the '10th August, 1850, from M.J. Verreaux' the then well-known ornithologist and dealer in Paris; and as to its habitat, only what the reverse of the label bears, in the handwriting presumably of M.J. Verreaux, the single word 'Madagascar'. . . . It is well known that M. Verreaux was often very inexact in the precise geographical data he inscribed on the labels of his specimens . . . 'Madagascar' . . . may, therefore . . . mean any part of the Mascarene region.

In other words, the skin could have come from the island or elsewhere; its provenance remains unknown. The grounds for linking the skin with *F. rodericanus* are not, therefore, entirely convincing. Perhaps the strongest indication of a genuine connection lies in the fact that both bones and skin appear to belong to otherwise unknown starlings.

According to taste, the three strands of evidence can be taken as indication that one, two, or three otherwise unrecorded species formerly existed. In his book *The Dodo and Kindred Birds* (1953), M. Hachisuka, for instance, named *Testudophaga bicolor*, the Bicoloured Chough, from the anonymous description, then listed *Necropsar rodericanus* on account of the bones and *Orphanopsar leguati* on the basis of the skin.

In whatever way the information available is classified, it is certain that at least one population of birds with starling-like affinities has been lost from Rodrigues during historical times. Reasons for extinction are not clear and neither is the timing of it, but if the anonymous author's account can be relied upon it would appear that the birds were rare even by 1730.

HUIA

Heteralocha acutirostris

PLATE LVII; FIGURES 78–9

Heteralocha acutirostris
(Gould, 1836)

Neomorpha acutirostris Gould, 1836 (North Island, New Zealand)
Neomorpha crassirostris Gould, 1836
Neomorpha gouldi Gray, 1841

Length 48 cm (19 in)

Description
Adult: plumage completely black with metallic green gloss apart from broad white terminal tail band; large rounded facial wattles rich orange; iris brown; legs and feet bluish grey. Sexes alike except for marked difference in shape of bill (ivory white darkening to blackish grey at base in both sexes) – medium length and sturdy in male; long, delicate and curved in female.
Immature: similar to adult but plumage dull black suffused with brown; white tail bar washed rufous.

Measurements
Male: wing 210–220 mm; tail 200 mm; culmen 57–60 mm; tarsus 84–86 mm.
Female: wing 200–205 mm; tail 195–200 mm; culmen 85–88 mm; tarsus 78 mm.

Prized by the Maori for its tail feathers, sought after by European skin collectors on account of an anomalous contrast in shape between bill of male and female, the Huia – caught between hammer and anvil – was doomed, perhaps from the hour Captain Cook first dropped anchor in New Zealand waters.

To the Maori it held an almost sacred significance; their wonder of the Huia is expressed in beautiful artefacts – pendants, amulets and the carved wooden boxes known as *waka huias* especially made for housing feathers. These feathers were commonly worn in battle, a war plume of twelve being called a *mare-reko*. Even more often, they were given as tokens of friendship or bestowed as marks of respect, holding a significance deeper than that conveyed by a purely decorative item. Most particularly, feathers were used during funeral rites, having somehow acquired a peculiar association with death. They are often found attached to the dried, tattooed Maori heads in which there was such an unpleasant trade during the nineteenth century.

Until the coming of Europeans chiefs of great rank were alone permitted to wear feathers (Figure 77) but this right was assumed by all who could acquire them as tribal traditions and culture were rapidly worn down; hunting pressure undoubtedly increased. An instance is on record of a haul of 646 skins taken by just eleven Maoris within a single month from woods between the Manawatu Gorge and Akitio. No species could long remain in a locality subject to persecution of this intensity. At the century's turn, a Maori guide placed feathers in the hatband of the visiting Duke of York during his tour of Rotorua. Trade in Huia tails, new or old, reached a climax!

The flavour of the ambivalent European attitude during the last quarter of the nineteenth century is captured by Walter Buller writing in 1888:

> Whilst we were looking at and admiring this little picture of bird-life, a pair of Huias, without uttering a sound, appeared in a tree overhead, and as they were caressing each other with their beautiful bills, a charge of No. 6 brought both to the ground together. The incident was rather touching, and I felt almost glad that the shot was not mine, although by no means loth to appropriate the two fine specimens.

Stuffed birds were needed to decorate colonial drawing rooms and many more were exported from New Zealand to museums and collectors' cabinets around the world. J.C. Greenway (1958) believed that hunting, either for tail feathers or for cabinet specimens, contributed little towards the species' disappearance; his research was able to reveal only sixty-five skins and skeletons in collections and the Maori had, after all, hunted Huias for centuries without exterminating them. His assumption is wrong. It overlooks the change in outlook and social ranking caused by the arrival of the colonialist *pakeha* and the sweeping changes this brought about. Geenway's estimate of Huia specimens falls well short of reality for W.J. Phillipps (1963) records 119 skins from the major New Zealand museums alone. Many more examples exist in private hands (a series of 20 splendidly preserved specimens was until recently in the possession of a single collector); mounted pairs are still quite regularly offered in London

77 *The Last of the Cannibals*, an oil painting (*c.* 1913) by the great New Zealand artist C.F. Goldie of a tattooed Maori chief wearing Huia feathers. Collection of the Aigantighe Art Gallery, Timaru, New Zealand.

LVII Huias: male (front); female (behind). Chromolithograph after a painting by J.G. Keulemans from W.L. Buller's *History of the Birds of New Zealand*, Vol. 1 (London, 1888), Pl.2.

salerooms. The total number of birds taken as trophies or as objects for scientific study would surely be surprisingly large if it could be ascertained. The European invasion caused devastation to the native birds of New Zealand and the Huia, it seems, was quite simply swept away before this onslaught.

In addition to the radical changes brought to Maori culture and the Europeans' own instinct to hunt whenever opportunity arose, the introduced mammalian predators probably caused telling losses to these fragile birds. Introduced birds may also have contributed to the Huia's disappearance; it is possible that some were direct competitors and perhaps carried avian diseases to which native birds had no natural immunity. Ticks of the species *Haemaphysalis leachi* and *Hyalomma aegyptium*, probably passed on by introduced Mynahs (*Acridotheres tristis*), have been found on Huia skins.

According to Maori tradition, the Huia was never a widespread bird; its home grounds were among the Rimutaka, Tararua, Ruahine and Kaimanawa ranges together with adjacent stretches of lowland forest – all areas lying within the southern half of New Zealand's North Island. Territory a little more to the north may also have been occupied and Huias are alleged to have lived on the South Island in the Marlborough and Nelson districts; Buller (1888) felt little doubt this was so, although the claim aroused a good deal of controversy.

As settlement, cultivation and grazing changed the face of large tracts of country, terrain wherein Huias could flourish became rapidly more restricted: habitat destruction must have played a significant role in the species' extinction.

It is from Buller's magnificent account of the species that almost everything now known about the living Huia is derived and the bird's decline can be traced through the editions of *A History of the Birds of New Zealand* and the *Supplement* to it. In 1872, when the first issue was circulated, Huias seem not to have been particularly well-known birds and were considered rather rare. By 1888, the date of publication of the second edition, opening up of the countryside had created the impression that they were more plentiful than previously supposed. Even so, Buller drew attention to the savage persecution then being waged against these birds. Seventeen years later, writing in the *Supplement*, he acknowledged that the species was facing inevitable extinction.

The year of the last fully authenticated sighting was, in fact, fast approaching. The *Supplement* became available in 1905 and in 1907, a few days after Christmas, W.W. Smith saw three birds, two males and a female, during an observation that has come to be regarded as the final indubitable record of the species.

There is little doubt, however, that a few pairs survived considerably after this date. Into the 1920s reports of black birds with orange facial wattles and white tips to the tail feathers occurred with regularity, many of them detailed by Phillipps (1963). Rather more recent unsubstantiated sightings have been made and perhaps there is a foundation of truth to some of these claims; the rediscovery of the Takahe (*Notornis mantelli*) (see page 71) in 1948 is proof enough that populations of birds may remain undetected in New Zealand. More recently, during 1977, examples of the very rare Kakapo (*Strigops habroptilus*) (see page 133) were found in southeastern Stewart Island where they were thought to be quite extinct. Similar instances concerning the recent discovery of colonies of rare New Zealand birds might be cited.

It is sometimes pointed out that although lowland parts of the Huia's former range were rapidly deforested to make way for farmland, many of the more mountainous areas once occupied remain covered to this day with forest. Undoubtedly this is true, but Buller drew attention to the need of the Huia to descend from its mountainous refuges to escape the cold of winter. As the years pass without authenticated sightings, it becomes increasingly certain that the species has failed to survive.

It seems a pity that Huias were not more often kept in captivity as they might have become great favourites with aviculturalists and breeding stocks could, conceivably, have been built up and maintained. Buller kept several, gradually weaning them from their natural foods, inducing them to eat, among other things, rice, potatoes and minced meat. He remarked (1888) in typically picturesque style:

> The captive Eagle frets in his sulky pride; the Bittern refuses food and dies untamable; the fluttering little Humming-bird beats itself to death against the tiny bars of its prison in its futile efforts to escape; and many species that appear to submit readily . . . ultimately pine, sicken, and die. There are other species, again, which cheerfully adapt themselves to their new life . . . Parrots, for example, are easily tamed . . . This character of tamability was exemplified to perfection in the Huias.

78 Two drawings of the Huia made by Laura Buller at the London Zoo in 1870.

Although in many respects so fragile, individuals withstood the rigours of transportation; at least one example was exhibited at the Gardens of the Zoological Society where it was drawn by Laura Buller during a stay in London in 1870 (Figure 78). A curious emotion must have stirred as the writer and his family watched – in its cage – this bird so familiar to them from the wooded ranges of their distant homeland.

One of the Huia's most intriguing aspects was the marked divergence in shape between the bill of male and female, an anomaly not recorded in any other bird species. John Gould, on receiving specimens in 1836, was so misled by the variance that he described two distinct species – *Neomorpha acutirostris* from the female, *Neomorpha crassirostris* from the male.

The beak of the male – thick, fairly straight and comparatively short – contrasts with the much longer, delicately formed and gracefully curved bill of the female.

Such a singular adaptation suggests cooperation between sexes in food gathering. Males used their powerful bills to break up rotting logs, attacking the wood almost in the manner of woodpeckers. Meanwhile, the following female extracted grubs and insects with her more delicate probe. Huhus, succulent grubs of a nocturnal beetle (*Prionoplus reticularis*), were especially favourite food items. These grubs, as large and plump as a man's little finger, were seized about the middle and transferred to a perch whereon the victim could be secured by the bird's foot. Hard parts were ripped off, the grub thrown upwards and caught lengthways in the bill – whereupon it would be swallowed whole. Another tasty dainty, the Weta (*Hemideina megacephala*), could be caught after the male had stripped the bark from saplings.

Despite this observed cooperation, females were never seen to share items they obtained – a female trait not completely without parallel, it might be argued. Possibly, the peculiarity in feeding technique allowed a more efficient use of territory. If the sexes, working in conjunction, could tap differing sources of food, then the feeding grounds they had to defend could be reduced.

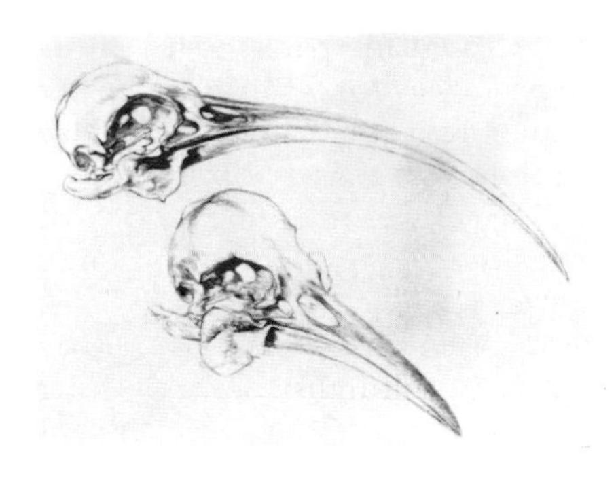

79 Huia skulls. Pencil drawing by Raymond Ching.

Clearly, some bond of mutual dependence was built by pairs. After the shooting or capture of a bird it was frequently noticed that its mate would show very evident signs of distress. Doubtless, many individuals suffered through staying to search for a captured or fallen mate, rather than making off as danger threatened.

The soft, fluting call of the Huia was something never forgotten by those who heard it. Often, it provided the first indication of the bird's presence. If within hearing distance, Huias could easily be lured by a decoy call, a device made great use of by Maori hunters. An imitation of the cry has been preserved in musical notation and perhaps this is the only extinct bird for which such a record exists. J.C. Andersen (1926) gives a distress call from which, incidentally, the species received its name,

and also a copy of the notes sung while searching for food.

According to Buller (1888):

> The Huia never leaves the shade of the forest. It moves along the ground, or from tree to tree, with surprising celerity by a series of bounds or jumps. In its flight it never rises, like other birds, above the tree-tops, except in the depth of the woods, when it happens to fly from one high tree to another . . . They are generally met with in pairs, but sometimes a party of four or more are found consorting together . . . This species builds its nest in hollow trees, forming it of dry grass, leaves, and the withered stems of herbaceous plants, carefully twined together in a circular form, and lined with softer materials of a similar kind.

The clutch consisted of two, three or four eggs; a supposed example, stone grey in colour with brown and dusty purple spots and blotches, is preserved in the Dominion Museum, Wellington. Nests containing young were located, all records apparently relating to the month of November.

The last word on the Huia is left to Buller whose writing evokes the spirit of an age and an avifauna now largely gone and forms a poignant memorial to a bird passed beyond our reach:

> I do not know of any more picturesque sight in the New Zealand woods – now, alas! the opportunities are becoming few and far between – than that of a small party of these handsome birds, playfully disporting themselves among the branches, in the intervals between their customary feeding times. Take for our purpose a dense piece of native vegetation . . . and furnish it, in imagination, with two pairs . . . They are hopping actively from branch to branch, and at short intervals balance themselves and spread to their full extent their broad white-tipped tails, as if in sheer delight; then the sexes meet for a moment to caress each other with their beautiful ivory bills, while they utter a low, whimpering love-note; and then, without any warning, as if moved by a sudden inspiration, they bound off in company, flying and leaping in succession, to some favourite feeding-place, far away in the silent depths of the forest.

HYPOTHETICAL SPECIES AND MYSTERY BIRDS

Among the thousands of scientifically described species of birds are a surprisingly large number about which virtually nothing is known. While more familiar species have been studied in depth over and over again, it might seem remarkable that information is quite inadequate for so many others. Of some species only a few details of habit and lifestyle are recorded and there are others still, known from nothing more than a handful of museum specimens, which, in a few instances, even lack precise locality data.

Reasons for these blanks in knowledge are not difficult to identify. If a species occurs only in a particularly inaccessible or inhospitable area, or even if its home territory just happens to be one seldom frequented by ornithologists, then, obviously, it is likely to receive less attention than others. If the species happens also to be excessively rare or localized, it may be overlooked almost entirely. Although such a bird may have a quite precise identity as a species and its affinities with other kinds be well understood, nothing more may be known of it. There are cases where it is likely that the knowledge gleaned from skins will never be added to, for exceptional rarity can easily give way to extinction.

Of this, the Choiseul Crested Pigeon (*Microgoura meeki*), known from little more than a series of skins taken by Walter Rothschild's collector A.S. Meek during January of 1904, provides a good example. Although the island of Choiseul in the Solomon group has been searched several times since the bird's discovery, no population of this large and handsome pigeon has been located; the species is now almost certainly extinct.

A species that until recently could have been written of in similar terms is the Yellow-fronted Bowerbird (*Amblyornis flavifrons*). Known from three – perhaps four – trade skins that came to notice during the nineteenth century, absolutely nothing was known of this bird apart from the appearance of the adult male. Even the exact place of origin was in doubt, the only certainty being that the skins came from New Guinea. The home grounds of this mysterious bird were searched for many times without success until in recent years they were located by an expedition to the

Table 22

NAMES UNDER WHICH BIRDS OF PARADISE, PRESUMED TO BE HYBRIDS, HAVE BEEN DESCRIBED

(1) Forms almost certainly of hybrid origin
Astrarchia barnesi
Paradisea apoda luptoni
Paradisea mixta
Paradisea bloodi
Paradisea maria
Paradisea duivenbodei
Rhipidornis gulielmi-tertii
Epimachus astrapioides

(2) Forms likely to be of hybrid origin
Parotia duivenbodei
Janthothorax mirabilis
Heteroptilorhis mantoui
Cicinnurus lyogyrus
Cicinnurus goodfellowi

(3) Forms possibly of hybrid origin but ones that may constitute valid species
Epimachus ellioti
Pseudastrapia lobata
Loborhamphus ptilorhis
Loborhamphus nobilis
Paryphephorus duivenbodei
Janthothorax bensbachi
Lamprothorax wilhelminae
Neoparadisea ruysi

Gauttier (Foja) Mountains. Despite the mystery surrounding them for so many years, *Amblyornis flavifrons* skins were so distinctive that the identity as a species was never seriously challenged; this applies to many other species known from similarly inadequate material.

Mystery birds

Skins of birds occasionally turn up that are even more problematical. These skins, while not directly referable to any known species, cannot be unquestionably accepted as a basis upon which to found new ones and, for a variety of reasons, the actual status of a considerable number of forms has never been satisfactorily determined. Almost certainly, many of these rather mysterious forms are the results of hybridization, but if a putative hybrid is a particularly uncommon phenomenon and also happens to be without marks incontrovertibly characteristic of two likely parent species, then positive designation is virtually impossible. In cases where a form is represented by just a single skin or a very small series of skins it is by no means easy to determine whether such material is referable to a very rare species, to the effects of hybridization or merely to some freak of nature. If the specimen or specimens are many years old and nothing similar has come to light, an additional possibility may be considered – the material might represent the only remains of a now extinct species.

Perhaps no family of birds so well highlights the difficulties associated with the interpretation of excessivly rare forms as the Paradiseidae – the Birds of Paradise. Because problematical bird skins are relevant to the subject of extinct birds some detail is necessary here.

There are just over forty generally accepted species within the family, but there remain twenty-four additional forms, the exact status of each of which is obscure. These constitute what have come to be known as the 'hybrid' or 'rare' birds of paradise (see Table 22). Almost all are known just from isolated and usually ageing museum specimens that as a rule lack precise locality data. Some forms are represented by only the type specimen (e.g. *Loborhamphus ptilorhis* and *Pseudastrapia lobata*); others, such as *Rhipidornis gulielmi-terti*, exist in several museum collections.

The circumstances under which many of these strange skins came to light are in themselves remarkable. The skins of certain kinds of birds were much in demand for use in the millinery and fashion industry during the latter part of the nineteenth century and the early years of the twentieth. Vast quantities arrived in Europe from all over the world and none were more in demand than birds of paradise. Among the crate-loads of commoner skins, anomalous specimens now and then turned up. Occasionally, these were brought to the attention of naturalists but usually they were not; many rare plumes perhaps became discarded as a hat aged or fashion changed. Most of the 'rare' birds of paradise have been known since the heyday of plume hunting and because they came to light exclusively as a result of that trade, these forms have never been observed by ornithologists upon their own home grounds.

Although almost all were originally described as distinct and valid species, the mystery surrounding them led commentators to speculate that hybridization may have been responsible for some. Undoubtedly this is correct. Many of the Paradiseidae show behavioural and morphological characteristics that render the production of occasional hybrids not unlikely; a number of species are highly polygynous and although male plumage may be quite divergent, close relationship is betrayed by the females, which are often remarkably alike. Erwin Stresemann (1930) postulated a hybrid origin for all of the eighteen 'rare' forms then known and since the time of his writing six additional kinds have come to notice.

Some of Stresemann's designations are clearly correct, others are more doubful but the likelihood of his speculation being accurate is high. This probability, taken together with Stresemann's enormous prestige as an ornithologist, led to all his assumptions being accepted almost without question, even though a number of forms exist in which the marks of hybridization he lists are by no means clear enough to justify positive designations. It is fair to say that several 'rare' forms appear to have been relegated to hybrid status largely as a result of others being demonstrably of such an origin.

Although these enigmatic bird forms might very well be hybrids, they might with equal validity be regarded as species in their own right – as indeed they were for many years before the Stresemann pronouncement. If this is so, and bearing in mind the long periods that have elapsed since they were last taken, such species might easily be extinct. Nor is it too strong to say that several of the 'rare' forms make highly unlikely hybrids for the parent species Stresemann proposed. To account for three of the most cryptic of the forms, Stresemann suggested that birds of the genus *Paradigalla* had mated with birds from highly polygynous genera (*Epimachus, Lophorina* and *Parotia*), an assumption that can be ruled out on ethological grounds alone, for the sexes of *Paradigalla* are alike, a trait generally assumed to be indicative of monogamous association. Whatever else they may be, these three rare forms cannot be accounted for in the generally accepted way.

As an instance of the cryptic nature of some of the 'rare' birds of paradise, the history of one particular form may be cited as fairly representative of others.

An example of a previously unrecorded long-tailed sickle-bill was received in 1872 by a London taxidermist, E. Ward, among a collection of newly imported bird skins. Typically, the specimen came without locality data. As one might expect from an unknown sickle-bill the bird's appearance was so singular that Ward realized he had acquired something special and exhibited and described it before the Zoological Society of London, naming it *Epimachus ellioti*. Apart from striking differences in plumage coloration, what really distinguishes *ellioti* from other long-tailed sickle-bills is its comparatively small size – perhaps one-third less than the more familiar Black Sickle-bill (*Epimachus fastuosus*).

Ward's bird was used as the model for an illustration by Joseph Wolf (Plate LVIII) reproduced in D.G. Elliot's *Monograph of the Paradiseidae* (1873) and then passed into the collection of John Gould, who later caused it to be figured for his own, partly posthumous, publication *The Birds of New Guinea* (1875–88). At Gould's death in 1881 this then-unique skin was acquired by the British Museum (Natural History) but 9 years later an additional example was obtained by the Dresden Museum from Dutch merchants in Ternate. This bird also came without locality data, its place of origin given, vaguely, as northwestern New Guinea. The English ornithologist R. Bowdler Sharpe recorded what may have been yet another specimen. This has now vanished but could in any case be the Dresden example.

By the close of the nineteenth century the form had acquired a very precise identity as a species and none of the ornithologists of this period considered it in any other terms; and these were people handling comparatively fresh specimens.

When Stresemann was critically evaluating 'rare' birds of paradise in 1930, it was natural that he should review the status of *E. ellioti*, but he designated it a hybrid for reasons that are not altogether clear. Stresemann's hypothesis seems to be founded upon the extreme rarity of specimens coupled with certain plumage characteristics alleged to be significant, but a variety of factors combine to make his conclusion questionable. The parent birds suggested are the Black Sickle-bill (*Epimachus fastuosus*) and the Arfak Astrapia (*Astrapia nigra*), both long-tailed species but otherwise quite different in form. A relationship with sickle-bills being apparent, the grounds for associating *ellioti* with *Astrapia* are crucial if it is to be accepted as an intergeneric hybrid; Stresemann connected the two largely on account of plumage coloration of the underparts – some red and some green is shown in both birds.

Taken at face value, this hybrid designation may or may not be considered conclusive but when critically examined it appears to be founded upon very tenuous connections. Although the lower breast of *ellioti* is certainly green this is in the nature of a wash, whereas *A. nigra* carries a very striking metallic green breast shield. Similarly, a faded red that shades broadly across the green underparts of *ellioti* is quite unlike the vivid, narrow, metallic red band that divides the breast of the proposed parent. In fact, the mysterious specimens show several characteristics not shared with either putative parent, the most telling of which are a distinct lobing of the gape and considerably smaller size.

What then can be made of *E. ellioti*? Is it a hybrid or an exceptionally rare species? A very localized population could conceivably remain hidden in such a formidable wilderness as New Guinea, but if *ellioti* actually constitutes a species it might easily be now extinct having perhaps failed to withstand competition from a close relative. The problems posed by this mysterious bird are fairly typical of those involved in the assessment of many excessively rare forms. The birds of paradise provide possibly the most spectacular examples of problematical birds, but similar cases exist in other bird groups. Assigned to the order Charadriiformes, for instance, are two distinct kinds of bird, both originally described as full species from unique specimens and named accordingly. These are the Auckland Island Shore Plover (*Thinornis rossi*) and Cooper's Sandpiper (*Pisobia cooperi*). These birds were collected during the first half of the nineteenth century and doubts about the validity of specific rank have been expressed ever since, especially as no similar examples have been taken. Either or both may constitute freaks, hybrids (this seems unlikely in the case of *T. rossi*) or valid species, which, bearing in mind the early date of collection, would probably now be extinct.

Among the diurnal raptors (order Falconiformes) is listed an enigmatic form of a rather different kind, Kleinschmidt's Falcon. Variously regarded as a full species (*Falco kreyenborgi*), a valid subspecies of the Peregrine (*F. peregrinus kreyenborgi*) or just a rare colour morph of the South African race (*F. peregrinus cassini*), the form is known not just from skins but from living birds kept at the Münster Zoo. Although there may be only small reason to suppose it an extinct species, Kleinschmidt's Falcon provides an indication of how difficult it can be to evaluate even living individuals of little-known forms.

Many of the problems presented by these birds and other similar examples will never be satisfactorily resolved. Under the headings of appropriate bird groups in the main part of the book are listed other enigmatic forms that may have relevance to the subject of extinct birds.

LVIII Elliot's Bird of Paradise, a hybrid or a vanished species? Hand-coloured lithograph by Joseph Wolf and Joseph Smit from D.G. Elliot's *Monograph of the Paradiseidae* (London, 1873), Pl.20.

Hypothetical species

Rather different are the difficulties posed by those described species that can most usefully be called 'hypothetical extinct birds'. These are birds that have gained some place in ornithological literature but of which no satisfactory specimen exists. Many are forms resurrected by nineteenth- or twentieth-century naturalists from the pages of early travel books, and by these later authors given names to which either less or more detailed scientific descriptions were appended.

When these descriptions are based upon such narratives, dependence on their accuracy must be weighed against the fact that primary sources were often put together by writers lacking ornithological expertise and whose chief interest in birds lay in whether or not they were good to eat. The problems associated with them are therefore self-evident. Travellers may have completely made up birds in order to fill out their narratives, or may have written of birds that they knew only through hearsay. Even assuming a description to be truthful in spirit, it may nevertheless be thoroughly inaccurate and any conclusions drawn from it irrelevant. On the other hand, a recorded observation may be accurate enough yet still misleading or liable to misinterpretation. A vague description can quite easily apply to some well-known species with which the author of the original source was unfamiliar.

Hypothetical birds are almost always assumed to be extinct but it is clear that the validity of their definition as species should be viewed with a certain amount of scepticism. It seems surprising that so many have been formally named because the names proposed have no more meaning than those that have been given to the Loch Ness Monster, the Great Serpent of the Sea or the Abominable Snowman. This is not to say that such creatures do not exist or have not existed, but is recognition that little purpose is served by the attachment of precise scientific labels to what are at present unknown quantities.

Species described from fragments of bone or based on very slight variation among skeletal remains present different but connected problems and some of these are perhaps also best considered as hypothetical. It is not necessarily an easy matter to assign successfully a single bone to a known species – and far more risky to found a new one upon such material. Palaeontologists have to cope regularly with these kind of problems, and make whatever they can from insubstantial remains; but forms or populations that seem to have become recently extinct and have been described from very incomplete matter are especially difficult to assess.

Without question, spectacular deductions have been made from very flimsy material. Richard Owen's celebrated statement in 1839 that 'struthious' birds were living, or had at one time lived, in New Zealand, was given on the strength of an examination of a remarkably small amount of bone (see page 20). But subsequent and more complete finds of moas proved the deduction to have been amazingly accurate. For all the other faults that seem to have afflicted Owen, he was a very great comparative anatomist and few others have been endowed with similar talent. The formal naming of a new species from a single bone is not necessarily an exercise of value, and indeed Owen himself made no attempt to do this in the case of the moas. One problem is that remains belonging to an undescribed species might be assigned to the wrong group. Another is that a fragment assumed to come from an otherwise unknown form may in reality be from a familiar one.

Extensive collections of skeletons are uncommon and if osseous remains are not compared with a wide range of material, the true identity may stay hidden. Even with a part as distinctive as a skull, mistakes of this kind can be made. W.R.B. Oliver (1955) described from a skull found in New Zealand *Cereopsis novaezealandiae*, a putative species, ostensibly extinct, closely related to the Australian Cape Barren Goose (*C. novaehollandiae*). Later, it was discovered that this skull, found during the nineteenth century, was identical to skulls of Australian birds. Since individuals belonging to the Australian species were taken across the Tasman Sea to New Zealand and liberated on various occasions, it is not difficult to understand how such a find could come to be made there. If mistakes are made with skulls, it follows that they are even more likely to occur when less-distinct fragments of bone are considered.

A surprisingly large number of hypothetical species have been proposed, by various authors, as former inhabitants of the Mascarene Islands of Mauritius, Réunion and Rodrigues. Some are based upon travellers' tales or supposed variations among them, others are founded upon inconclusive osteological evidence.

A. Newton and H. Gadow in 1893 identified a single right ulna found in the Mare aux Songes, Mauritius as that of a grebe. M. Hachisuka (1953) considered this provided enough evidence to warrant naming a species and this he called *Popiceps gadowi*. Whether or not there was ever a distinct form of grebe on the island cannot be said. The length of the bone, 82 mm ($3\frac{1}{4}$ in), is claimed by J.C. Greenway (1958) to fall within the range of an African species; but even Greenway's meaning is unclear for the species he cites, *Podiceps auritus*, is not an African bird at all but a grebe of North America and northern Asia and Europe. However, unless more evidence is forthcoming, it cannot be assumed that Mauritius had its own distinct grebe, and the validity of *gadowi* is open to question.

Another perhaps invalid species founded, like *Podiceps gadowi*, upon remains that may prove attributable to an extant African species, was named *Astur alphonsi* by Newton and Gadow in 1893. As speculated, the remains could conceivably be those of an otherwise unknown extinct species, but can with equal justification be considered the bones of the Great Sparrow Hawk (*Accipiter malanoleucus*), a quite widespread African bird of prey.

Skeletal remains and the records of naturalists indicate that at least two species of owl were once endemic to Mauritius but are now extinct. These have been given the names *Scops commersoni* and *Strix sauzieri*. Newton and Gadow described *sauzieri* in 1893 from a series of metatarsi, tibiae and humeri. To account for certain size differentials among the metatarsi, Walter Rothschild (1907) proposed the additional name *Strix newtoni* but Greenway (1958) considered that differences in proportion would be quite within the range of expected individual variation.

Two species of owl have been proposed as former inhabitants of Rodrigues. The existence of *Athene murivora* in the quite recent past is supported by skeletal evidence and, perhaps, a description written during the eighteenth century. The suggestion that a second species of owl was endemic to the island is much less certain. A tibiotarsus from Rodrigues, once part of the Rothschild collection, was used to found the supposed species *Bubo leguati*. Rothschild proposed this name with the claim that François Leguat, an early Huguenot castaway upon the island (see page 127), was the first to mention owls on Rodrigues. Both Hachisuka (1953) and Greenway (1958) agree that Leguat's written account of his stay contains no mention of any such birds and Greenway added that the single bone upon which this supposed species is based cannot with confidence be assigned to any particular genus.

More than one species of heron alleged to have inhabited the Mascarenes has been described scientifically upon the basis of flimsy material and the merits of these are discussed in the account for *Nycticorax megacephalus* (page 42).

A very large and spectacular bird may have lived on Mauritius until the end of the seventeenth century. In 1858, H. Schlegel named it *Leguatia gigantea*, since which time a considerable ornithological literature has become attached to it. Schlegel's description is based upon Leguat's narrative published in 1707. Many of Leguat's observations seem reliable so there may be some foundation of truth to his description and illustration of the Géan, a long-legged, long-necked bird, 1.8 m (6 ft) tall, which allegedly inhabited the marshes of Mauritius (Figure 80). Difficulty arises because other early travellers made lists and gave accounts of birds that they saw upon the island but none mentions any unknown bird that tallies with Leguat's description.

It seems unlikely that all could have overlooked a bird so large and striking as this surely was. K. Lambrecht (1933) called *Leguatia gigantea*

80 A large bird supposedly seen in Mauritius in the late 1600s by the Huguenot refugee François Leguat. Engraving from Leguat's *New Voyage to the East Indies* (London, 1708).

'Ein erdachter Vogel' – a dreamed-up bird (Plate LIX). This, of course, it may be, but it is not impossible that Leguat saw something quite familiar – a flamingo perhaps – and failed to observe or describe it with accuracy. Hachisuka (1953) advanced reasons why this could not be so but none seems particularly convincing and flamingos were certainly recorded in the Mascarenes by early travellers. Greenway (1958) rejected *L. gigantea* as a valid species but other commentators, including Hachisuka (1953) and Rothschild (1907), have assumed that Leguat really did observe an otherwise unknown bird. Generally, those who have upheld *L. gigantea* as a valid species, identify it as a gigantic long-legged rail or some sort of crane.

Several coloured illustrations of this alleged species were made but although these may have worth as works of the imagination their scientific value is questionable. The spirit in which two of these pictures were put together highlights some of the confusion surrounding hypothetical species. For plate 31 of *Extinct Birds* Walter Rothschild commissioned F.W. Frohawk to produce a picture based upon the early record. One of the features included is black tipping to the otherwise white primary feathers. As grounds for this particular piece of decoration seem to be lacking, Hachisuka in *The Dodo and Kindred Birds* complained, 'I am at a loss to know upon what basis Lord Rothschild has coloured the tip of the wings black, especially since Leguat specifically informs us that the bird is altogether white'. At the bottom of the page on which this sentence occurs (p. 165), Hachisuka was obliged to make certain confessions regarding the plate (pl. 18) in his own book. A footnote reads: 'Through the carelessness of my supervision the wing-tip was coloured black; it should be entirely white.'

Another hypothetical species, *Kuina mundyi*, which would if valid probably be a rail, was proposed by Hachisuka in 1937 because of differences in old accounts, generally supposed to refer to the Mauritius Red Hen (*Aphanapteryx bonasia*) (see page 87). The early traveller Peter Mundy wrote a description at some variance with those written by his near contemporaries, but it is difficult to interpret what significance might lie in these differences. Hachisuka makes much of the colour Mundy mentioned, a 'yellowish wheaten', whereas the general impression had come down of a reddish-brown bird. Assessments of colour in old writings are notoriously unreliable and the colour described might fall within a considerable range. Although it is apparent from the tone of his work that Mundy was a truthful man, there is evidence to suggest that his remarks on colour are particularly misleading. If real differences did exist among these Mauritian rails, it seems likely that they could be accounted for by age or sex and Mundy, in fact, saw only one individual. Two large, closely related rails may have inhabited Mauritius but this seems very doubtful.

Pezocrex (Didus) herberti is in many respects a comparable form. This doubtful species is based upon a seventeenth-century drawing by Sir Thomas Herbert (see Figure 24), but it does not exactly correspond to any other early drawing of a Mauritius bird. Since the picture is extremely primitive, this is hardly surprising. Schlegel (1854) believed it represented a kind of dodo, Hachisuka (1953) an unknown rail; but it seems almost certain that the illustration is meant to show the Mauritius Red Hen.

The name *Apterornis coerulescens*, applied to a hypothetical species from Réunion, often occurs in lists of extinct birds (Figure 81). The form was first given a scientific identity by Baron de Sélys Longchamps in 1848, who based his description upon an account contained in the manuscript of

LIX *Leguatia gigantea*, a dreamed-up bird? Chromolithograph after a painting by F.W. Frohawk from W. Rothschild's *Extinct Birds* (London, 1907), Pl.31. Courtesy of The Hon. Miriam Rothschild.

81 A reconstruction of a hypothetical species from Réunion, *Apterornis coerulescens*. Chromolithograph after a painting by J.G. Keulemans from W. Rothschild's *Extinct Birds* (London, 1907), Pl.32. Courtesy of The Hon. Miriam Rothschild.

the traveller Dubois relating to the years 1669–72. 'Le Sieur' Dubois wrote of flightless blue birds with red bills and feet that attained large size and were good to eat. From this very spare entry has grown a considerable literature.

It is difficult to understand how conclusions have been drawn from such flimsy material, but researchers seem to agree that, if Dubois's description actually does refer to an otherwise unknown species, such a bird would have been a rail, most probably related or belonging to the genus *Porphyrio*. Following such an assumption, Walter Rothschild commissioned J.G. Keulemans to illustrate the supposed species as plate 32 of *Extinct Birds*. The painting depicts a creature looking very much like a slimmed-down Takahe (*Notornis mantelli*), but, although a well-produced piece of work, it is almost totally conjectural.

A brief description by Dubois of russet pigeons inhabiting Réunion is the basis of Rothschild's *Nesoenas duboisi*, and another from this source provoked the same author into naming a parrot *Necropsittacus borbonicus*, also allegedly an inhabitant of the island. The original descriptions may be perfectly accurate, or they may not. Greenway (1958) pointed out that the parrot, said to have been green with the head and upper parts of the wings the colour of fire, might have been an escaped lory. Another parrot named by Rothschild *Necropsittacus francicus* is based on even less certain material. The species is supposed to have inhabited Mauritius and, according to Rothschild, was mentioned by early travellers. It is not certain which original sources he had in mind.

In E.L. Daubenton's *Planches Enluminées* (1765–81) is a picture of a bird resembling the Madagascan Red Fody (*Foudia madagascariensis*) but showing some conspicuous differences. Associated with this illustration, supposed to depict a bird found in Réunion, is the name *Foudia* or *Fringilla bruante*. Whether there ever was a population of birds inhabiting Réunion and distinct from those of Madagascar cannot be said. Daubenton's plates are not always to be relied upon for their accuracy and it seems at least as likely that the mysterious *Foudia* of Réunion was introduced from Madagascar, as that a distinct species occurred there.

The number of hypothetical forms from the Mascarenes is out of all proportion to those described from the rest of the world. Since these islands have suffered such appalling losses in recent times, the Mascarenes have attracted attention from ornithologists fascinated by the rare and the curious. Presumably, it is the very nature of this interest that has provoked so many into naming doubtful species.

Undoubtedly, hypothetical species with just as much claim to validity could be described from elsewhere and, occasionally, have been. Greenway (1958), for instance, lists *Dysmorodrepanis munroi* from Lanai in the Hawaiian Islands and *Pterodroma hindwoodi* from Norfolk Island as hypothetical. To these might be added the Painted Vulture (*Sarcorhamphus sacra*), alleged to have been a close but distinct relative of the King Vulture (*S. papa*), and once to have been an inhabitant of Florida.

For the most part, however, ornithologists studying avifaunas other than that of the Mascarenes have been more restrained in naming species. The *roa roa*, the mysterious Great Kiwi of New Zealand, might easily have been named by an over-zealous researcher, but this appears never to have happened; other similar instances might be cited. Only from the West Indies have extinct hypothetical birds been described on anything like the same scale; the validity of these is assessed together with other mysterious parrots in the introduction to the order Psittaciformes (page 131).

BIOGRAPHICAL NOTES

Audubon, John James (1785–1851)
Painter and backwoodsman, great romantic hero of American natural history. Audubon was born in New Orleans of a French father and Creole mother. He spent most of his childhood in France (during the turbulent years of the revolution) where he is said to have studied briefly under the great painter Jacques Louis David. He returned to America in 1803, destined to revolutionize the art of bird painting and produce, in four volumes, the most valuable of all natural history books – the double-elephant folio *Birds of America*, sets of which have sold for more than £1,000,000. Audubon spent much of his life in the wild places of America searching for specimens from which to paint the designs for the 435 plates of his masterwork.

Banks, Sir Joseph (1743–1820)
English naturalist, disciple of Linnaeus, collector, wealthy landowner and President of the Royal Society for 43 years. Banks travelled in HMS *Endeavour* on the first of Captain Cook's epic circumnavigations of the globe in 1768–71 and it is for this voyage that he is chiefly remembered. Later, he took control of the Royal Botanic Gardens at Kew and raised them to their position of worldwide pre-eminence. He was made a baronet in 1781. His library and collections passed to the British Museum.

Buller, Sir Walter Lawry (1838–1906)
Son of a Cornish missionary, Buller was born in New Zealand and by the age of 17 was fluent enough in Maori to be appointed Goverment Interpreter at Wellington. He was made a Justice of the Peace in 1862 and had a lifelong involvement with negotiations over Maori land claims. He was decorated for gallantry in the Maori wars and then began, increasingly, to devote himself to natural history studies; both the first and the second editions of his *A History of the Birds of New Zealand* (these can, with justice, be considered quite separate works) are among the finest bird books of the nineteenth century. In 1885, he was appointed New Zealand Commissioner at the Colonial and Indian Exhibition and was knighted a year later. Buller came to London in 1898 and spent the last 8 years of his life in England.

Cook, Captain James (1728–79)
English navigator, born in Whitby, Yorkshire. Ran away to sea at an early age and in 1755 joined the Royal Navy. Quickly rising through the ranks, he was appointed to lead the expedition to chart the transit of the planet Venus. Circumnavigated the globe aboard the *Endeavour* (1768–71), then again in the *Resolution*. During what would have been his third voyage of discovery, begun in 1776, Cook was treacherously murdered by Hawaiian natives at Karakakoa Bay, Hawaii on 14 February 1779.

Erxleben, James (*c*.1830–80)
A little-known but phenomenally gifted scientific illustrator. His forté was the production of beautiful lithographs showing the skeletal anatomy of all kinds of rare and curious creatures and to this work he brought an almost unbelievable technical skill. He enjoyed a long association with Sir Richard Owen and produced hundreds of lithographed plates for Owen's various natural history monographs.

Keulemans, John Gerrard (1842–1912)
Probably the most skilled, almost certainly the most prolific, ornithological illustrator of his day. Born in Holland, he left his home country at the age of 27 and settled in England where he spent the rest of his life. He contributed the plates for over 100 major ornithological volumes and produced many hundreds of other illustrations for scientific journals of the period.

Latham, John (1740–1837)
Long-lived British ornithologist who wrote a *General Synopsis of Birds* (1781–5) in which he attempted to describe every bird species then known. He gave his birds English names rather than scientific ones in Latin and this allowed the German zoologist J.F. Gmelin to profit from Latham's work; Gmelin simply copied Latham's descriptions and attached to them Latin names of his own. Latham etched all the copper plates for the engravings in his original work.

Owen, Sir Richard (1804–92)
The greatest comparative anatomist of his time. Among his more celebrated achievements were the introduction of the Moa to the scientific world and the coining of the word 'dinosaur'. He was knighted in 1884 having, apparently, turned down the honour in 1845. During his life he wrote a vast number of papers and learned treatises on all kinds of living and fossil creatures. He was instrumental in the setting up of the British Museum of Natural History in London's South Kensington area, having been Superintendent of the Department of Natural History at the British Museum in Bloomsbury. A great favourite of Queen Victoria.

Rothschild, Lionel Walter (2nd Baron Rothschild) (1868–1937)
Collector, naturalist, politician, great patron of natural history, who was instrumental in the production of the Balfour Declaration. Walter Rothschild succeeded to the baronetcy on the death of his father in 1915, turned his back on a career in banking and devoted his life to the love of natural history. Used his immense financial resources to commission collectors and agents to scour the world in search of specimens and knowledge. Commissioned or inspired many important papers and works, and himself contributed hundreds of papers to the leading scientific journals of his day; also produced several important and beautiful books. Assembled a vast library and an equally important collection of natural history specimens; much was bequeathed to the British people at his death, together with the building that housed it.

BIBLIOGRAPHY

Abbott, C.G. 1933. Closing history of the Guadalupe Caracara. *Condor* **35**, 10–14.

Ali, S. and Ripley, S.D. 1968–74. *Handbook of the Birds of India and Pakistan*, 10 vols. Oxford University Press: Bombay and London.

Allen, J.A. 1876. Extinction of the Great Auk on Funk Island. *American Naturalist* **10**, 48.

Allingham, E.G. 1924. *A Romance of the Rostrum; being the Business Life of Henry Stevens, and the History of Thirty-eight King Street.* H.F. and G. Witherby: London.

Amadon, D. 1947. An estimated weight of the largest known bird. *Condor* **49**, 159–64.

1950. The Hawaiian Honeycreepers. *Bulletin of the American Museum of Natural History* **95**, 157–257.

Andersen, J.C. 1926. *Bird Song and New Zealand Song Birds.* Whitcombe and Tombs: Auckland.

Anon. 1789. *The Voyage of Governor Phillip to Botany Bay.* London.

1891. *Relation de l'Île Rodrigue.* Hakluyt Society: London.

Anthony, A.W. 1898. Petrels of southern California. *Auk* **15**, 140–4.

1925. Birds and Mammals of Guadalupe Island. *Proceedings of the California Academy of Sciences*, ser. 4, **14**, 277–320.

Archey, G. 1941. The Moa, a study of the Dinornithiformes. *Bulletin of the Auckland Institute and Museum* No. 1, 5–101.

Audubon, J.J. 1827–38. *The Birds of America*, 4 vols. London.

1831–9 *Ornithological Biography*, 5 vols. A. & C. Black: Edinburgh.

Austin, O.L. and Singer, A. 1963. *Birds of the World.* Hamlyn: London.

Bailey, A.M. 1956. *Birds of Midway and Laysan Islands.* Museum of Natural History: Denver.

Baker, E.C.S. 1908. *Indian Ducks and their Allies.* Bombay Natural History Society: London.

1922–30. *The Fauna of British India*, 8 vols. Taylor and Francis: London.

Baker, R.H. 1951. The avifauna of Micronesia, its origin, evolution and distribution. *University of Kansas Publications of the Museum of Natural History* No.3, 1–359.

Baldwin, P.H. 1949. The life history of the Laysan Rail. *Condor* **51**, 14–21.

Bannerman, D.A. 1930–51. *The Birds of Tropical West Africa*, 8 vols. Oliver and Boyd: London.

1958. *The Birds of Cyprus.* Oliver and Boyd: London.

1963–8. *The Birds of the Atlantic Islands*, 4 vols. Oliver and Boyd: London.

Bannerman, D.A. and Lodge, G.E. 1953–63. *The Birds of the British Isles*, 12 vols. Oliver and Boyd: London.

Barbour, T. 1943. *Cuban Ornithology.* Nuttall Ornithological Club: Cambridge, Mass.

Beebe, W. 1918–22. *A Monograph of the Pheasants*, 4 vols. H.F. and G. Witherby: London.

Beechey, F.W. 1839. *The Zoology of Captain Beechey's Voyage to the Pacific . . . in the Blossom . . . 1825–1828.* London.

Bendire, C.E. 1892–5. Life histories of North American birds, 2 vols. *US National Museum Special Bulletin* No. 1.

Bent, A.C. 1922. Life histories of North American petrels and pelicans and their allies. *US National Museum Bulletin* No. 121.

1923–5. Life histories of North American Wildfowl, 2 vols. *US National Museum Bulletin* Nos. 126 and 130.

1932. Life histories of North American gallinaceous birds. *US National Museum Bulletin* No. 162.

1937–8. Life histories of North American birds of prey, 2 vols. *US National Museum Bulletin* Nos. 167 and 170.

1940. Life histories of North American cuckoos, goatsuckers, hummingbirds and their allies. *US National Museum Bulletin* No. 176.

Berger, A.J. 1972. *Hawaiian Birdlife.* University Press of Hawaii: Honolulu.

Berlioz, J. 1935. Notice sur les specimens naturalisés d'oiseaux éteint existant dans les collections du Muséum. *Archives du Muséum d'Histoire Naturelle, Paris*, ser. 6, **4**, 485–95.

1946. Oiseaux de la Réunion. *Faune de l'Empire Français* **55**, 1–81.

Blanford, W.T. 1898. *The Fauna of British India including Ceylon, and Burma: Birds.* Taylor and Francis: London.

Blasius, W. 1884. Zur geschichte der uberreste von *Alca impennis. Journal für Ornithologie* **4**, 58–176.

Bonaparte, C.L. and Schlegel, H. 1850. *Monographie des Loxiens.* Paris.

Bond, J. 1936. *Birds of the West Indies.* Academy of Natural Sciences: Philadelphia.

Bontekoe, V.I. 1646. *Iovrnael ofte Gedenckwaerdige beschrijvinghe vande Oost-Indische Reyse van Villem Ijsbrantszoon Bontekoe van Hoorn.* Hoorn.

Bouillard, M. 1875. Relation de l'Île Rodrigues. *Proceedings of the Zoological Society of London,* 39–42.

Bowman, R. 1961. Morphological differentiation and adaptation in the Galapagos finches. *University of California Zoological Publications* No. 58.

Brewster, W. 1889a. Present status of the wild pigeon. *Auk* **6**, 285–91.

1889b. Nesting habits of the Carolina Parakeet. *Auk* **6**, 336–7.

Brown, L. and Amadon, D. 1968. *Eagles, Hawks and Falcons of the World.* Country Life: London.

Brush, A.H. 1976. Waterfowl feather proteins: analysis of use in taxonomic studies. *Journal of Zoology* **179**, 467–98.

Bry, T. and Bry, J.J. 1601. *Variorum Navigationis.* Amsterdam.

Bryan, W.A. 1901. *A Key to the Birds of the Hawaiian Group.* Bishop Museum Press: Honolulu.

1908. Some birds of Molokai. *Occasional Papers of the Bernice P. Bishop Museum, Honolulu* No. 4, 133–76.

Bryant, W.E. 1887. Additions to the ornithology of Guadalupe Island. *Bulletin of the California Academy of Science* **6**, 269–318.

Bucknill, J.A. 1924. The disappearance of the Pink-headed Duck. *Ibis*, ser. 11, **6**, 146–51.

Buick, T.L. 1931. *The Mystery of the Moa.* Thomas Avery: New Plymouth.

1936. *The Discovery of Dinornis.* Thomas Avery: New Plymouth.

1937. *The Moa-hunters of New Zealand.* Thomas Avery: New Plymouth.

Buller, W.L. 1872–3. *A History of the Birds of New Zealand.* John van Voorst: London.

1882. *Manual of the Birds of New Zealand.* Colonial Museum and Geological Survey Department: Wellington.

1887–8. *A History of the Birds of New Zealand*, 2 vols. Second edn. Published (for the subscribers) by the author: London.
1895. Stephen Island Wren. *Ibis*, ser. 7, **1**, 236.
1904. The Laughing Owl. *Ibis*, ser. 8, **4**, 639.
1905. *Supplement to the 'Birds of New Zealand'*, 2 vols. Published (for the subscribers) by the author: London.
Burger, R., Ducate, K., Robinson, K. and Walter, H. 1975. Radiocarbon date for the largest extinct bird. *Nature* **258**, 709.
Burton, J.A. 1973. *Owls of the World*. Peter Lowe: London.
Burton, P.J.K. 1974. Anatomy of the head and neck of the Huia with comparative notes on other Callaeidae. *Bulletin of the British Museum (Natural History) Zoology* **27**(1), 1–48.
Byron, Lord. 1827. *Voyage of H.M.S. Blonde to the Sandwich Islands in the Years 1824–1825*. John Murray: London.

Cade, T.J. and Digby, R.D. 1982. *Falcons of the World*. Collins. London.
Caldwell, J. 1876. Notes on the zoology of Rodriguez. *Proceedings of the Zoological Society of London* [1875], 644–7.
Carré, M. 1669. *Voyage des Indes Orientales*. Paris.
Cartier, J. (ed. Michelant, M.H.) 1865. *Voyage de Jacques Cartier au Canada en 1534*. Paris.
Cauche, F. 1651. *Relations véritables et curieuses de l'Île de Madagascar et Brésil*. Paris.
Chalmers-Hunt, J.M. 1976. *Natural History Auctions, 1700–1972*. Sotheby Parke Bernet: London.
Chilton, C. (ed.) 1909. *The Sub-antarctic Islands of New Zealand*, 2 vols. Philosophical Institution of Canterbury: Canterbury.
Chisholm, A.H. 1922. *Mateship with Birds*. Whitcombe and Tombs: Melbourne.
Clark, A.H. 1905. Lesser Antillean Macaws. *Auk* **22**, 266–73.
1908. The Macaw of Dominica. *Auk* **25**, 309–11.
Colenso, W. 1843. Some enormous fossil bones lately discovered in New Zealand. *Tasmanian Journal of Science* **2**, 31–105.
1879. On the Moa. *Transactions of the New Zealand Institute* **12**, 63–108.
Cory, C.B. 1880–3. *The Beautiful and Curious Birds of the World*. Boston.
Coues, E. 1874. Birds of the Northwest. *US Geological Survey Miscellaneous Publications* No. 3.
Cracraft, J. 1974. Phylogeny and evolution of the ratite birds. *Ibis* **115**, 494–521.
1976. The species of moas. *Smithsonian Contributions to Paleobiology* No. 27, 189–205.

Darwin, C. 1839. *Narrative of the Surveying Voyages of Her Majesty's Ships Adventure and Beagle between the years 1826 and 1836*. London.
Daubenton, E.L. 1765–81. *Planches Enluminée*. Paris.
Day, D. 1981. *The Doomsday Book of Animals*. Ebury Press: London.
Delacour, J. 1932. Les oiseaux de la mission zoologique . . . à Madagascar. *Revue Française d'Ornithologie* **2**, 1–96.
1977. *The Pheasants of the World*. Spur Publications: Hindhead, Surrey.
Delacour J. and Amadon, D. 1973. *Curassows and Related Birds*. New York: American Museum of Natural History.
Delacour, J. and Scott, P. 1954–64. *The Waterfowl of the World*, 4 vols. Country Life: London:
Des Murs, M.A.P.O. 1849. *Iconographie Ornithologique*. Paris.
D'Essling, M. le Prince, Duc de Rivoli. 1846. *Catalogue de la Magnifique Collection d'Oiseaux de M. le Prince D'Essling, Duc de Rivoli*. Paris.
De Vis, C.W. 1884. The Moa in Australia. *Proceedings of the Royal Society of Queensland* **1**, 23–8.
Dickerman, R.W. 1965. The Slender-billed Grackle *Scaphidurus palustris*. *Auk* **82**, 268.
Dill, H.R. and Bryan, W.A. 1912. Report of an expedition to Laysan Island in 1911. *US Biological Survey Bulletin* No. 42, 1–30.
Dole, S.B. 1879. List of birds of the Hawaiian Islands. In: Thrum, T.G., *Hawaiian Almanac and Annual*. Honolulu.
Dresser, H.E. 1871–96. *A History of the Birds of Europe*, 8 vols and supplement. London.
Dubois, 'Le Sieur'. 1674. *Les voyages faits par le Sieur D.B. aux Îsles Dauphine ou Madagascar et Bourbon ou Mascarenne, és années 1669, 1670, 1671, 1672*. Paris.
Duff, R.S. 1949. Moas and Man. *Antiquity* **23**, 172–9.
Dumont D'Urville, J.S.C. 1830–5. *Voyage de décôuvertes de l'Astrolabe exécuté par ordre du Roi, pendant les années 1826, 1827, 1828 et 1829, sous le commandement de M.J. Dumont D'Urville*, 4 vols. J. Tastu: Paris
1842–54. *Voyage au Pole Sud et dans l'océanie sur les corvettes l'Astrolabe et la Zelée; exécuté par ordre du Roi pendant les années 1837, 1838, 1839 et 1840 sous le commandement de M.J. Dumont D'Urville*, 7 vols. Paris.
Dutcher, W. 1891. The Labrador Duck, a revised list of extant specimens in North America. *Auk* **8**, 301–16.
Du Tertre, J.B. 1667. *Histoire générale des Antilles habitées par les François*. Paris.

Eaton, E.H. 1910–14. The Birds of New York State. *New York State Museum Memoir* No. 12.
Elliot, D.G. 1866–9. *The New and Heretofore Unfigured Species of the Birds of North America*, 2 vols. New York.
1870–2. *A Monograph of the Phasianidae*, 2 vols. New York.
1873. *A Monograph of the Paradiseidae*. London.
1895. *North American Shorebirds*. New York.

Falla, R.A., Sibson, R.B. and Turbott, E.G. 1979. *The New Guide to the Birds of New Zealand and Outlying Islands*. Collins: Auckland.
Feduccia, A. 1980. *The Age of Birds*. Harvard University Press: Cambridge, Mass.
Finn, F. 1915. *Indian Sporting Birds*. F. Edwards: London.
Finsch. O. 1867–8. *Die Papageien*, 2 vols. Leiden.
1881. Ornithological letters from the Pacific. *Ibis*, ser. 4, **5**, 102–14.
Finsch, O. and Hartlaub, G. 1867. *Beitrag zur Fauna Centralpolynesiens*. Halle.
Fisher, H.I. and Baldwin, P.H. 1945, A recent trip to Midway Islands, Pacific Ocean. *Elepaio, Journal of the Honolulu Audubon Society* **6**, 11–16.
1946. War and the birds of Midway Atoll. *Condor* **48**, 3–15.
Fisher, J., Simon, N. and Vincent, J. 1969. *The Red Book: Wildlife in Danger*. Collins: London.
Fisher, W.K. 1906. Birds of Laysan and the Leeward Islands. *Bulletin of the US Fish Commission* No. 23 (pt 3), 767–807.
Flacourt, Étienne de. 1658. *Histoire de la Grande Isle Madagascar*. Paris.
Fleming, C.A. and Lodge, G.E. 1983. *George Edward Lodge, Unpublished Bird Paintings*. Michael Joseph: London.
Fleming, J.H. 1915. On the Piopio. *Proceedings of the Biological Society, Washington* **28**, 121.
Forbes, H.O. 1893. Birds inhabiting the Chatham Islands. *Ibis*, ser. 6, **5**, 521–45.

1898. On a species from the Mascarene Islands, provisionally referred to *Necropsar*. *Bulletin of the Liverpool Museum* **1**, 29–35.

Forshaw, J. and Cooper, W.T. 1973 (1978). *Parrots of the World*. Lansdowne: Melbourne.

1981. *Australian Parrots*. Second edn. Lansdowne: Melbourne.

Forster, J.R. (ed. H. Lichtenstein) 1844. *Descriptiones animalium quae in itinere ad Maris Australis Terras per annos 1772, 1773 et 1774 suscepto*. Berlin Academy: Berlin.

(ed. M. Hoare) 1982. *The Resolution Journal of Johann Reinhold Forster*. London: Hakluyt Society.

Frauenfeld, G.R. von. 1868. *Neu aufgefundene Abbildung des Dronte und eines zweiten kurzflugeligen Vogels, wahrscheinlich des poule rouge au bec de bécasse der Maskarenen in der Privatbibliothek S.M. des verstorbenen Kaisers Franz*. Wien.

Frith, H.J. 1967. *Waterfowl in Australia*. Angus and Robertson: Sydney.

1982. *Pigeons and Doves of Australia*. Rigby: Adelaide.

Frohawk, F. 1892. Description of a new species of rail from Laysan Island (North Pacific). *Annals and Magazine of Natural History* **9**, 247–9.

Fuller, E. 1979. Hybridization amongst the Paradisaeidae. *Bulletin of the British Ornithologists' Club* **99** (4), 145–52.

Galbraith, I.C.J. and Galbraith, E.H. 1962. Land birds of Guadalcanal and the San Cristoval Group, eastern Solomon Islands. *Bulletin of the British Museum (Natural History) Zoology* **9** (1).

Garrod, A.H. 1875. On the form of the lower larynx in certain species of ducks. *Proceedings of the Zoological Society of London*, 151–6.

Gill, F.B. 1967. Birds of Rodrigues Island. *Ibis* **109**, 383–90.

Gilliard, E.T. 1969. *Birds of Paradise and Bower Birds*. Weidenfeld and Nicolson: London.

Gmelin, J.F. 1789. *Systema Naturae*. Leipzig.

Godman, F. du C. 1907–10. *A Monograph of the Petrels*. H.F. & G. Witherby: London.

Goodwin, D. 1970. *Pigeons and Doves of the World*. London: British Museum (Natural History).

Gosse, P.H. 1847. *The Birds of Jamaica*. Van Voorst: London.

Gould, J. 1832–7. *The Birds of Europe*, 5 vols. London.

1840–8. *The Birds of Australia*, 7 vols. London.

1843–4. *See* Hinds, R.B., *The Zoology of the Voyage of H.M.S. Sulphur*.

1851–69. *Supplement to the Birds of Australia*. London.

Gould, J. and Sharpe, R.B. 1875–88. *The Birds of New Guinea*, 5 vols. London.

Grandidier, A. 1876–85. *Histoire physique, naturelle et politique de Madagascar*, 15 vols. Paris.

Gray, G.R. 1844. *See* Richardson, J., *The Zoology of the Voyage of H.M.S. Erebus and Terror*.

Gray, G.R. 1862. List of the birds of New Zealand. *Ibis* **4**, 214.

Gray, J.E. 1846–50. *Gleanings from the Menagerie and Aviary at Knowsley Hall*. Privately printed.

Greenway, J.C. 1958. *Extinct and Vanishing Birds of the World*. American Committee for International Wildlife Protection: New York.

Grieve, S. 1885. *The Great Auk or Garefowl*. Edinburgh.

Groen, H. 1966. *Australian Parakeets: Their Maintenance and Breeding in Europe*. Privately published by the author: Haren.

Grossman, M.L. and Hamlet, J. 1964. *Birds of Prey of the World*. Bonanza Books: New York.

Gundlach, J. 1876. *Contribución a la Ornitología Cubana*. Havana.

Günther, A. and Newton, E. 1879. Extinct birds of Rodriguez. *Philosophical Transactions of the Royal Society* (extra) **168**, 423–37.

Hachisuka, M. 1931–5. *The Birds of the Philippine Islands*, 2 vols. H.F. and G. Witherby: London.

1937a. Kuina mundyi. *Bulletin of the British Ornithologists' Club* **57**, 156.

1937b. Revisional note on the didine birds of Réunion. *Proceedings of the Biological Society of Washington* **1**, 69–72.

1953. *The Dodo and Kindred Birds, or the Extinct Birds of the Mascarene Islands*. H.F. and G. Witherby: London.

Hadden, F.C. 1941. Midway Islands. *Hawaiian Planters' Record* **45**, 179–221.

Hahn, C.W. 1834–41. *Ornithologischer Atlas*. Nurnberg.

Hahn, P. 1963. *Where is that Vanished Bird?* Royal Ontario Museum: Toronto.

Halliday, T. 1978, *Vanishing Birds*. Sidgwick and Jackson: London.

Hancock, J. 1874. Catalogue of the Birds of Northumberland and Durham. *Transactions of the Natural History Society of Northumbland and Durham* **6**, 1–174.

Hancock, J. and Elliott, A. 1978. *Herons of the World*. London Editions: London.

Harrison, M. 1970. The Orange-fronted Parakeet. *Notornis* **17**, 115–25.

Harry, B. A coppey of Mr Benj. Harry's Journall when he was chief mate of the shippe *Berkley Castle*, Capt. William Talbot then Commander, on a voyage to the Coste and Bay, 1679, which voyage they wintered at the Maurisshes. *British Museum Additional Manuscript* No. 3668, 11 D.

Hartert, E. 1927. Types of birds in the Tring Museum. *Novitates Zoologicae* **34**, 1–38.

Hartlaub, G. 1893. Vier seltene Rallen. *Abhandlungen der Naturwissenschaftiche Verein zu Bremen* **12**, 389–402.

Hartlaub, G. and Finsch, O. 1871. The Samoan Wood Rail. *Proceedings of the Zoological Society of London*, 25.

Hasbrouck, E.M. 1891. The Carolina Parakeet. *Auk* **8**, 369–79.

Henshaw, H.W. 1901–4. Birds of the Hawaiian Islands, being a complete list of the birds of the Hawaiian possessions with notes on their habits. In: Thrum, T.G., *Hawaiian Almanac and Annual*. Honolulu.

Herbert, Sir T. 1634. *A Relation of some yeares' Travaile, begunne Anno 1626, into Afrique and the greater Asia, especially the territories of the Persian Monarchie, and some parts of the Oriental Indies and Isles adiacent*. London.

Heuvelmans, B. 1958. *On the Track of Unknown Animals*. Rupert Hart-Davies: London.

1966. Le Dodo. *Sandorama* (Paris), May, 26–9.

Hinds, R.B. 1843–4. *The Zoology of the Voyage of H.M.S. Sulphur, under the command of Captain Sir Edward Belcher . . . during the years 1836–42*. London.

Hindwood, K.A. 1940. The birds of Lord Howe Island. *Emu* **40**, 1–86.

1965. John Hunter: a naturalist and artist of the First Fleet. *Emu* **65**, 83–95.

Hochstetter, F. von. 1867. *New Zealand, its Physical Geography, Geology and Natural History*. J.G. Cotta: Stuttgart.

Holyoak, D.T. 1971. Comments on the extinct parrot *Lophopsittacus mauritianus*. *Ardea* **59**, 50–1.

1974a. *Cyanoramphus malherbi*, is it a colour morph of *C. auriceps*? *Bulletin of the British Ornithologists' Club* **94**, 4–9.

1974b. Les oiseaux des Îles de la Société. *Oiseaux et la Revue Française d'Ornithologie* **44**.

Holyoak, D.T. and Thibault, J.C. 1978. Undescribed *Acrocephalus* warblers from Pacific Ocean islands. *Bulletin of the British Ornithologists' Club* **98**, 122.

Howard, R. and Moore, A. 1980. *A Complete Checklist of the Birds of the World.* Oxford University Press: Oxford.

Hugel, Baron von. 1875. A new specimen of the Auckland Islands Merganser. *Ibis*, ser. 3, **5**, 392.

Hull, A.F. Bassett. 1909. The birds of Lord Howe and Norfolk Islands. *Proceedings of the Linnean Society of New South Wales* **34**, 636–93.

Hume, A.O. and Marshall, C.H.T. 1879–81. *The Game Birds of India, Burmah and Ceylon*, 3 vols. A. Acton: Calcutta.

Hutton, F.W. and Drumond, J. 1904. *The Animals of New Zealand.* Whitcombe and Tombs: Wellington.

Jehl, J.R. 1972. On the cold trail of an extinct petrel. *Pacific Discovery* **25**, 24–9.

Jerdon, T.C. 1864. *Birds of India*, 3 vols. Calcutta.

Johnsgard, P.A. 1961. The trachial anatomy of the Anatidae and its taxonomic significance. *Wildfowl Trust 12th Annual Report*, 58–69.

1978. *Ducks, Geese and Swans of the World.* University of Nebraska: Lincoln.

1981. *The Plovers, Sandpipers and Snipes of the World.* University of Nebraska: Lincoln.

Kalm, P. 1759. (1911). A description of wild pigeons which visit the southern English colonies in North America during certain years in incredible multitudes. Reproduced in *Auk* **28**, 53–66.

Kear, J. and Scarlett, R.J. 1970. The Auckland Islands Merganser. *Wildfowl* **21**, 78–86.

King, W.B. 1981. *Endangered Birds of the World: the ICBP Bird Red Data Book.* Smithsonian Institution Press in cooperation with the International Council for Bird Preservation: Washington.

Kittlitz, F. von. 1830. Kittlitz Thrush. *Memoir of the Academy of Science of St. Petersburg* **1**, 245.

1832–3. *Kupfertafeln zur Naturgeschichte der Vogel.* Frankfurt am Main.

1858. *Denkwürdigkeiten einer Reise nach dem russichen Amerika, nach Mikronesien und durch Kamtschatka.* Gotha.

Knip, Madame and Temminck, C.J. 1809–11. *Les Pigeons.* Paris.

Kuroda, N. 1917. [Korean Crested Shelduck.] *Tori* **1** [in Japanese].

1924. [Korean Crested Shelduck.] *Tori* **4** [in Japanese].

1925. *A Contribution to the Knowledge of the Avifauna of the Riu Kiu Islands.* Tokyo.

1940. [Korean Crested Shelduck.] *Tori* **10** [in Japanese]

Labat, J.B. 1742. *Nouveau voyage aux isles de l'Amerique.* Paris.

Lack, D. 1947. *Darwin's Finches.* Cambridge University Press: Cambridge.

Lambrecht, K. 1933. *Handbuch der Palaeornithologie.* Gebrüder Borntraeger: Berlin.

Latham, J. 1781–5. *A General Synopsis of Birds*, 3 vols. London.

1787. *Supplement to the General Synopsis of Birds.* London.

Lavauden, L. 1932. Étude d'une petit collection d'oiseaux de Madagascar. *Bulletin du Muséum d'Histoire Naturelle, Paris* **4**, 629–40.

Laycock, G. 1969. The last parakeet. *Audubon* **71**, 21–5.

Leguat, F. 1707. *Voyages et avantures en deux isles désertes des Indes Orientales.* Paris.

1708. *A New Voyage to the East Indies.* London.

Lendon, A.H. 1973. *Australian Parrots in Field and Aviary.* Angus and Robertson: Sydney.

L'Estrange, Sir Hamon. In: Browne, Sir Thomas 1836. *Works* (Wilkin edition). London. *See* Vol. 1, p. 369: Vol.2, p. 173.

Levaillant, F. 1799–1808. *Histoire Naturelle des Oiseaux d'Afrique*, 6 vols. Paris.

1801–5. *Histoire Naturelle des Perroquets*, 2 vols. Paris.

Lever, Sir Christopher. 1985. *Naturalized Mammals of the World.* Longman: Harlow.

Linnaeus, C. 1758. *Systema Naturae.* Stockholm.

Lysaght, A.M. 1953. A rail from Tonga *Rallus philippensis ecaudata* Miller 1785. *Bulletin of the British Ornithologists' Club* **73**, 74–5.

Mathew, M.A. 1866. Great Auks on Lundy. *Zoologist* **1**, 100.

Mathews, G.M. 1910–28. *Birds of Australia*, 12 vols. H.T. and G. Witherby: London.

1928. *The Birds of Norfolk and Lord Howe Islands.* H.F. and G. Witherby: London.

1936. *A Supplement to the Birds of Norfolk and Lord Howe Islands.* H.F. and G. Witherby: London.

Mayr, E. 1945. *Birds of the South-west Pacific.* MacMillan: New York.

Medway, D.G. 1979. Some ornithological results of Cook's third voyage. *Journal of the Society for the Bibliography of Natural History* **9**, 315–51.

Meek, A.S. 1913, *A Naturalist in Cannibal Land.* Fisher Unwin: London.

Mees, G.F. 1977. Enige gegevens over de uitgestorven rail *Pareudiastes pacifica. Zoologische Mededeelingen Leiden* **50**, 230–42.

Mershon, W.B. 1907. *The Passenger Pigeon.* Outing Publishing: New York.

Milne-Edwards, A. 1866–73. *Recherches sur la Faune Ornithologique Éteinte des îles Mascareignes et de Madagascar.* Paris.

1869. Researches into the zoological affinities of . . . *Aphanapteryx imperialis. Ibis*, ser. 2, **5**, 256–75.

1874. Recherches sur la faune ancienne des îles Mascareignes. *Annales des Sciences Naturelles Zoologie*, ser. 5, **19**, 1–31.

Milne-Edwards, A. and Oustalet, E. 1893. Notice sur quelques espèces d'oiseaux actuallement éteintes. *Centenaire de la fondation du Muséum d'Histoire Naturelle, Paris*, 190–252.

Mitchell, M. 1935. The Passenger Pigeon in Ontario. *Contributions of the Royal Ontario Museum* No. 7, 1–181.

Morton, T. 1637 (1883). *The New English Canaan.* Boston: Prince Society.

Mundy, P. *See* Temple, R.C. and Anstey, L.M.

Munro, G.C. 1944. *Birds of Hawaii.* Tuttle: Honolulu.

Murphy, R.C. 1936. *The Oceanic Birds of South America*, 2 vols. American Museum of Natural History: New York.

Neck, J.C. van. 1601. *Het tweede Boek, Journal oft Dagh-register, inhoudende een warachtig verhael.* Middleburg.

Newton, A. 1861. Abstract of J. Wolley's researches in Iceland respecting the Garefowl or Great Auk. *Ibis* **3**, 374–99.

1865. The Gare-fowl and its historians. *Natural History Review* **5**, 467–88.

1867. On a picture supposed to represent the didine bird of the island of Bourbon (Réunion). *Transactions of the Zoological Society of London* **6**, 373–6.

1875. Additional evidence as to the original fauna of Rodriguez. *Proceedings of the Zoological Society of London*, 39–40.

1875. Note on *Palaeornis exsul*. *Ibis*, ser. 3, **5**, 342–3.
1879. A specimen of *Alectroenas nitidissima* in Edinburgh. *Proceedings of the Zoological Society of London*, 2–4.
Newton, A. and Gadow, H. 1893. On additional bones of the dodo and other extinct birds of Mauritius. *Transactions of the Zoological Society of London* **13**, 281–302.
Newton, A. and Newton, E. 1868. On the osteology of the Solitaire or didine bird of the Island of Rodriguez, *Pezophaps solitaria* (Gmel.). *Philosophical Transactions of the Royal Society of London* **159**, 327–62.
1876. On the Psittaci of the Mascarene Islands. *Ibis*, ser. 3, **6**, 281–9.
Newton, E. 1861. Ornithological notes from Mauritius. *Ibis* **3**, 270–7.
Nichols, R.A. 1943. The breeding birds of St. Thomas and St. John, Virgin Islands. *Memorias Sociedad Cubana de Historia Natural* **17**, 23–37.
Nixon, A.J. 1981. The external morphology and taxonomic status of the Orange-fronted Parakeet. *Notornis* **28**, 292–300.

Oliver, W.R.B. 1930. *New Zealand Birds*. Fine Arts (N.Z.): Wellington.
1949. The Moas of New Zealand and Australia. *Dominion Museum Bulletin* No. 15, 1–195.
1955. *New Zealand Birds*. A.H. and A.W. Reed: Wellington.
Olson, S.L. 1973. Evolution of the rails of the South Atlantic islands. *Smithsonian Contributions to Zoology* **152**, 7.
1977. A synopsis of the fossil Rallidae. In: Ripley, S.D. *Rails of the World*, 339–73. Godine: Boston.
1986. An early account of some birds from Mauke, Cook Islands, and the origin of the 'Mysterious Starling' *Aplonis mavornata* Buller. *Notornis* **33**, 197–208.
Oudemans, A.C. 1917. *Dodo-studien*. Johannes Müller: Amsterdam.
Oustalet, E. 1896. Faune des îles Mascareignes. *Annales des Sciences Naturelles, Zoologie* **3**, 1 128.
Owen, M. 1977. *Wildfowl of Europe*. Macmillan: London.
Owen, R. 1839. Exhibited bone of an unknown struthious bird from New Zealand. *Proceedings of the Zoological Society of London* **7**, 169–71.
1842. Notice of a fragment of the femur of a gigantic bird from New Zealand. *Transactions of the Zoological Society of London* **3**, 29–33.
1864. Description of the skeleton of the Great Auk or Garefowl. *Transactions of the Zoological Society of London* **5**, 317–35.
1866a. *Memoir on the Dodo* (with an historical introduction by William John Broderip). London.
1866b. Evidence of a species, perhaps extinct, of a large parrot *Psittacus mauritianus*, contemporary with the Dodo, on the island of Mauritius. *Ibis*, ser. 2, **3**, 168–71.
1871. Notes on the articulated skeleton of the Dodo in the British Museum. *Transactions of the Zoological Society of London* **7**, 513–25.
1879. *Memoirs on the Extinct Wingless Birds of New Zealand, with an appendix on those of England, Australia, Newfoundland, Mauritius and Rodriguez*, 2 vols. London.

Palmer, R.S. (ed.) 1976. *Handbook of North American Birds*, 3 vols. Yale University Press: New Haven.
Parker, S. 1967. New information on the Solomon Islands Crowned Pigeon *Microgoura meeki* Rothschild. *Bulletin of the British Ornithologists' Club* **87**, 86–9.
1972. An unsuccessful search for the Solomon Islands Crowned Pigeon. *Emu* **72**, 24–6.
1982. On a new species of sandpiper. *South Australian Naturalist* **56** (4), 63.
Parker, T.J. 1893. On the presence of a crest of feathers in certain species of moa. *Transactions of the New Zealand Institute* **25**, 3–6.
Parkin, T. 1911. The Great Auk. A record of sales of birds and eggs by public auction in Great Britain, 1806–1910. *Hastings and East Sussex Naturalist* (extra) **1** (6), 1–36.
Peale, T.R. 1848. *United States Exploring Expedition, during the years 1838, 1839, 1840, 1842*. Philadelphia.
Pelzeln, A. von. 1860. Zur ornithologie der Insel Norfolk. *Sitzungsberichte der Mathematisch, Naturwissenschafliche Klasse der Akademie der Wissenschaften, Wien* **41** (15), 319–32.
1873. On the birds in the Imperial Collection at Vienna obtained from the Leverian Museum. *Ibis*, ser. 3, **3**, 105–24.
Perkins, R.C.L. 1893. The Koa Finch. *Ibis*, ser. 6, **5**, 103–4.
1903. *Vertebrata*. In: *Fauna Hawaiiensis, or the Zoology of the Sandwich Islands*, Vol. 1, Pt 4, pp. 365–466. Cambridge University Press: Cambridge.
Peters, J.L. (and others) 1931–86. *A Check-list of the Birds of the World*, 15 vols. Museum of Comparative Zoology: Cambridge, Mass.
Phillipps, W.J. 1963. *The Book of the Huia*. Whitcombe and Tombs: Christchurch.
Phillips, J.C. 1922–6. *A Natural History of the Ducks*, 4 vols. Houghton Mifflin: Boston.
Piveteau, J. 1945. Étude sur l'Aphanapteryx oiseau éteint de l'Île Maurice. *Annales de Paléontologie* **31**, 31–7.
Polack, J.S. 1838. *New Zealand: Being a Narrative of Travels and Adventures During a Residence in that Country between the years 1831 and 1837*. London.
Potts, T.H. 1870–4. On the birds of New Zealand, Parts 1–4. *Transactions and Proceedings of the New Zealand Institute* Vols 2, 3, 5 and 6.
1882. *Out in the Open: a Budget of Scraps of Natural History gathered in New Zealand*. Christchurch.
Pratt, H.D. 1979. *A Systematic Analysis of the Endemic Avifauna of the Hawaiian Islands*. Published on demand by University Microfilms International: Ann Arbor, Michigan.
1981. The Drepanididae. In: *The Living Bird, 19th Annual*, pp.73–90. Cornell University: Ithaca, New York.

Quoy, J.R.C. and Gaimard, J.P. 1830. Zoologie, Tome Premier. In: *Voyage de Découvertes de l'Astrolabe*. Paris. *See* Dumont D'Urville, J.S.C.

Rand, A.L. 1936. Distribution of Madagascan birds. *Bulletin of the American Museum of Natural History* **72** (5), 143–499.
Reischek. A. 1930. *Yesterdays in Maoriland*. Whitcombe and Tombs: Wellington.
Renshaw, G. 1931. The Mauritius Dodo. *Journal of the Society for the Preservation of the Fauna of the Empire* **15**, 14–21.
Richardson, J. and Gray G.R. 1844–75. *The Zoology of the voyage of H.M.S. Erebus and Terror, under command of Captain Sir James Clark Ross, R.N., F.R.S., during the years 1839 to 1843*, 2 vols. London.
Ridgway, R. 1876a. Studies of the American Falconidae: monograph of the Polybori. *Bulletin of the US Geological and Geographical Survey of the Territories*, ser. 2, No. 6, 451–73.
1876b. Ornithology of Guadalupe Island. *Bulletin of the US Geological and Geographical Survey of the Territories*, ser. 2, No. 7, 483–95.

Ripley, S.D. 1977. *Rails of the World*. Godine: Boston.
Rothschild, M. 1983. *Dear Lord Rothschild*. Hutchinson: London.
Rothschild, W. 1892. Descriptions of seven new species of birds from the Sandwich Islands. *Annals and Magazine of Natural History* **6**, 108–12.
1893–1900. *The Avifauna of Laysan and the Neighbouring Islands*, 2 vols. R.H. Porter: London.
1894. A new species from Stephens Island. *Bulletin of the British Ornithologists' Club* **4**, 19.
1903. *Rallus wakensis*. *Bulletin of the British Ornithologists' Club* **13**, 78.
1904. *Microgoura meeki*. *Bulletin of the British Ornithologists' Club* **14**, 78.
1905a. On extinct and living parrots of the West Indies. *Bulletin of the British Ornithologists' Club* **16**, 13–14.
1905b (1907). On extinct and vanishing birds. *Proceedings of the 4th International Ornithological Congress, London, 1905*, 191–217. Dulau: London.
1907. *Extinct Birds*. Hutchinson: London.
1919. On one of the four original pictures from life of the Réunion or white dodo. *Ibis*, ser. 11, **1**, 78–9.
Rowley, D.G. 1875–8. *Ornithological Miscellany*, 3 vols. Trübner: London.

Salvin, O. 1873. Note on the *Fulica alba* of White. *Ibis*, ser. 6, **5**, 295.
Schauensee, R. Meyer de. 1957. On some avian types, principally Gould's, in the collection of the Academy. *Proceedings of the Academy of Natural Science, Philadelphia* **109**, 123–46.
Schlegel, H. 1854. Ook een woordje over den Dodo, *Didus ineptus*, en zijne verwanten. *Verslagen en Mededeelingen der Koninklijke Nederlandsch Akademie van Wetenschappen te Amsterdam* **7**, 232.
Schorger, A.W. 1955. *The Passenger Pigeon*. University of Wisconsin: Madison.
Sclater, W.L. 1915. The 'Mauritius Hen' of Peter Mundy. *Ibis*, ser. 10, **3**, 316–9.
Scott, H.H. 1924. Note on the King Island Emu. *Papers and Proceedings of the Royal Society of Tasmania, 1923*, 103–7.
1932. The extinct Tasmanian Emu. *Papers and Proceedings of the Royal Society of Tasmania, 1931*, 108–10.
Seebohm, H. 1888. *The Geographical Distribution of the Charadriidae*. Henry Southeran: London.
1890. Birds of the Bonin Islands. *Ibis*, ser. 6, **2**, 95–108.
1898–1902. *Monograph of the Turdidae*, 2 vols. Henry Southeran: London.
Sélys Longchamps, Baron E. de. 1910. *Collections Zoologique*. Fasc. 31, *Oiseaux*. Hayez: Brussels.
Seth-Smith, D. 1932. Note on *Rhodonessa*. *Avicultural Magazine*, ser. 4, **10**, 118.
Sharpe, R.B., Sclater, P.L. and others. 1874–98. *Catalogue of the Birds in the British Museum*, 27 vols. London.
Sharpe, R.B. 1891–8. *Monograph of the Paradiseidae and Ptilonorhynchidae*, 2 vols. London.
1906a. A note on the White-winged Sandpiper. *Bulletin of the British Ornithologists' Club* **16**, 86.
1906b. *History of the Collections contained in the Natural History Departments, British Museum*. London: British Museum.
Shelley, G.K. 1883. On the Columbidae of the Ethiopian region. *Ibis*, ser. 5, **1**, 258–330.
Simson, F.B. 1884. Note on the Pink-headed Duck. *Ibis*, ser. 54, **2**, 271–5.
Slater, P. 1978. *Rare and Vanishing Australian Birds*. Rigby: Sydney.
1980. *Masterpieces of Australian Bird Photography*. Rigby: Sydney.
Sonnerat, P. 1782. *Voyage aux Indes Orientales et à la Chine*, 2 vols. Paris.
Steenstrup. J.J.S. 1855. Et bidrag til gierfuglens, *Alca impennis*. *Videnskabelige Meddelelser fra Dansk naturhistorik Forening i Kjøbenhavn* Nos. 3–7, 33–115.
Stejneger, L. 1883. The status of the Spectacled, or Pallas', Cormorant. *Proceedings of the US National Museum* **6**, 65.
1885. Results of ornithological exploration in the Commander Islands and in Kamchatka. *US National Museum Bulletin* No. 29, 1–382.
1890. Contributions to the history of Pallas' Cormorant. *Proceedings of the US National Museum* **12**, 83–8.
1936. *Georg Wilhelm Steller*. Harvard University Press: Cambridge, Mass.
Stresemann, E. 1930. Welche Paradiesvogelarten der literatur sind hybriden ursprungs? *Novitates Zoologicae* **36**, 3–13.
1949. Birds collected in the North Pacific area during Capt. Cook's last voyage. *Ibis*, **89**, 244–55.
1950. Birds collected during Captain Cook's last expedition. *Auk* **67**, 66–88.
Strickland, H.E. 1849. Supposed existence of a giant bird in Madagascar. *Annals and Magazine of Natural History* ser. 2, **4**, 338–9.
Strickland, H.E. and Melville, A.G. 1848. *The Dodo and its Kindred*. Reeve, Benham & Reeve: London.
Swainson, W. 1827. [*Scaphidurus palustris*.] *Philosophical Magazine* **1**, 437.
Swann, H.K. 1924–36. *A Monograph of the Birds of Prey*, 2 vols. Wheldon and Wesley: London.

Tatton, J. 1625. Voyages of Castleton. In: *Purchas's Pilgrimage*. London.
Taylor, R. 1870. *Te Ika a Maui*. Second edn. William Macintosh: London.
1873. An account of the first discovery of moa remains. *Transactions of the New Zealand Institute* **5**, 97–101.
Temminck, C.J. and Laugier de Chartrouse, Baron M. 1820–39. *Nouveau Recueil de Planches colorées d'Oiseaux*, 5 vols. Paris.
Temple, R.C. and Anstey, L.M. (eds). 1919–36. *The Travels of Peter Mundy in Europe and Asia, 1608–1667*. Haklyut Society: London.
Thayer, J.E. and Bangs, O. 1908. The present state of the ornis of Guadalupe Island. *Condor* **10**, 101–6.
Thibault, J.C. 1974. *Le Peuplement Avian des Îles de la Société*. Les Editions du Pacifique: Papeete, Tahiti.
Todd, F.S. 1979. *Waterfowl: Ducks, Geese and Swans of the World*. Sea World Press: London.
Tomkinson, P.M.L. and Tomkinson, J.W. 1966. Eggs of the Great Auk. *Bulletin of the British Museum (Natural History), Historical Series* **3** (4), 1–97.
Tristram, H.B. 1880. Description of a new species and genus of owl from the Seychelles Islands. *Ibis*, ser. 4, **4**, 456–9.
Tweeddale, A., Marquis of. 1881. *The Ornithological Works of Arthur, Ninth Marquis of Tweeddale*. London.

Uchida, S. 1918. [Korean Crested Shelduck.] *Tori* **2** [in Japanese].

Walkinshaw, L. 1973. *Cranes of the World*. Winchester Press: New York.

Watling, D. 1982. *Birds of Fiji, Tonga and Samoa*. Millwood Press: Wellington.
Wetmore, A. 1925. Bird life among lava rock and coral sand. *National Geographical Magazine* **48**, 76–108.
1937. [Cuban Red Macaw.] *Journal of Agriculture, University of Puerto Rico* **21**, 12.
1967. Recreating Madagascar's giant bird. *National Geographical Magazine* **132**, 488–93.
Wheaton. J.M. 1882. Report on the birds of Ohio. *Ohio Geological Survey* **4**, 187–628.
White, J. 1790. *Journal of a Voyage to New South Wales*. London.
Whitmee, S.J. 1874. Letter to P.L. Sclater on birds of Samoa. *Proceedings of the Zoological Society of London*, 183–6.
Wiglesworth, L.W. 1891. Aves Polynesiae. A catalogue of birds of the Polynesian subregion. *Abhandlungen des Zoologisches und Anthropologisch-Ethnographisches Museum zu Dresden* No. 6.
Williams, G.R. 1956. The Kakapo: a review and re-appraisal of a near-extinct species. *Notornis* **7**, 29–56.
Williams, G. and Harrison, M. 1972. The Laughing Owl. *Notornis* **19**, 4–19.
Wilson, A. 1808–14. *American Ornithology*, 7 vols. Philadelphia.
Wilson, A. and Bonaparte, C.L. 1832. *American Ornithology*, 3 vols. London.
Wilson, S.B. and Evans, A.H. 1890–9. *Aves Hawaiiensis. The Birds of the Sandwich Islands*. R.H. Porter: London.
Wilson, S.B. 1888. The Kona Grosbeak. *Proceedings of the Zoological Society of London*, 218.
Wood, J. 1871. [*Argusianus bipunctatus*.] *Annals and Magazine of Natural History*, ser. 4, **8**, 67.
Woolfenden, G.E. 1961. Postcranial osteology of the waterfowl. *Bulletin of the Florida State Museum for Biological Science* **6**, 1–29.

ACKNOWLEDGEMENTS

The publishers have made every attempt to contact the owners of the photographs appearing in this book. In the few instances where they have been unsuccessful, they invite the copyright holders to contact them direct.

British Museum (Natural History): 110, 198.
Collection of Aigantighe Art Gallery, Timaru, New Zealand: 230.
Denver Museum of Natural History Photo Archives: 81, 82, 179, 187 (top and bottom). (All rights reserved.)
National Photographic Index of Australian Wildlife, 137.
Teylers Museum, Haarlem: 126.
University Library, Utrecht, R.fol.41: 123.
Zoological Society of London/photos Martin Lyster: 35, 38, 66, 107, 110, 143, 147, 155, 199, 206, 210, 214, 215, 218, (left and right), 219, 222 (left and right).

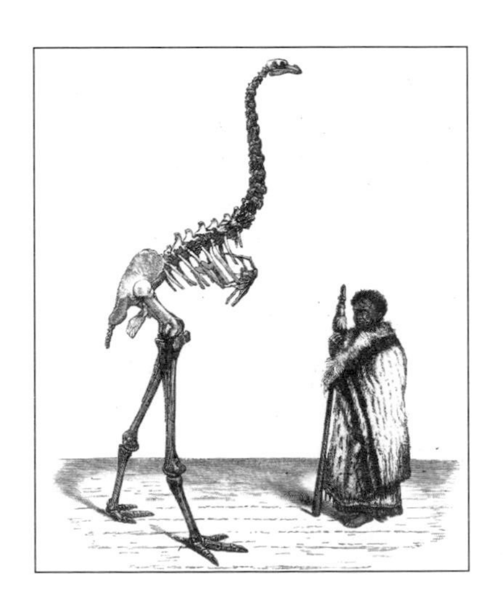

INDEX

Page numbers in *italics* refer to illustrations

'They will whirl abou
times together on the
ſpace of four or five M
their … kes the
of a R … d one m
Paces … Bone of
ter tow … Extrer
round … der the
Muske … hat and
fence … rd. 'T
in the … but eaſ
cauſe we run faſter tha
we approach them withc